Medicinal Plants of Laos

This book provides a description of medicinal plants of Laos, including their role in maintaining healthcare among the population, their potential as a source for new medicinal compounds, their preservation, and their importance for the well-being of the communities for present and future generations. The focus of this book is to draw on the rich culture, folklore, and environment of medicinal plants in the country. This is an opportunity to describe medicinal plants from a scientifically underrepresented area, with the hope of making an important contribution to the knowledge of the region for academics, scientists, and anyone who has interest in Laos.

Features

- Describes terrestrial medicinal plants from a scientifically underrepresented region.
- Includes a wider variety of plants found growing in Laos than has previously been published.
- Discusses past and present research on medicinal plants that may lead to the discovery of new medicines.
- Describes efforts in the preservation of these medicinal plants for present and future generations.
- Focuses on the rich culture, folklore, and environment of medicinal plants in Laos.
- Provides an important contribution to knowledge of the region and will benefit anyone interested in the medicinal plants of Laos.

Natural Products Chemistry of Global Plants

Editor: *Raymond Cooper*

This unique book series focuses on the natural products chemistry of botanical medicines from different countries such as Sri Lanka, Cambodia, Brazil, China, Africa, Borneo, Thailand, and Silk Road Countries. These fascinating volumes are written by experts from their respective countries. The series will focus on the pharmacognosy covering recognized areas rich in folklore as well as botanical medicinal uses as a platform to present the natural products and organic chemistry. Where possible, the authors will link these molecules to pharmacological modes of action. The series intends to trace a route through history from ancient civilizations to the modern day showing the importance to man of natural products in medicines, foods, and a variety of other ways.

Recent Titles in this Series:

Medicinal Plants and Mushrooms of Yunnan Province of China
Clara Lau and Chun-Lin Long

Medicinal Plants of Borneo
Simon Gibbons and Stephen P. Teo

Natural Products and Botanical Medicines of Iran
Reza Eddin Owfi

Natural Products of Silk Road Plants
Raymond Cooper and Jeffrey John Deakin

Brazilian Medicinal Plants
Luzia Modolo and Mary Ann Foglio

Medicinal Plants of Bangladesh and West Bengal: Botany, Natural Products, and Ethnopharmacology
Christophe Wiart

Traditional Herbal Remedies of Sri Lanka
Viduranga Y. Waisundara

Medicinal Plants of Ecuador
Pablo A. Chong Aguirre, Patricia Manzano Santana, Migdalia Miranda Martínez (Eds)

Medicinal Plants of Laos
Djaja Djendoel Soejarto, Kongmany Sydara, and Bethany Gwen Elkington

Medicinal Plants of Laos

Edited by
Djaja Djendoel Soejarto

Professor Emeritus, University of Illinois at Chicago and Adjunct
Curator, Science and Education, Field Museum of Natural History,
Chicago, USA

Kongmany Sydara

Former Director General, Institute of Traditional Medicine,
Ministry of Health, Vientiane, Laos and Invited Lecturer,
Faculty of Pharmacy, University of Health Sciences,
Vientiane, Laos

Bethany Gwen Elkington

Research Assistant Professor, University of Illinois at Chicago
and Research Associate, Science and Education, Field Museum of
Natural History, Chicago, USA

CRC Press
Taylor & Francis Group
Boca Raton New York London

CRC Press is an imprint of the
Taylor & Francis Group, an **informa** business

First edition published 2023
by CRC Press
6000 Broken Sound Parkway NW, Suite 300, Boca Raton, FL 33487-2742

and by CRC Press
4 Park Square, Milton Park, Abingdon, Oxon, OX14 4RN

© 2023 selection and editorial matter, **Djaja Djendoel Soejarto, Kongmany Sydara, and Bethany Gwen Elkington**; individual chapters, the contributors

CRC Press is an imprint of Taylor & Francis Group, LLC

Library of Congress Cataloging-in-Publication Data
Names: Soejarto, Djaja D., editor. | Elkington, Bethany Gwen, editor. |
Kongmany Sydara, editor.
Title: Medicinal plants of Laos / edited by Djaja Djendoel Soejarto,
Department of Pharmaceutical Sciences, College of Pharmacy, University
of Illinois at Chicago, Chicago, Illinois and Adjunct Curator, Science
and Education, Field Museum of Natural History, Chicago, USA, Kongmany
Sydara, Institute of Traditional Medicine, Ministry of Health,
Vientiane, Lao People's Democratic Republic, and Invited Lecturer,
Faculty of Pharmacy, University of Health Sciences, Vientiane, Lao PDR,
Bethany G. Elkington, Department of Pharmaceutical Sciences, College of
Pharmacy, University of Illinois at Chicago, Chicago, Illinois, and
Research Associate, Science and Education, Field Museum of Natural
History, Chicago, USA.
Description: First edition. | Boca Raton, FL : CRC Press, 2023. | Includes
bibliographical references and index. | Identifiers: LCCN 2022052278 (print) | LCCN 2022052279 (ebook) |
ISBN 9781032107028 (HB) | ISBN 9781032077772 (PB) | ISBN 9781003216636 (EB)
Subjects: LCSH: Medicinal plants--Laos.
Classification: LCC QK99.L26 M43 2023 (print) | LCC QK99.L26 (ebook) |
DDC 581.6/3409594--dc23/eng/20221222
LC record available at https://lccn.loc.gov/2022052278
LC ebook record available at https://lccn.loc.gov/2022052279

ISBN: 978-1-032-10702-8 (hbk)
ISBN: 978-1-032-07777-2 (pbk)
ISBN: 978-1-003-21663-6 (ebk)

DOI: 10.1201/9781003216636

Typeset in Times
by SPi Technologies India Pvt Ltd (Straive)

Dedication

Firstly, this book is dedicated to the people of Laos. The authors learned much about medicinal plants from them, especially from the healers and members of the various communities around Laos with whom we worked.

Secondly, this book is dedicated to our good friend and mentor who provided the inspiration and guidance to write this book, the late Dr. Raymond Cooper, Visiting Professor in the Department of Applied Biology and Chemical Technology at The Hong Kong Polytechnic University, Hong Kong. We are deeply saddened that Ray will not see the final copy of this book, as he passed away while we were finishing the last chapter. We are grateful for Ray's many contributions to the study of medicinal plants and share our deepest sympathies with his family.

Contents

*James G. Graham, Bethany Gwen Elkington, Jonathan Bisson,
Charlotte Gyllenhaal, Yulin Ren, A. Douglas Kinghorn,
Kongmany Sydara, and Djaja Djendoel Soejarto*

*Djaja Djendoel Soejarto, Kongmany Sydara,
Mouachanh Xayvue, Onevilay Souliya,
Bethany Gwen Elkington, Mary Riley, and Charlotte Gyllenhaal*

*Djaja Djendoel Soejarto, Kongmany Sydara,
Bethany Gwen Elkington, Chun-Tao Che, Guido F. Pauli,
Bounleuane Douangdeuane, Charlotte Gyllenhaal, and
Mary Riley*

Foreword

Out of the estimated 374,000 species[1] of plants known to exist on Earth, an astounding 9% of these have been documented for their application in medicine. Over the span of millennia, this body of traditional knowledge concerning the collection, preparation, and application of medicinal plants for the treatment of myriad diseases pertinent to both human and veterinary medicine has emerged from different cultures, ecosystems, and languages across the globe. Scientific investigation of these traditions has revealed new insights and led to advances in modern medicine. Indeed, some of the world's most essential medicines for the treatment of cancer, pain, heart disease, infection, and more are based on compounds originally discovered in medicinal plants.

While the importance of plants to advances in modern medicine as we know it today is irrefutable, there remains much work to be done concerning the chemical characterization and pharmacological evaluation of these species. Indeed, the vast majority of the estimated 34,000 medicinal plants[2] on Earth have never been examined with the tools of modern science. Moreover, beyond their incredible potential as drivers of drug discovery, medicinal plants play a fundamental role in supporting the health and livelihood of billions of people across the globe. The World Health Organization estimates that up to 80% of people living in developing countries rely on traditional medicine as their main form of health care,[3] with plants serving as the foundation for their pharmacopoeia. Tragically, today these resources are under threat due to changes in land use practices, environmental stressors caused by global climate change, and overharvesting of wild plant populations. The 2020 Kew Report on the State of the World's Plants and Fungi[4] makes the stark reality that we face perfectly clear, highlighting that human activities are accelerating biodiversity loss and now two in five plants are estimated to be threatened with extinction. As a result, we are facing a dual threat to the future of human health: 1) the loss of biodiversity as a resource for future therapeutic discoveries and innovations for global use under the umbrella of modern medicine; and 2) the loss of species critical to the health and well-being of billions of people, endangering access to traditional medicines that serve as the backbone of community health care.

In the face of these challenges, one may wonder if there is anything that can be done to stem the tide of loss both for biodiversity and the traditional medical knowledge linked to these species. There absolutely is, and this is where the disciplines of ethnobotany, pharmacognosy, and conservation biology can unite, bringing together strengths from the social and natural sciences.

This book is a testament to such an approach. In it, we find insights into the incredible biodiversity of Laos, with special emphasis on both cultivated and wild-crafted medicinal species of importance to local people and community health. Going further, the authors explore the pharmacological properties of these species, covering a broad swathe of bioactive plant secondary metabolites of medical interest. Conservation measures and recommendations are also presented; importantly, these highlight the importance of community engagement in the conservation process.

The future of planetary health and human health are inextricably intertwined. Without one, we cannot have the other. Through documenting traditional knowledge concerning medicinal plants, evaluating these species through the lens of modern science, and working closely with communities to foster sustainable and economically beneficial conservation strategies, we can make real progress in securing health for all.

Cassandra L. Quave, Ph.D.
Emory University School of Medicine
Associate Professor and Herbarium Curator
Atlanta, GA, USA

NOTES

1 Christenhusz, M.J.M. and J.W. Byng. 2016. The number of known plants species in the world and its annual increase. *Phytotaxa* 261(3): 201–217.
2 *Medicinal Plant Names Services*, Version 11. https://mpns.science.kew.org/mpns-portal/version.
3 Bodeker C., G. Bodeker, C.K. Ong, C.K. Grundy, G. Burford, K. Shein. 2005. *WHO Global Atlas of Traditional, Complementary and Alternative Medicine*. Geneva, Switzerland: World Health Organization.
4 Antonelli, A. et al. 2020. *State of the World's Plants and Fungi*. Royal Botanic Gardens, Kew.

Preface

This book is part of the series on "Natural Products Chemistry of Global Plants" and presents one approach to studying medicinal plants by ethnobotanists and natural product researchers. It is the authors' intent to positively contribute to the knowledge of the Lao People's Democratic Republic (Lao PDR).

The book is to be used by the people of Laos and those who have interest in medicinal plants, but especially scientists, botanists, pharmacognosists, pharmacists, chemists, and students, as well as researchers in the fields of natural products, herbal medicines, ethnobotany, pharmacology, and biology, who are searching for new discoveries about medicinal plants of Laos.

This book provides a description of the medicinal plants of Laos, their role in improving human health, their potential as a source for new medicinal compounds, and their preservation. The focus of this book is to draw on the rich culture, folklore, and environment of medicinal plants in Laos, which is an important region for the study of medicinal plants.

The Editors

Acknowledgements

We express our deepest thanks and appreciation first and foremost to the healers of Laos who have trusted us with their intellectual property. To the healers of Somsavath village in Laos, where we have spent the most time for this research, we express special thanks for their gracious collaboration. To members of all the communities where we carried out our fieldwork, we express our thanks and appreciation for their support and collaboration.

We hold immeasurable gratitude for the late Dr. Raymond Cooper, who graciously invited the first Editor, Dr. Djaja D. Soejarto, to write a book as part of the Series on "Natural Products Chemistry of Global Plants," which he and Dr. Clara Bik-san Lau, Associate Director, Institute of Chinese Medicine at the Chinese University of Hong Kong, edited. Dr. Cooper was the founding Editor of this series.

To the former Director and the present Director of the Institute of Traditional Medicine, Ministry of Health, Lao PDR, we express our thanks for their support and collaboration during our field research in Laos, and in the writing of this book. To the staff of this Institute, we owe our thanks for their support and collaboration given to the research process that resulted in materials for the writing of this book. We especially want to thank Mouachanh Xayvue and Onevilay Souliya. They have been exemplary colleagues for many years and serve as co-authors of several chapters of this book.

We also express thanks to our other co-authors for their gracious support and collaboration in the writing of various chapters of the book: Chun-Tao Che, Charlotte Gyllenhaal, Mary Riley, Doug Kinghorn, Yulin Ren, James Graham, Jonathan Bisson, Josh Henkin, and Guido Pauli. These professional scientists have, in one period or another, been a part of the research on the medicinal plants of Laos. Special thanks to James Graham, Research Assistant Professor at the College of Pharmacy, University of Illinois at Chicago (UIC), USA, for his leadership in the writing of Chapter 6.

Our heartfelt thanks to Deena Wolfson, a Research Volunteer at the Field Museum of Natural History, Chicago, Illinois, USA, who provided excellent proofreading and edits for all the chapters.

To the Field Museum of Natural History and its John G. Searle Herbarium, we express our thanks for its support and for providing access to botany resources for research that resulted in this book.

Our sincere thanks and appreciation go to Dr. Cassandra Leah Quave, ethnobotanist, herbarium curator, and Associate Professor at Emory University, Atlanta, Georgia, USA, for writing the Foreword to this book.

We thank Hilary LaFoe, Senior Acquisitions Editor at CRC Press/Taylor and Francis, USA and UK, Abingdon, England, United Kingdom, who kindly accepted our book proposal and, together with her staff, provided guidance during the process of writing this book.

To the Dean and staff of the College of Pharmacy, UIC, we express our thanks for their support during the research and writing that led to this book.

The research for this book was supported, directly and indirectly, by grants from the California Community Foundation (CCF), Los Angeles, California, USA (grant numbers 103861 and 155689; Co-Principal Investigators: Chun-Tao Che and Djaja D. Soejarto, 2013–2014; Bethany G. Elkington, Djaja D. Soejarto, Guido F. Pauli and Chun-Tao Che, 2019–2023), awarded to UIC; and the Anticancer Drug discovery grant (P01 grant) from the United States National Cancer Institute (NCI), National Institutes of Health (NIH), Bethesda, MD, USA. (grant number CA125066; Principal Investigator: A. Douglas Kinghorn) awarded to Ohio State University (OSU), Columbus, Ohio, USA.

And last, but certainly not least, we thank our families and friends for their support and patience during the writing of this book, along with all of the research that has gone into it.

The Editors

Editors

Professor Djaja Djendoel Soejarto is Professor Emeritus at UIC, College of Pharmacy and Adjunct Curator in Science and Education, The Field Museum. He earned his Ph.D. degree (Biology/Botany) in 1969 from Harvard University. He served as Fellow of the Latin American Teaching Fellowships, Fletcher School of Law and Diplomacy, Tufts University, Medford, Massachusetts, in a mission assignment to the Department of Biology, University of Antioquia, Medellin, Colombia (1969–1972). In 1969, as a LATF Fellow, he founded the Herbarium of the University of Antioquia (HUA; http://sweetgum.nybg.org/science/ih/herbarium-details/?irn=126256). He joined the Department of Biology, first as an Assistant Professor, then Associate Professor in Biology (1972–1976). He served as a postdoctoral Fellow in Ethnopharmacology at Harvard University (1972–1973) under the mentorship of Professor Richard Evans Schultes. After serving as a Consultant to the World Health Organization (WHO), Geneva, Switzerland (1976–1978), he became an Adjunct Associate Professor (1979–1983) in the Department of Pharmaceutical Sciences of the College of Pharmacy, UIC, an Associate Professor (1983–1989), then a Full Professor (1989–2015). From 1979 to 2020, he was involved in research on plants and medicinal plants under a multidisciplinary setting at UIC, traveling extensively to many countries to collect plants as part of his collaborative research. He was the Principal Investigator of an NCI contract on Plant Exploration and Collection with UIC (1986–2004) and was the Principal Investigator of a multinational, multidisciplinary program called the International Cooperative Biodiversity Groups (ICBG) program of the Fogarty International Center, NIH (1998–2010). The research focus of this ICBG group was drug discovery, biodiversity conservation, and economic development in Vietnam and Laos. Under grants from the CCF (2013–2014; 2019–2023) he carried out research on medicinal plants of Laos and on medicinal plant conservation. He is the author and co-author of more than 200 peer reviewed papers with a focus on plant taxonomy, medicinal plant studies, plant-derived sweetening agents, drug discovery from plants, and conservation of medicinal plants; 22 book chapters; and 26 symposium proceedings.

Dr. Kongmany Sydara is a former Director General of the Institute of Traditional Medicine (ITM). He has been an invited lecturer at the University of Health Sciences, Vientiane since 2006. From 2014 to 2020 he served as the Director General of the ITM, Ministry of Health, Vientiane, Lao PDR. Before that, he served as the Deputy Director of the Traditional Medicine Research Center (TMRC), Ministry of Health, Vientiane, Laos, from 1991 to 2014. He completed his Bachelor of Science degree and his Master of Science degree in Biochemical Engineering at the Technical University of Budapest, Hungary. Aside from administrative duties, he has been active in carrying out research on medicinal plants since 1984. As part of the advancement of his work and scientific research, he has also received additional training opportunities in the Republic of Korea, Hungary, Japan, and the United States. Professor Sydara is the author and co-author of many books and peer-reviewed papers focusing on the uses and conservation of medicinal plants of Laos in both English and Lao languages.

Dr. Bethany Gwen Elkington is a Research Assistant Professor at the Department of Pharmaceutical Sciences, UIC, and a scientific/research affiliate of the Pharmacognosy Institute. She has studied the medicinal plants of Laos since 2005. She earned her Ph.D. degree in Pharmacognosy-Medical Ethnobotany from UIC in 2013, with her dissertation titled "Herbal Treatments for Tuberculosis in Laos: Ethnobotany and Pharmacognosy Studies." She was a Fulbright Fellow in Laos (2008–2009), and an NIH National Center for Complementary and Alternative Medicine (NCCAM) grantee (2010–2013). She also holds a Bachelor of Science degree in Biology from Purdue University, West Lafayette, Indiana, USA. She is the author and co-author of more than 20 peer-reviewed papers, books and book chapters, and plant-identification field guides on medicinal plants of Southeast Asia and West Africa. Dr. Elkington has been a scientific/research affiliate of the Field Museum of Natural History as a student and professional researcher since 2005. She is also a biology instructor at City Colleges of Chicago – Malcolm X College, Chicago, Illinois, USA.

Contributors

PRIMARY CONTRIBUTORS

Bethany Gwen Elkington
Research Assistant Professor,
   University of Illinois at Chicago
   (UIC)
Chicago, USA
and
Research Affiliate, Field Museum of
   Natural History
Chicago, USA

Djaja Djendoel Soejarto
Professor Emeritus, University of Illinois
   at Chicago (UIC)
Chicago, USA

and
Adjunct Curator, Field Museum of
   Natural History
Chicago, USA

Kongmany Sydara
Former Director General,
   Institute of Traditional Medicine,
   Ministry of Health
Vientiane, Laos

CHAPTER CONTRIBUTORS

Jonathan Bisson
Adjunct Research Assistant Professor,
   University of Illinois at Chicago
   (UIC)
Chicago, USA

Chun-Tao Che
Professor, University of Illinois at
   Chicago (UIC)
Chicago, USA

Bounleuane Douangdeuane
Director General, Institute of
   Traditional Medicine (ITM),
   Ministry of Health
Vientiane, Laos

Bethany Gwen Elkington
Research Assistant Professor, University
   of Illinois at Chicago (UIC)
Chicago, USA
and
Research Affiliate, Field Museum of
   Natural History
Chicago, USA

James G. Graham
Research Assistant Professor, University
   of Illinois at Chicago (UIC)
Chicago, USA
Research Affiliate, Field Museum of
   Natural History
Chicago, USA

Charlotte Gyllenhaal
Adjunct Research Assistant Professor,
  University of Illinois at Chicago
  (UIC)
Chicago, USA

Joshua Matthew Henkin
University of Illinois at Chicago,
  Chicago, and Research Affiliate,
  Field Museum of Natural History
Chicago, USA

A. Douglas Kinghorn
Distinguished University Professor,
  The Ohio State University (OSU)
Columbus, Ohio, USA

Guido F. Pauli
Distinguished University Professor,
  University of Illinois at Chicago
  (UIC)
Chicago, USA

Yulin Ren
Research Professor, The Ohio State
  University (OSU)
Columbus, Ohio, USA

Mary Riley
University of Illinois at Chicago (UIC)
Chicago, USA

Djaja Djendoel Soejarto
Professor Emeritus, University of
  Illinois at Chicago (UIC)
Chicago, USA
and
Adjunct Curator, Field Museum of
  Natural History
Chicago, USA

Onevilay Souliya
Deputy Director General,
  Institute of Traditional Medicine,
  Ministry of Health
Vientiane, Laos

Kongmany Sydara
Former Director General,
  Institute of Traditional Medicine,
  Ministry of Health, Vientiane,
  Laos
and
Invited Lecturer, Faculty of Pharmacy,
  University of Health Sciences
Vientiane, Laos

Mouachanh Xayvue
Head of Pharmacognosy Division,
  Institute of Traditional Medicine,
  Ministry of Health
Vientiane, Laos

1 Introduction

The Country of Laos and Research Leading to This Book

Djaja Djendoel Soejarto
University of Illinois at Chicago, Chicago, IL, USA
Field Museum of Natural History, Chicago, IL, USA

Kongmany Sydara
Institute of Traditional Medicine, Ministry of Health, Vientiane, Laos
University of Health Sciences, Vientiane, Laos

Bethany Gwen Elkington
University of Illinois at Chicago, Chicago, IL, USA
Field Museum of Natural History, Chicago, IL, USA

Mary Riley
University of Illinois at Chicago, Chicago, IL, USA

Onevilay Souliya, Mouachanh Xayvue, and Bounleuane Douangdeuane
Institute of Traditional Medicine, Ministry of Health, Vientiane, Laos

CONTENTS

DOI: 10.1201/9781003216636-1

1.1   INTRODUCTION

Laos occupies a vital place in Southeast Asia, as well as globally, because of its unique position geographically, culturally, politically, and especially ecologically. Due to various factors, including its geographic position and history, Laos boasts many relatively healthy and intact forested areas. Within Laos, the geological and climate variations contribute to a wide variety of forest types. Likewise, it holds an incredibly diverse population of people coming from numerous ethnic backgrounds, languages, and belief systems. Consequently, there is an immeasurable wealth of cultural perspectives on and knowledge about the natural environment, including the use of plants for medicine.

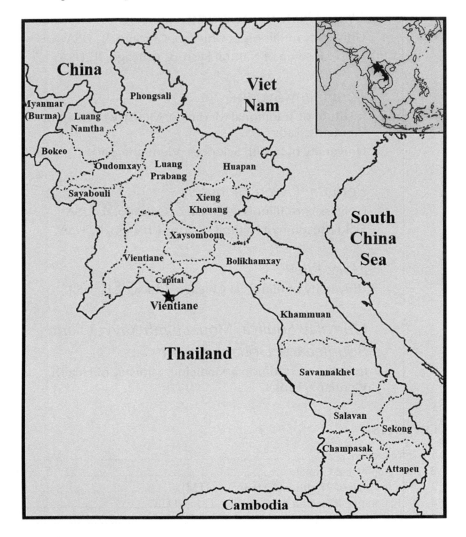

FIGURE 1.1   Map of provinces in Laos. The Vientiane capital city is marked with a star.

Flanked by Thailand to the west, Myanmar (Burma) to the northwest, China to the north, Vietnam to the east and Cambodia to the south, Laos is a small country in the heart of Southeast Asia. It has a surface area of 236,800 square kilometers (91,429 square miles), of which 70–80% of the land is covered by forest (Inoguchi and Bounthapandid 2021), though recent changes in this estimate are unclear (FAO 2020). Its population, estimated to be approximately 7.6 million (United Nations Statistics Division 2021; Kemp 2021), is often subdivided according to the altitude at which people live (Lowland, Midland, and Upland Laotians) (WPR 2021). Approximately 60% currently reside in rural areas, though this has been declining since the 1960s (World Bank Group 2021).

The country of Laos is officially recognized as "Lao People's Democratic Republic" or "Lao PDR," a name that reflects its socialist republic status. Laos is divided into 18 provinces: Attapeu, Bokeo, Bolikhamxay, Champasak, Huapan, Khammuan, Luang Namtha, Luang Prabang, Oudomxay, Phongsali, Salavan, Savannakhet, Sayabouli, Sekong, Vientiane (pronounced Wieng Chan), Vientiane Capital, Xaysomboun, and Xieng Khouang (Wikipedia 2022). Due to a multitude of transliteration systems, many of the province names (Figure 1.1), along with countless other words in the Lao language, have multiple accepted spellings with the Roman script (Lewis 2009). In Figure 1.1, the capital city of Vientiane (Wieng Chan) is marked with a star.

According to the Human Development Indicators from the United Nations, as of the writing of this book, Laos ranks 140 out of 191 countries, with nearly 25% of the population living in poverty (United Nations Development Programme 2020). The proportion of people living in rural vs. urban areas is higher in Laos than most other countries. Most of the population lives in or near forested areas and relatively far from health care centers using Western-based medical approaches. This makes it most convenient and cost-effective for people to go to the forest when medicines are needed.

1.2 GEOGRAPHY AND CLIMATE

Laos lies in a tropical monsoon region with a rainy season from May to November and a dry season from December to April. It is frequently subject to floods and droughts. The terrain includes many rugged mountains, and the tallest, Phou Bia, is more than 2,800 meters above sea level. In 2018, it was estimated that between 6% and 10% of the land was used for growing crops (The World Bank 2018; The World Factbook 2018).

While Laos is surrounded by five other countries, it remains isolated due to its mountainous geography, which has historically made transportation difficult. The Annamite Mountains run along the eastern border between Laos and Vietnam. In the northwest, the Luang Prabang mountain range creates obstacles for road travel to and from China, Thailand, and Myanmar. In addition, the Mekong River runs along the western border, separating much of Laos from Thailand and Myanmar.

The Mekong River is vital to Laos, and has been important for travel, fishing, and farming for centuries (Viravong 1964). Running most of the length of Laos' western border with Thailand, the Mekong River is a freshwater system that has supported a high biodiversity of aquatic life in the past (Máiz-Tomé 2019), but its ability to support aquatic life has decreased in recent years (RFA's Lao Service 2022). The word "Mekong" actually comes from the name "maenam kong" in Thailand and Laos, meaning "Khong River" (Campbell 2009, page 1). It originates on the Tibetan Plateau and runs through China, Myanmar, Laos, Thailand, Cambodia, and Vietnam, before emptying into the South China Sea. It is the earth's tenth longest river, with an estimated length of almost 5,000 km, releasing around 475 km^3 of water each year (MRC 2022).

1.3 THE INSTITUTE OF TRADITIONAL MEDICINE (ITM)

The majority of the people of Laos use and rely on traditional medicines, primarily in the form of medicinal plants. Due in part to the large percentage of the population that relies on traditional medicines, one of the Government of Laos'primary tasks is to safeguard public health care services by ensuring proper medicinal plant production and use of these traditional resources. To enforce these tasks, on April 27, 1976, the Ministry of Health established the Institute of Traditional Medicine (ITM). The ITM has made great strides in gathering and cataloging traditional medical treatments—the greater part of which consists of herbal remedies—and with publishing results of their research for use by contemporary healers and the wider public (see Chapter 2). In addition, it adapts *materia medica* for wider use by producing traditional medicine preparations and products for the population in its laboratory and manufacturing facilities. The ITM is a primary player in training future health care students, especially pharmacy students, to be knowledgeable about herbal medicines (Elliott 2021).

As part of its function and directive, the ITM has established smaller traditional medicine centers, referred to here as provincial Traditional Medicine Stations (TMS). Each TMS is affiliated with a traditional medicine hospital located in the capital of the province. The staff of a TMS works in direct contact with rural traditional healers. As a result, the traditional healers are then able to both give and receive information on treatment methods currently in use around the country.

Experts at the ITM have been trained in plant collection and identification (taxonomy), and its herbarium houses more than 13,000 plant collections, most of which are medicinal plants (Sydara et al. 2014). The Institute has a facility to produce herbal products, as well as a phytochemistry laboratory equipped to extract plant samples. It also holds a collection of medical palm leaf manuscripts, which were used in the past by Buddhist monks to keep records of traditional medicines (see Chapter 2).

1.4   COLLABORATIVE RESEARCH BETWEEN THE ITM AND UIC

The ITM has collaborated with UIC in various research efforts on medicinal plants since 1996. This culminated in the implementation of a multidisciplinary and multi-institutional project, the International Cooperative Biodiversity Groups (ICBG) program with funding from the Fogarty International Center of the National Institute of Health (NIH) in the United States (US) (Soejarto et al. 1999, 2002, 2006, 2007, 2009; Gyllenhaal et al. 2012; Zhang et al. 2016) (see Chapters 3, 4, 5, 7). The ICBG Program had multiple goals that included: (1) discovery of lead compounds as candidates for pharmaceutical development for therapies against tuberculosis, malaria, HIV-AIDS, and cancer; (2) conservation of biodiversity, with specific focus on seed plants and medicinal plants; (3) economic development to strengthen institutional infrastructure and community support and improvement. Chapter 3 describes the ICBG's drug discovery results; Chapters 4 and 5 describe the processes and approaches the ICBG used to promote conservation efforts in Laos; and Chapter 7 describes the ICBG's contributions to institutional strengthening, capacity building, and the promotion of economic development within participating communities.

After termination of the ICBG Program in 2010, research collaboration between the ITM and UIC has continued. One project has a goal of discovering anticancer lead compounds as part of a multi-year project based at Ohio State University (OSU). Funding has been through the US National Cancer Institute (NCI), NIH, Bethesda, MD, USA (Henkin et al. 2017; Henkin 2019; Ren et al. 2021) (see Chapter 3). This research cooperation, known as the P01 Anticancer Drug Discovery Project, is ongoing as of the writing of this book.

Research collaboration between the ITM and UIC for 2013–2014 also continued through funding from the California Community Foundation (CCF). This project is titled "Medicinal Plants of Laos: Discovery, Conservation and Community Engagement," and its mission was "… to further the study of medicinal/pharmacological use of wildlife products, particularly in relation to local communities." This funding was re-awarded for 2019–2020, but its period of operation has been extended through 2023 due to the Covid-19 pandemic inhibiting field research.

1.5   ABOUT THIS BOOK

As a result of the long-standing collaboration between the ITM and UIC, a considerable amount of data has been amassed, much of which appeared in publications throughout 2000–2021, as mentioned above. Except for some recent papers (Sydara et al. 2014; Henkin et al. 2017; Henkin 2019; Ren et al. 2021), the major portion of data from the CCF grants has not been published. There have been and currently are also many highly valuable contributions to similar ethnobotanical and medical anthropology studies in Laos, which the authors acknowledge throughout the book. In an effort to remain focused, the main purpose of this work is to provide an update on the results of our most recent collaborative research effort between the ITM and UIC.

REFERENCES

Campbell, I.C. 2009. "Introduction." *The Mekong: Biophysical Environment of an International River Basin*, edited by I.C. Campbell. New York: Elsevier. 464 pages.

Elliott, E.M. 2021. *Potent Plants, Cool Hearts: A Landscape of Healing in Laos* (PhD thesis, University College London): University College London. Accessed on January 8, 2023. https://discovery.ucl.ac.uk/id/eprint/10126896/.

FAO 2020. "FRA Platform." Food and Agriculture Organization of the United Nations – Global Forest Resources Assessment. Accessed on January 8, 2023. https://fra-data.fao.org/LAO/fra2020/home/.

Gyllenhaal, C., M.R. Kadushin, B. Southavong, K. Sydara, S. Bouamanivong, M. Xayveu, L.T. Xuan, et al. 2012. "Ethnobotanical approach versus random approach in the search for new bioactive compounds: Support of a hypothesis." *Pharmaceutical Biology* 50 (1): 30–41. doi:10.3109/13880209.2011.634424.

Henkin, J.M. 2019. *Medical Ethnobotany And The Search For New Anticancer Agents From Plants Of Laos And Vietnam*. PhD Dissertation. Chicago, IL: University of Illinois at Chicago.

Henkin, J.M., K. Sydara, M. Xayvue, O. Souliya, A.D. Kinghorn, J.E. Burdette, W.-L. Chen, B.G. Elkington, D.D. Soejarto. 2017. "Revisiting the linkage between ethnomedical use and development of new medicines: A novel plant collection strategy towards the discovery of anticancer agents." *Journal of Medicinal Plants Research Research* 11 (40): 621–34. doi:10.5897/jmpr2017.6485.

Inoguchi, A. and S. Bounthapandit. 2021. "What is the Forest Area of Lao People's Democratic Republic?" Vientiane. Accessed on January 8, 2023. https://www.fao.org/3/cb3000en/cb3000en.pdf.

Kemp, S. 2021. "Digital 2021: Laos." *All the Statistics You Need in 2021 – DataReportal – Global Digital Insights*. Accessed on January 8, 2023. https://datareportal.com/reports/digital-2021-laos.

Lewis, P.M. 2009. *Ethnologue: Languages of the World*. Dallas, TX: SIL International.

Máiz-Tomé, L., 2019. "Freshwater Key Biodiversity Areas in the Lower Mekong River Basin – IUCN." Accessed on January 8, 2023. https://docslib.org/doc/13115613/freshwater-key-biodiversity-areas-in-the-lower-mekong-river-basin-informing-species-conservation-and-investment-planning-in-freshwater-ecosystems-june-2019

MRC. 2022. "Mekong Basin: Geography." *Mekong River Commission For Sustainable Development*. Vientiane: Mekong River Commission. Accessed on January 8, 2023. https://www.mrcmekong.org/about/mekong-basin/geography/.

Ren, Y., B.G. Elkington, J.M. Henkin, K. Sydara, A.D. Kinghorn, D.D. Soejarto. 2021. "Bioactive small-molecule constituents of Lao plants." *Journal of Medicinal Plants Research* 15 (12): 540–59. doi:10.5897/JMPR2021.7137.

RFA's Lao Service. 2022. "Worries in Laos and Thailand as Upstream Dams Drain Mekong River — Radio Free Asia." Radio Free Asia. Accessed on January 8, 2023. https://www.rfa.org/english/news/laos/mekong-water-01212022141419.html.

Soejarto, D.D., C. Gyllenhaal, J.A. Tarzian Sorensen, H.S. Fong, L.T. Xuan, L.T. Binh, N.V. Hung, et al. 2007. "Bioprospecting arrangements: Cooperation between the north and the south." In: Proceedings: *The 2nd International Conference on Medicinal Mushroom and the International Conference on Biodiversity and Bioactive Compounds*, edited by Biotec, NSTDA, and Ministry of Science and Technology, pp. 5–22. Pattaya.

Soejarto, D.D., C. Gyllenhaal, J.C. Regalado, J.M. Pezzuto, H.S. Fong, G.T. Tan, N.T. Hiep, et al. 2002. "An international collaborative program to discover new drugs from tropical biodiversity of Vietnam and Laos." *Natural Product Sciences* 8: 1–15.

Soejarto, D.D., C. Gyllenhaal, J. Regalado, J. Pezzuto, H. Fong, G.T. Tan, N.T. Hiep, et al. 1999. "Studies on biodiversity of Vietnam and Laos: The UIC-based ICBG program." *Pharmaceutical Biology* 37 (4): 100–13.

Soejarto, D.D., B. Southavong, K. Sydara, S. Bouamanivong, M.C. Riley, A.S. Libman, M.R. Kadushin, C. Gyllenhaal. 2009. "Studies on medicinal plants of Laos – A collaborative program between the University of Illinois at Chicago (UIC) and the Traditional Medicine Research Center (TMRC), Lao PDR: Accomplishments 1998–2005." In: *Contemporary Lao Studies: Research on Development, Culture, Language, and Traditional Medicine*, edited by C.J. Compton, J.F. Hartmann, V. Sysamouth, pp. 307–23. San Fransisco, De Kalb: Center for Lao Studies and the Center for Southeast Asian Studies, Northern Illinois University.

Soejarto, D.D., H.J. Zhang, H.H. Fong, G.T. Tan, C.Y. Ma, C. Gyllenhaal, M.C. Riley, et al. 2006. "'Studies on biodiversity of Vietnam and Laos' 1998–2005: Examining the impact." *Journal of Natural Products* 69 (3): 473–81. doi:10.1021/np058107t.

Sydara, K., M. Xayvue, O. Souliya, B.G. Elkington, D.D. Soejarto. 2014. "Inventory of medicinal plants of the Lao peoples democratic republic: A mini review." *Journal of Medicinal Plants Research* 8 (43): 1262–74. doi:10.5897/JMPR2014.5534.

United Nations Development Programme 2020. "Human Development Insights." *Human Development Reports.* Accessed on January 8, 2023. http://hdr.undp.org/en/countries/profiles/LAO.

United Nations Statistics Division 2021. "UNData: Lao People's Democratic Republic." UNData: A World of Information. Accessed on January 8, 2023. http://data.un.org/en/iso/la.html.

Viravong, M.S. 1964. *History of Laos*. New York: Paragon Book Reprint Corp.

Wikipedia 2022. "Laos." Accessed on January 8, 2023. https://en.wikipedia.org/wiki/Laos.

World Bank Group 2021. "Lao PDR." The World Bank – Data: Rural Popluation (% of Total Population). Accessed on January 8, 2023. https://data.worldbank.org/indicator/SP.RUR.TOTL.ZS?end=2019&locations=LA&start=1960&view=chart.

The World Bank 2018. "Arable Land (% of Land Area) – Data." Arable Land (% of Land Area). Accessed on January 8, 2023. https://data.worldbank.org/indicator/AG.LND.ARBL.ZS?view=map.

The World Factbook 2018. "Land Use." Field Listing – Land Use. Accessed on January 8, 2023. https://www.cia.gov/the-world-factbook/field/land-use/.

WPR 2021. "Laos Population 2021 (Demographics, Maps, Graphs)." World Population Review. Accessed on January 8, 2023. https://worldpopulationreview.com/countries/laos-population.

Zhang, H.-J., W.-F. Li, H.H.S. Fong, D.D. Soejarto. 2016. "Discovery of bioactive compounds by the UIC-ICBG drug discovery program in the 18 years since 1998." *Molecules* 21 (11): 1448. doi:10.3390/molecules21111448.

2 Role of Medicinal Plants in The Lao People's Democratic Republic

Kongmany Sydara
Institute of Traditional Medicine, Ministry of Health,
Vientiane, Laos
University of Health Sciences, Vientiane, Laos

*Onevilay Souliya, Mouachanh Xayvue, and
Bounleuane Douangdeuane*
Institute of Traditional Medicine, Ministry of Health,
Vientiane, Laos

Bethany Gwen Elkington
University of Illinois at Chicago, Chicago, IL, USA
Field Museum of Natural History, Chicago, IL, USA

Mary Riley
University of Illinois at Chicago, Chicago, IL, USA

Djaja Djendoel Soejarto
University of Illinois at Chicago, Chicago, IL, USA
Field Museum of Natural History, Chicago, IL, USA

CONTENTS

DOI: 10.1201/9781003216636-2

9

2.1   INTRODUCTION

Laos, the only landlocked country in Southeast Asia, has a total surface area of 236,800 square kilometers (91,429 square miles), of which approximately 80% is covered by forests (Inoguchi and Bounthapandid 2021). Its population is estimated at around 7.6 million, with 49.8% reported as female and 36.6% male (Kemp 2021), and who are often subdivided according to the altitude at which they live (Lowland, Midland, and Upland Laotians) (WPR 2021).

The people of this region have had a close relationship with agriculture, forestry, and hunting for at least 6,000 years, since wet-rice and millet farming techniques were introduced from the Yangtze River valley of southern China (Higham 2013). Even today, particularly in rural and remote areas, the population turns primarily to the forest for their many needs, including dwelling, construction materials, food, medicine, and fuel. The rural population still has limited access to modern health care facilities, thus relying on the use of medicinal plants and traditional medicines for the prevention and treatment of diseases (MoNRE and IUCN 2016).

Laos is one of the few countries in the world to boast university-level training about traditional medicines (WHO 2019; Elliott 2021). Oversight of herbal medicine has transferred through various government entities, with research today taking place at the Institute of Traditional Medicine (ITM). There is a threat to medicinal plants as deforestation increases (Hughes 2017). The medical potential yet to be discovered from plants underscores the need for laws governing conservation and the sustainable use of undeveloped forested areas.

Even though the Lao PDR has undergone rapid economic progress in recent decades (The World Bank 2011), people living in rural areas today still have less access to modern health care facilities than urban populations. Therefore, they rely on medicinal plants for the prevention and treatment of diseases. In past decades, medicinal plants served as a primary component of health care, which is documented in old palm leaf manuscripts kept in Buddhist temples in various pagodas, as well as in the National Library of Laos (Elkington et al. 2013). The use of medicinal plants in past eras reflects the community-based knowledge of traditional medicine by the people of Laos.

2.2 HISTORY OF THE USE OF MEDICINAL PLANTS AND TRADITIONAL MEDICINE IN THE LAO PDR

Around 4,000 years ago, the Lao nation was one of the four biggest nations in Asia (Higham 2013). Agriculture was the main livelihood for most of the population, who had a close relationship with nature (Yan et al. 2014). The use of medicinal plants and traditional medicines has a long history in the region (Pottier 2007), and traditional medical knowledge about the use of medicinal plants has been passed down from one generation to the next and preserved by many healers today (Elkington et al. 2013). At present, the Government of Laos acknowledges the importance of medicinal plants in primary health care. An accounting of the historical uses of medicinal plants in different eras is presented below.

2.2.1 TRADITIONAL MEDICINE DURING THE KINGDOM OF LAN XANG

The Kingdom of Lan Xang was established by the first King of the nation, Chao Fa Ngum, in the 14th century (Evans 2002, 9–10; Lorrillard 2006; Stuart-Fox 1997; Viravong 1964). At that time, pagodas were constructed and seen in many communities (meuang). The use of medicinal plants and traditional medicine was practiced in various districts and records kept. These districts are the equivalent of what are currently referred to as "provinces."

In the Laotian traditional medicine system, three major categories of healers are recognized: (1) healers that treat patients physically; (2) healers that treat patients based on ritual ceremonies; and (3) healers that treat patients psychologically. A person is healthy when the four elements (earth, water, wind, fire) in the body are in balance.

The transfer of knowledge from generation to generation has been through different methods, including orally and in the form of documents recorded in various scripts and on diverse materials, such as silver or gold sheets, tree bark or leaves, animal skin, and/or stone. The languages used in these documents include the "old Lao" languages, including Tham ("Dhamma"), Leu, and Khom (Khmer). Since these languages became difficult to understand over time, old texts have been translated into the contemporary Lao language, and also transcribed onto palm leaves, paper made from *Streblus asper* Lour. (Moraceae; common names: sandpaper tree, Siamese rough bush) and/or paper made from *Broussonetia papyrifera* (L.) Ventenat (Moraceae; common name: paper mulberry) (Elkington et al. 2013).

Previous to the Lan Xang Kingdom, when the Lao people lived near the Yangtze River, a note about soldiers mentioned that soldiers wounded on the front of the body received medical treatment and care (Viravong 1964). One may infer from this that treatment of the injured soldier was performed using traditional medicines.

During the Muang Sua Era (698–1353), natural disasters and wars resulted in sparse and vague written records. Under the reign of Chao Fa Ngum (1316–1393), society became more developed and organized. Knowledge—including information about traditional medicines—was collected, recorded in written form, and the records kept in pagodas. From this, we know there were royal healers to serve various kings' families. These records also serve as evidence of the country's ancient knowledge of traditional medicine (Viravong 1964).

In approximately 1350 AD, Chao Fa Kham Hiew, the uncle of Chao Fa Ngum, committed suicide by using *Gelsemium elegans* (Gardner & Chapm.) Benth. (Gelsemiaceae; common name: heartbreak grass) to redeem his shame because he lost the war to his nephew, demonstrating another use for a medicinal plant (Viravong 1964). During the 15th and the 16th centuries there were many pagodas in the Kingdom. In the 17th century, the mission led by Father Manili recorded information about Lao society at that time. This record states that "... they have also other weapons, which is the poison, that they sometimes used. They have perfect knowledge how to prepare this kind of weapon...." (Viravong 1964).

The national heritage (including documents containing information on traditional medicines) was destroyed in Vientiane several times: (1) In 1778, the monarchy of Siam (Thailand's former name) destroyed Vientiane by burning it for the first time; (2) Siamese armies burned Vientiane for the second time in 1828 following Anouvong's war for independence; (3) In 1882, Siamese troops used seven elephants to carry the important palm leaf manuscripts from Vientiane to Bangkok, and more than twenty elephants carried other important documents to Yasothone (now in Thailand) (Viravong 1964).

Another major event that destroyed the national heritage was the fire that burned Luang Prabang during the Lao New Year in 1774. In 1887, the wars of the Black and Red Flag Hor (Chinese minorities) destroyed Luang Prabang, which included the destruction of important pagodas and palm leaf manuscripts. These events are significant as they represent a massive loss of important information about the Lao nation (Viravong 1964).

Anthropologist Elizabeth Elliott (2021) provides an overview of the history of traditional medicine in Laos, noting that written accounts are relatively sparse prior to the French colonial period. One important distinction Elliott makes between the development of medicinal traditions in Laos, as compared to the surrounding areas, is that Laotian traditional medicinal knowledge did not evolve into a formal, standardized national system of medicine such as Traditional Chinese Medicine, Traditional Thai Medicine, or Ayurvedic Medicine. While traditional medicine does have healers and specialists, traditional medicinal knowledge is folk medicine and more heterogeneous in form, in contrast to formal systems of medicine that have centralized knowledge sources and require greater degrees of specialization and training (Elliott 2021).

2.2.2 TRADITIONAL MEDICINE DURING THE KINGDOM OF LAOS (BEFORE 1975)

After 1893, the Kingdom of Laos was under the protectorate of France. In 1900, there was only one hospital in Vientiane, which served the French community, monarchial families, and high-ranking Lao officers. The majority of the Lao people had to treat themselves by using spiritual means and traditional medicine—the old heritage of the country. Several French language historical resources included accounts of traditional medicine and medicinal plant knowledge during the colonial period (1893–1953), and additional published accounts provide information on the status of traditional medicine in the following decades (Elliott 2021).

2.2.3 TRADITIONAL MEDICINE IN THE CURRENT LAO PDR (AFTER 2 DECEMBER 1975)

Traditional medicine today, especially in the form of medicinal plants, forms part of the backbone of health care (Libman et al. 2008). Medicinal plants have been used since time immemorial and continue to be used today in various forms: raw material, extracts, and formulated medicines. They are offered for sale by roadside vendors, in every marketplace (Figure 2.1), and in pharmacies throughout the country (Figure 2.2).

FIGURE 2.1 A traditional medicine stand with a buyer in a marketplace in Vientiane. Photo credit: D. Soejarto.

FIGURE 2.2 Traditional medicine products and modern pharmaceutical products in a contemporary pharmacy in Vientiane. Photo credit: D. Soejarto.

One reason for the heavy use of medicinal plants is the limited access to Western health care systems for most of the population. A 2005 survey of 600 households from lowland and mountainous districts found that 77% of families interviewed reported using traditional medicine in their day-to-day health maintenance practices (Sydara et al. 2005). Subsequent research showed similar findings among 1600 participants surveyed in four rural and four urban areas (Peltzer et al. 2016). The heavy use of medicinal plants as part of traditional medicine practices is in line with findings of the World Health Organization (WHO). In 2019, the WHO reported 88% of its Member States acknowledged use of traditional and complementary medicine (WHO 2019), with the greater part consisting of medicinal plants.

The Government of Laos recognizes the significance of medicinal plants and traditional medicine and acknowledges that traditional medicine is an important component of the health care sector. The government promotes the integration of traditional medicine with conventional medicine (Zhang 2018). As evidence of this effort and policy, the Ministry of Health established the Institute of Traditional Medicine on the 27th of April, 1976 (about four months after the establishment of the Lao PDR). In 1990, the name was changed to the Research Institute of Medicinal Plants (RIMP), which was then renamed as the Traditional Medicine Research Center (TMRC) in 1999, and then again renamed as the Institute of Traditional Medicine (ITM) in 2010. The mission of the ITM is to carry out research to adapt traditional medicinal uses of *materia medica* for wider use by the population, including through the production of traditional medicine preparations and products in its laboratory and manufacturing facilities.

2.3 STATUS OF MEDICINAL PLANTS IN THE PRESENT LAO PDR

2.3.1 USES OF MEDICINAL PLANTS

Many people in Laos prefer modern medicine for treating their health conditions (Lundberg 2008). However, many people in rural and urban communities continue to rely on medicinal plants from the wild, from cultivated sources, from individuals, familial factories, as well as from domestic and international commercial companies.

In addition to harvesting from the wild, some domestic and foreign companies also cultivate select medicinal plant species (Table 2.1). It should be noted that some species listed in Table 2.1 are also used as health foods. Some foreign companies focus their operation on the processing of select medicinal and aromatic plants, such as *Dracaena cambodiana* Pierre ex Gagnep. (Asparagaceae), *Cinnamomum cassia* (L.) J. Presl (Lauraceae), *Coscinium fenestratum* (Gaertn.) Colebr. (Menispermaceae), *Zingiber officinale* Roscoe (Zingiberaceae), and *Codonopsis pilosula* Franch. (Campanulaceae). Since 1975, government institutions have carried out surveys on medicinal plants in various districts throughout the country in collaboration with foreign institutions. This chapter focuses specifically on plants noted through the ITM–UIC (University of Illinois at Chicago) collaboration.

A summary of examples of the medicinal importance of each plant species listed in Table 2.1 (above) is provided below.

(Acanthaceae) *Andrographis paniculata* (La Sa Bii or Sam Pan Bii). Also known as "King of bitters," this plant is native to Assam, Bangladesh, India, and Sri Lanka, and has been introduced and become naturalized in many other tropical countries, including Laos (Kew 2022a). It has been used traditionally for centuries in the Asian, American, and African continents to treat many diseases such as cancer, diabetes, high blood pressure, ulcers, leprosy, bronchitis, skin diseases, flatulence, colic, influenza, dysentery, dyspepsia, and malaria (Okhuarobo et al. 2014; Najib Nik A Rahman et al. 1999). Diterpenes, flavonoids, xanthones, noriridoides, and other miscellaneous compounds have been isolated from the plant. Both the extract and pure compounds have been reported to have biological activities, including antimicrobial, cytotoxicity, anti-protozoan, anti-inflammatory, antioxidant, immunostimulant, antidiabetic, anti-infective, antiangiogenic, hepato-renal protection, sex hormone/sexual function modulation, liver enzyme modulation, insecticidal, and toxicity (Okhuarobo et al. 2014).

(Apiaceae) *Centella asiatica* (Pak Nohk). This plant is also known as "gotu kola," and is a traditional Chinese medicine with extensive medicinal value. It is commonly used in Southeast Asian countries. The plant's native range is Caucasus, Tropical & Subtropical Old World to New Zealand and the W. Pacific (Kew 2022b). Its main effective components are a group of triterpenoids, including

TABLE 2.1
A List of Selected Species of Cultivated Medicinal Plants[a]

Common Name in Laos	Scientific Name	Medicinal Use (Disease Treated)	Part Used
La Sa Bii, Sam Pan Bii	Acanthaceae *Andrographis paniculata* (Burm.f.) Nees	Tonic, antiviral, fever	Herb
Pak Nohk	Apiaceae *Centella asiatica* (L.) Urb.	High fever, cough	Whole plant
Van Hang Khae	Asphodelaceae *Aloe vera* (L.) Burm.f.	Minor burns	Leaf gel
Sohm Dark Cheung	Campanulaceae *Codonopsis pilosula* Franch.	Tonic	Rhizome
Pak Bua Leuat	Iridaceae *Eleutherine subaphylla* Gagnep.	Stomach ache, bleeding	Bulb (rhizome)
Nya Nouat Meo	Lamiaceae *Orthosiphon stamineus* Benth.	Kidney stone, diuresis, diabetes	Leaf
Khae Hohm	Lauraceae *Cinnamomum cassia* (L.) J. Presl	Tonic, heart tonic, flatulence	Stem bark
Mak Chong Ban	Malvaceae *Sterculia lychnophora* Hance	Cold medicine, heat inside the body	Fruit
Mohn	Moraceae *Morus alba* L.	Fever, chronic cough	Leaf
Mak Deuy	Poaceae *Coix lacrymal-jobi* L.	Warts, cancers	Fruit
Khi Hout	Rutaceae *Citrus hystrix* DC.	Indigestion, toothache	Leaf (leaf oil)
Mak Khen	Rutaceae *Zanthoxylum nitidum* (Roxb.) DC.	Stomach ache, toothache	Root
Nhane	Styracaceae *Styrax tonkinensis* (Pierre) Craib ex Hartwich	Cough	Resin
Kha	Zingiberaceae *Alpinia galanga* (L.) Willd.	Stomach ache, diarrhea	Rhizome
Mak Neng	Zingiberaceae *Amomum ovoideum* Pierre ex Gagn.	Tonic	Seed
Van Hoa Deo	Zingiberaceae *Curcuma xanthorrhiza* Roxb.	Leucorrhea, uterine pain	Rhizome
Khing Dam	Zingiberaceae *Kaempferia parviflora* Wall. ex Baker	Tonic	Rhizome
Van Pai	Zingiberaceae *Zingiber cassumunar* Roxb.	Flatulence	Rhizome
Khing	Zingiberaceae *Zingiber officinale* Roscoe	Cough, sore throat	Rhizome

[a] Updated from Sydara 2007; unpublished reports at ITM.

asiaticoside, asiatic acid, madecassoside, and madecassic acid. In vivo and in vitro studies have shown that its triterpenoids exert therapeutic and relieving effects on multi-system diseases, including relief from sleep deprivation, Alzheimer disease, type 2 diabetes mellitus (T2DM), hyperlipidemia, gestational diabetes, baldness, atopic dermatitis, wounds, drug-induced liver toxicity, liver injury, gastric mucosal injury, gastric ulcers, breast cancer, leukemia, oral submucous fibrosis, and migraines (Sun et al. 2020).

(Asphodelaceae) *Aloe vera* (Van Hang Khae). Aloe is a native of the Arabian Peninsula (Oman) but has been introduced and naturalized in hot and dry areas of North Africa, the Middle East, the Southern Mediterranean, the Canary Islands, and in the subtropical belt of Asia, Australia, Africa, and South America (Kew 2022c). The aloe plant has been used in traditional medicine to treat and to heal wounds and burns, and today it has found its place in the cosmetic industry (Gupta et al. 2018; Sánchez et al. 2020). More than 75 active and health-promoting ingredients from the inner gel (containing polysaccharides) have been identified, including vitamins (A, C, E, B_{12}), minerals (zinc, copper, selenium, and calcium), enzymes (amylase, catalase, and peroxidase), sugars (monosaccharides such as mannose-6-phosphate and polysaccharides such as glucomannans), anthraquinones (aloin and emodin) and phenolic compounds, lignin, saponins, sterols, amino acids, and salicylic acid (Gupta et al. 2018; Sánchez et al. 2020). In vitro studies have shown antimicrobial, anti-inflammatory, cytotoxic, antitumor, anticancer, skin, and osteoporosis protection, while in vivo studies have been performed to evaluate cardioprotective effect, cytotoxic, antitumor and anticancer activities, and skin protection activities. It should be noted that clinical trials (using aloe, not isolated compounds) have been limited and focus on digestive and skin protective effects (Sánchez et al. 2020).

(Campanulaceae) *Codonopsis pilosula* (Sohm Dark Cheung). This plant is the subject of an ongoing conversation about medicinal plant conservation. See Chapter 4 for more details.

(Iridaceae) *Eleutherine subaphylla* (Pak Bua Leuat). This species name is a synonym of *E. bulbosa* (Mill.) Urb., and this species is a native of the West Indies to Northern South America, south to Argentina through the Andes cordillera, and has been introduced and naturalized in a number of countries in Africa, the Indian subcontinent, and Indo-China (Kew 2022e). In Brazil the plant is known as "coquinho" and is widely used in folk medicine for the treatment of giardiasis, amoebiasis, and diarrhea (Couto et al. 2016). In Borneo, the Dayak community uses the bulb to treat several diseases like diabetes, breast cancer, nasal congestion, and fertility problems (Kamarudin et al. 2021). In Laos, pills are processed using the bulb of *Eleutherine subaphylla*, *Curcuma longa*, and *Zingiber montanum* rhizomes, and it is sold for gastritis and stomach ulcers. The treatment should be taken internally, and repeatedly over time to cure the problem (Southavong, Sydara, and Souliya 2014). The bulb contains exceptionally rich phytoconstituents, especially phenolic and flavonoid derivatives, naphthalene, anthraquinone, and naphthoquinone. Studies have demonstrated various pharmacological activities of the bulb that include anticancer, antidiabetic, antibacterial, antifungal, antiviral, anti-inflammatory, dermatological problems, antioxidant, and anti-fertility (Kamarudin et al. 2021).

(Lamiaceae) *Orthosiphon stamineus* (Nya Nouat Meo). This plant name is a synonym of *O. aristatus* (Bl.) Miq. The leaves are used traditionally in the form of a tea as a diuretic to remove kidney stones and other diseases, including gout, arthritis, and inflammatory-related conditions (Tabana et al. 2016; Almatar, Ekal, and Rahmat 2014). The plant contains various phytochemicals including flavonoids, terpenoids, and essential oils. Studies have shown diuretic, hypouricemic, renal protective, antioxidant, anti-inflammatory, hepatoprotective, gastroprotective, antihypertensive, antidiabetic, antihyperlipidemic, antimicrobial, and anorexic activities (Ashraf, Sultan, and Adam 2018). A study to evaluate the effects of a 50% ethanolic extract on suppressed acute and chronic inflammation induced in rats showed that it prevented and managed rheumatoid arthritis and other chronic inflammatory disorders (Tabana et al. 2016).

(Lauraceae) *Cinnamomum cassia* (Khae Hohm). The bark of young branches is the source of cinnamon, which is used worldwide for its fragrance and spicy flavor, and also as a raw material for medical products. More than 160 chemicals have been isolated and identified, with the main constituents being phenylpropanoids and glycosides, among others. Modern studies have confirmed that *C. cassia* has a wide range of pharmacological effects, including antitumor, anti-inflammatory, analgesic, antidiabetic, anti-obesity, antibacterial, antiviral, cardiovascular protective, cytoprotective, neuroprotective, and immunoregulatory (C. Zhang et al. 2019).

(Malvaceae) *Sterculia lychnophora* (Mak Chong Ban). See Chapter 4.

(Moraceae) *Morus alba*; mulberry (Mohn). Mulberry is a plant native to China, but it has been introduced and naturalized in many countries worldwide (Kew 2022g). For centuries, the leaves have served as the primary food for silkworms. The leaves have also been used as animal feed for livestock, while the fruits have been made into a variety of food products. With flavonoids as the major constituents, the leaves possess various biological activities, including antioxidant, antimicrobial, skin-whitening, cytotoxic, antidiabetic, glucosidase inhibition, anti-hyperlipidemic, anti-atherosclerotic, anti-obesity, cardioprotective, and cognitive enhancement. The fruits, rich in anthocyanins and alkaloids, exert antioxidant, antidiabetic, anti-atherosclerotic, anti-obesity, and hepatoprotective activities (Ma et al. 2022). The root bark, rich in root flavonoids, alkaloids and stilbenoids, has antimicrobial, skin-whitening, cytotoxic, anti-inflammatory, and anti-hyperlipidemic properties. Other pharmacological properties include anti-platelet, anxiolytic, anti-asthmatic, anthelminthic, antidepressant, cardioprotective, and immunomodulatory activities (Chan et al. 2016).

(Poaceae) *Coix lacryma-jobi*; Job's tears (Mak Deuy). This plant is native to the Indian subcontinent east to China, Indo-China, the Philippines, Malay Peninsula, and Indonesia, but it has been introduced to and is currently cultivated in many countries worldwide (Kew 2022d). Coix grain is popular in Chinese traditional medicine and used in soups and beverages. The grains contain phenols, flavonoids, polysaccharides, proteins, fibers, vitamins, and oils, and is responsible for antioxidant, anti-inflammatory, and anti-obesity effects, and in stimulating reproductive hormones, promoting uterine contraction, and modulating gut microbiota (Devaraj, Jeepipalli, and Xu 2020). Other compounds that have been isolated from the Coix grains are triglycerides and beta-sitosterol (Ragasa et al. 2014). More recently, Yu et al. isolated an anticancer compound (20R)-22E-24-ethylcholesta-4,22-dien-3-one from the leaves and stem of Coix (Yu et al. 2021).

(Rutaceae) *Citrus hystrix*; Kaffir lime (Khi Hout). Kaffir lime is a native of India east to Southeast Asia and dates back in ancient literature for its use as a medicinal plant. The fruit is used to prepare juice, pickles, and as an acidulant in curries, while the fruit peel and leaves are used to treat various inflammatory ailments in Ayurvedic and other systems of medicine. Various phytoconstituents that give the plant its intense odor include citronellal, L-linalool, 1,8-cineole, α-terpeneol, and δ-cadinene, while other compounds including glycerol, glycolipids, tannins, tocopherols, furanocoumarins, flavonoids, and alkaloids account for its medicinal uses. Crude extracts and phytochemicals isolated from various parts have shown many pharmacological activities such as antibacterial, antifungal, anti-inflammatory, anticancer, antioxidant, chemopreventive, anticholinesterase, cardio, and hepatoprotective effects (Abirami, Nagarani, and Siddhuraju 2014; Agouillal et al. 2017).

(Rutaceae) *Zanthoxylum nitidum*; Shiny-leaf prickly-ash (Mak Khen). Shiny-leaf prickly-ash is a medicinal plant distributed in the Indian subcontinent east to Papua New Guinea and Australia through Indo-China, southern China, the Philippines, Malay peninsula, and Indonesia (GBIF 2022a). It is known in China as Liang-Mian-Zhen and is traditionally used to treat stomachaches, toothaches, rheumatic arthralgia, traumatic injuries, venomous snake bites, treatment of inflammations, various types of cancer, bacterial and viral infections, gastric and oral ulcers, and liver damage (Lu et al. 2020). In other Asian countries it is used to treat toothache and oral pathogens (Thailand), to prevent intoxication, and to heal respiratory illnesses (Indonesia) (Okagu et al. 2021). Chemically, more than 150 constituents have been isolated and identified from this plant, most of which are alkaloids, which account for biological activities such as anti-inflammation, analgesia, hemostasis, anticancer, and antibacterial effects determined through both in vitro and in vivo studies (Lu et al. 2020).

(Styracaceae) *Styrax tonkinensis*; Siam benzoin (Nhane). Siam benzoin is a tree species native to Southern China, Laos, Vietnam, Cambodia, and Thailand (Huang 1987, 256). It is of great economic importance as a timber source, medicine, oil, and for its ornamental value. From the trunk bark, Siam benzoin resin is collected to serve the food industry for flavoring a range of products including ice cream, baked goods, chocolate glaze, and various beverages (Wu et al. 2020). For medicine, it is used as a source of antiseptic, inhalant for cough, laryngitis, bronchitis and upper respiratory tract disorders, and to heal wounds (tinctures and balsams), as well as a source of fragrances for the perfume industry (Venkatramna Industries 2022). The seeds, a source for biofuels, are rich in carboxylic acids and derivatives, flavonoids, fatty acyls, glycerophospholipids, organooxygen compounds, prenol lipids, steroids and steroid derivatives, and is the source of biofuels (Wu et al. 2020). Studies of Siam benzoin extract have shown antiproliferative and differentiation effects in human leukemia HL-60 cells and neuroprotective effects on cerebral ischemia (against ischemic stroke) (Wang et al. 2006; Chen et al. 2020).

(Zingiberaceae) *Alpinia galanga*; Greater galangal (Kha). The rhizome of greater galangal is a plant native to Southern China and Southeast Asia (WFO 2022a). It has a spicy flavor and aromatic odor and has long been used for flavoring ingredients and spices in Asia (Zhou et al. 2018). Different parts of the plant have been used to treat asthma, gastrointestinal diseases (stomachache, dyspepsia, vomiting), fungal infection, tumors, diuresis, heart disease, rheumatic pains, fever, and diabetes (Shetty and Monisha 2015; Zhou et al. 2018). In vitro and in vivo studies have shown that greater galangal has many biological activities, including effectiveness as anti-inflammatory, antitumor, antiviral, antimicrobial, antioxidant, antiallergic, anti-proliferative, apoptotic, antiangiogenic, anti-leishmanial, and as cytotoxic agents (Kaur et al. 2010; Shetty and Monisha 2015; Chouni and Paul 2018). It contains a number of bioactive molecules including diarylheptanoids, flavonoids, volatile oils, terpenes, phenylpropanoids, glycosides, phenolic compounds, and essential oils (Zhou et al. 2018).

(Zingiberaceae) *Amomum ovoideum* (Mak Neng). This is a synonym of *Amomum uliginosum* J. Koenig (WFO 2022b). It is a native of Myanmar, Indochina, Thailand, and south to western Indonesia (GBIF 2022b). The essential oil derived from the fruit of this plant (known in Thailand under the name of "Reo-Noi") is used in Thai herbal formulas for the treatment of gastrointestinal diseases. It has demonstrated antibacterial, cytotoxic, and antimutagenic activities (Pulbutr et al. 2012). Major chemical constituents in the plant include monoterpenoids (α-terpineol, isopinocamphone, 1,8-cineole, and p-cymene) and sesquiterpenoids (limonene and camphene), α-fenchyl alcohol, α-fenchyl acetate, ß-chamigrene, globulol, γ-terpinene and terpinolene (0.6%), myrcene, pinocarvylacetate, alloaromadendrene and epiglobulol (0.5%), camphor, α-humulene, and viridiflorol. Methyl thymyl ether contributed approximately 0.9% of the total oil (Mailina et al. 2007).

(Zingiberaceae) *Curcuma xanthorrhiza*; Javanese turmeric (Van Hoa Deo). Van Hoa Deo is a valuable medicinal plant, traditionally used to treat heartburn, diarrhea, hemorrhoids, coughs, asthma, and canker sores. Phytochemically, this plant has been reported to contain curcumin, essential oils, saponins, flavonoids, alkaloids, and tannins, which are responsible for activities including antibacterial, antimicrobial, anticancer, antifungal, anti-acne, and antioxidant (Kustina et al. 2020).

(Zingiberaceae) *Kaempferia parviflora* (Khing Dam). This species is native to Bangladesh east to Myanmar, Indo-China, and Thailand (Kew 2022f). The rhizome of this plant is used in folk medicine to improve blood flow and to treat inflammatory, allergic, and gastrointestinal disorders (Yoshino et al. 2018). Studies have shown beneficial activities, such as anti-inflammatory, spasmolytic, gastric ulcer amelioration, antioxidative, and vasodilatory effects (Yoshino et al. 2018), as well as anticancer, vascular relaxation, cardioprotective, sexual enhancement, neuroprotective, antiallergic, and anti-inflammatory activities (D. Chen et al. 2018). The primary plant constituents responsible for these activities are several methoxyflavones, including 5,7-dimethoxyflavone, 5,7,4′-trimethoxyflavone, and 3,5,7,3′,4′-pentamethoxyflavone (D. Chen et al. 2018). Continual ingestion has been shown to reduce abdominal fat in healthy humans with a BMI ≥ 24 and < 30 kg/m^2 (Yoshino et al. 2018).

(Zingiberaceae) *Zingiber cassumunar*; Cassumunar ginger (Van Pai). This species name is a synonym of *Zingiber purpureum* Roscoe. It is a native of Assam, Bangladesh, East Himalaya, India, Myanmar, but has been introduced and naturalized in many tropical countries (Kew 2022h; CABI 2022). The plant has been used in traditional medicine in Southeast Asia to treat inflammation, respiratory problems, rheumatism, muscular pain, wounds, and asthma. Chemically, a series of compounds that include phenylbutenoids, curcuminoids, terpenoids, flavonoids, alkaloids, steroids, and benzenoids are responsible for the traditional uses of the plant. A number of in vitro and in vivo studies, along with clinical evaluations on the extracts, solvent fractions, and constituents of Cassumunar ginger, have described their diverse medicinal properties, including antioxidant, anti-inflammatory, anticancer, neuroprotective/neurotrophic, cosmeceutical, and antifungal/antimicrobial, analgesic, and anti-malarial bioactivities (Singh et al. 2015; Han et al. 2021).

(Zingiberaceae) *Zingiber officinale*; ginger (Khing). Khing is a plant native to Assam, India, East Himalaya, and South-Central China, but it has been introduced to many other tropical countries where it has become naturalized (Kew 2022i). It is an important medicinal plant with a long history of use for more than 2000 years, both in Ayurvedic and Chinese medicine systems. It is used to cure heart problems, menstruation disorders, food poisoning, osteoarthritis, epilepsy, nausea, inflammation, cough and cold, motion sickness, cancer, and many more ailments (Kumar, Gupta and Sharma 2014; Prasad and Tyagi 2015). Furthermore, ginger also exhibits antimicrobial, antioxidant, antiviral, radioprotective, anti-inflammatory, and anticancer properties. The presence of gingerol, paradol, shogaols, and zerumbone compounds, among others, account for the medicinal importance of ginger (Kumar Gupta and Sharma 2014; Prasad and Tyagi 2015; Dissanayake et al. 2020).

2.3.2 GOVERNMENT POLICY AND REGULATIONS ON THE USE OF MEDICINAL PLANTS

The key policies and regulations on the collection, cultivation, and production of medicinal plants are outlined in the following sections. Note that Chapter 6 also mentions some plants cited in the International Union for Conservation of Nature (IUCN) Red List and/or in the Convention on International Trade of Endangered Species (CITES) Checklist.

2.3.2.1 The Health Sector (Ministry of Health)

2.3.2.1.1 Law on Drugs and Medical Products No. 01/NA, 8 April 2000

Article 1 of this law (President of the Lao PDR 2000) defines principles, rules, and measures related to the management, cultivation, protection, production, importation, exportation, distribution, possession, and use of drugs and medical products in order to ensure the availability of high-quality, safe, and appropriately priced drugs and medical products for preventing and treating diseases and ensuring good health for the population. Article 2 reiterates the role that the government should take to promote the development of medical resources by cultivation, growth, protection, purchase, research, preparation, and production of modernized drugs and traditional medicines to be used locally, to substitute imports, and to be exported. Article 3 states that the government should widely promote the production and use of modern drugs in combination with traditional medicines.

2.3.2.1.2 Prime Ministerial Decree No. 155, 30 September 2003

The objective of this Decree (Prime Minister of Laos 2003) is to define measures related to the promotion, management, production, cultivation, and use of natural resources to protect the country's medicinal natural resources and rich biodiversity, and to ensure the sustainable use of medicinal natural resources. This decree classifies medicinal plants into three categories: *Category 1*: Plants, animals, and minerals that are rare and endangered. *Category 2*: Plants, animals, and minerals that have high commercial value and can be used for domestic consumption and for exportation. *Category 3*: Plants, animals, and minerals that are available in abundance throughout the country.

For harvesting and collection of medicinal plants in Category 1, the approval of the Ministry of Health and other competent authorities is required. For Category 2, harvesting and exploitation are also to be certified by the Ministry of Health and concerned authorities by providing a management plan for harvesting and replanting. For Category 3, the exploitation of plants is not restricted since they are abundantly available in nature. However, these classifications are not static. The plants in Category 2 and Category 3 may be moved to Category 1 in the future if management is inadequate. Examples of medicinal plants belonging to the three categories are shown in Table 2.2.

2.3.2.2 The Forestry Sector (Ministry of Agriculture and Forestry)

The forestry sector, which is an integral part of the Medicinal Plants Sector, also provides important guidance to the sustainable growth of medicinal plants. Important goals of the forestry sector include alleviating poverty, increasing forest area from the current coverage of approximately 40% to 70% (MAF 2005), protecting the environment, and developing a process for sustainable use of the country's natural resources.

The Environmental Protection Law, No. 09/PO, defines the basic principles, rules, and measures related to the management, protection, and use of natural resources and forests. It promotes the rehabilitation, growth, and extension of

TABLE 2.2

Selected Examples of Medicinal Plants Classified under the Ministerial Decree[a]

Common Name in Laos	Scientific Name	Traditional Use	Part Used
Category 1 - Rare and endangered			
Fang Deng	Fabaceae-Caesalpinioideae *Caesalpinia sappan* L.	As a blood tonic; to treat diarrhea	Stem wood
Mak Seng Beua	Loganiaceae *Strychnos nux-vomica* L.	To treat rheumatism, joint pain, paralysis. See also Chapter 4	Seed
Kout Tin Houng	Ophioglossaceae *Helminthostachys zeylanica* (L.) Hook.	Used as a tonic; to treat fever See also Chapter 4	Root
Van Bai Lay	Orchidaceae *Anoectochilus* sp.	Used as a tonic	Whole plant
Category 2 - High commercial value			
Mak Tek (Kheua)	Celastraceae *Celastrus paniculatus* Willd.	Treatment of stomach aches	Stem
Si Khai Ton	Lauraceae *Litsea cubeba* (Lour.) Pers.	To treat smallpox	Stem, root
Kheua Hem	Menispermaceae *Coscinium fenestratum* (Gaertn.) Colebr.	To treat dysentery, diarrhea, diabetes See also Chapter 4	Liana
Kok Had Mee	Moraceae *Artocarpus lakoocha* Roxb.	To treat tape worm infection See also Chapter 4	Stem wood (without sap)
Tin Cham (Kieng)	Primulaceae *Ardisia humilis* Vahl	Cases of urine retention (as a diuresis)	Root
Category 3 - Available throughout the country			
Kha Nhom Phou	Apocynaceae *Rauvolfia serpentina* Benth.	To treat cases of high blood pressure	Root
Chan Dai	Asparagaceae *Dracaena cambodiana* Pierre ex Gagn.	Used as a blood tonic; blood purifier See also Chapter 4	Dried stem wood
Som Khi Mone	Primulaceae *Embelia ribes* Burm.	Used to treat stomach ache, diarrhea	Root
Yik Bo Thong	Simaroubaceae *Eurycoma longifolia* Jack	As a tonic; cases of high fever	Root

(Continued)

TABLE 2.2   (CONTINUED)
Selected Examples of Medicinal Plants Classified under the Ministerial Decree[a]

Common Name in Laos	Scientific Name	Traditional Use	Part Used
Hua Sam Sip	Stemonaceae *Stemona tuberosa* Lour.	To treat lung inflammation; tuberculosis	Rhizome
Nhane (Kok)	Styracaceae *Styrax tonkinensis* (Pierre) Craib ex Hartwich	Treatment of cough	Resin

[a] Sources: Prime Minister Lao PDR 2003 – a public document widely available in Laos and online; Updated, ITM document.

natural resources in the Lao PDR to ensure a balanced environment and sustainable forests, which help to protect water resources, prevent soil erosion, and protect biodiversity and the environment. All these measures contribute to the socioeconomic development of the country (President of the Lao PDR 1999).

Although the program was successful in raising awareness on agro-biodiversity and instrumental in developing a number of projects, it also had a number of shortcomings. Most importantly, there was a lack of broad stakeholder involvement, resulting in inadequate funding for support of the program. Consequently, a new 10-year program, 2015–2025, was put in place to support the national priorities on food security, poverty-reduction, and socioeconomic development (MAF 2016).

2.3.2.3   The Security Sector (Ministry of Public Security)

The main role and duty of this Ministry is dealing with cultivation and use of medicinal plants classified as narcotic based on the Single Convention on Narcotic Drugs (UN Office on Drugs and Crime 1961). Following this convention, the Law on Narcotic Drugs of the Lao PDR highlights two plants as controlled drugs within the list of narcotics, *Papaver somniferum* L. (Papaveraceae) and *Cannabis indica* Lam. (synonym of *Cannabis sativa* L.) (Cannabaceae) (FDD of Lao PDR 2003). In recent years, WHO's experts on drug dependence (ECDD) have twice formulated recommendations for *Cannabis*, cannabidiol (CBD), and related matters.

The 2018 meeting then recommended that preparations considered to be pure cannabidiol (CBD) should not be scheduled within the international drug conventions if they contain no more than 0.2% of delta-9-tetrahydrocannabinol (THC, tetrahydrocanabinol) (WHO 2018). Still, cultivation and use of poppy and marijuana remains illegal based on current law. Nevertheless, under the control of concerned authorities, those plants can be used in medically necessary situations. At the time of this writing, the government was drafting regulations to control the cultivation and processing of *Cannabis* for medicinal purposes (Sydara, perso communication, 2022).

2.3.3 ENVIRONMENTAL CONCERNS ON THE USE OF NATURAL MEDICINAL

Laos has tropical rainforests of broad-leaved evergreens in the north and monsoon forests of mixed evergreens and deciduous trees in the south. Roughly two-fifths of the country is forest and partly serves as a natural medicinal plant resource.

Current trade liberalization policies implemented by the Government of Laos are opening new possibilities for international trade of medicinal plants. Over the past few decades, several products have been exported to neighboring countries, and the growing demand in these foreign markets has triggered increased harvesting and processing of medicinal plant resources. Trade may provide many benefits, particularly in the economic sector. On the other hand, forests and natural resources, if not managed appropriately, can have significant negative impacts on the country's rich biodiversity, its natural medicinal resources, and ultimately, the economy that is largely dependent on natural resources. Unsustainable harvesting of medicinal plants leads to a loss of forest biodiversity, as well as associated deforestation effects such as increased production-related pollution (Manivong 2008).

Some of the commonly traded wild medicinal plants of Laos are shown in Table 2.3 below.

2.3.4 MEASURES TO BE TAKEN FOR SUSTAINABLE USE OF MEDICINAL PLANTS IN THE LAO PDR

In the coming years, trade in medicinal plants will continue to play an important role toward eradication of poverty in rural areas. For many people, particularly in remote areas, the only source of income is from selling medicinal plants and medicinal plant products. If the harvesting of medicinal plants and other non-timber forest products is unregulated, the rural population will eventually face a shortage of natural resources (see Chapter 4 for some measures to protect medicinal plant resources).

One goal of the Government of Laos is to ensure sustainable growth of medicinal plant production through cultivation and sustainable harvesting, while adding value to processing and marketing. Some important recommendations for improved sustainable use of medicinal plant resources have previously been put forward (Sydara 2007) and include:

- Collaborate with neighboring countries on mutually agreed upon conservation of protected areas at borders and control of illegal trade in wildlife and prohibited plant species. Such action not only helps protect rare and endangered species, but also contributes to implementation of principles of CITES (UNEP-WCMC 2011). See Chapter 6 for examples.
- Work towards more scientific and community-related management of forest resources in order to generate medicinal plants and non-timber forest products (NTFPs) at sustainable levels.

TABLE 2.3
Some Commonly Traded Wild Medicinal Plants[a]

Common Name in Laos	Family-species Name	Medicinal Importance & Conservation Status
Pak Nohk	Apiaceae *Centella asiatica* (L.) Urb.	See Table 2.3.1 above.
Chan Dai	Asparagaceae *Dracaena cambodiana* Pierre ex Gagnep.	See Chapter 4.
Khing Pa	Asparagaceae *Polygonatum* sp.	See Chapter 4.
Kheua Sai Xang	Fabaceae-Papilionoideae *Spatholobus suberectus* Dunn	Anti-inflammatory (Liu, Xu, and Wang 2020)
Kok Xi Din	Hypoxidaceae *Curculigo orchioides* Gaertn.	See Chapter 4.
Man On Ling	Lauraceae *Cinnamomum cassia* Bl.	See Table 2.1 above.
Mak Seng Beua	Loganiaceae *Strychnos nux-vomica* L.	See Chapter 4.
Kheua Hem	Menispermaceae *Coscinium fenestratum* (Gaertn.) Colebr.	See Chapter 4.
Kout Tin Houng	Ophioglossaceae *Helminthostachys zeylanica* (L.) Hook.	See Chapter 4.
Van Bai Lay	Orchidaceae *Anoectochilus* spp.	
Mak Kham Pome	Phyllanthaceae *Phyllanthus emblica* L.	Tonic (Krishnaveni and Mirunalini 2010)
Song Fa	Rutaceae *Clausena harmandiana* (Pierre) Guillaumin	Cytotoxicity (Jantamat, Weerapreeyakul, and Puthongking 2019)
Ian Don	Simaroubaceae *Eurycoma harmandiana* Pierre	Antiparasitic (Wiart 2021)
Yik Bo Thong	Simaroubaceae *Eurycoma longifolia* Jack	Tonic (Rehman, Choe, and Yoo 2016)
Ya Hua	Smilacaceae *Smilax glabra* Roxb.	Many (Hua et al. 2018; She et al. 2015)
Hua Sam Sip	Stemonaceae *Stemona tuberosa* Lour.	Anti-insect (Bharali et al. 2015)

[a] Source: (STEA & UNEP 2006) – a public document widely available in Laos and online; Updated, ITM document.

- Improve statistics on resource harvesting and exporting to guide future policies. The first steps in the process are to assemble responsible agencies, increase collaboration, and define roles and responsibilities.
- Develop and enforce laws and regulations related to the forestry sector as a whole, especially regarding NTFPs.

2.3.5 IMPORTANT ROLE OF MEDICINAL PLANTS IN THE ECONOMIC DEVELOPMENT OF THE COUNTRY

Although the abundance of natural resources has been recognized, the Lao PDR remains one of the least-developed countries in Southeast Asia, ranking 140 out of 189 countries on the United Nations Human Development Index (United Nations Development Programme 2020). When Laos joined the WTO in 2013, it was expected that the country would rise above the "least-developed country" status by 2020, with trade being the mechanism for changing this ranking, including the trade of medicinal plants and spices (WTO 2019). According to recent reports, Lao PDR is on track to graduate from least-developed country status by 2026, after an extended preparatory period of five years to support the country with planning for a post-COVID-19 economic recovery by implementing policies and strategies to reverse COVID-19-related economic shocks (UNDESA 2021).

The International Institute for Sustainable Development's 2007 report listed medicinal plants and spices of Laos as "Prioritized for export in the National Export Strategy" (Shaw et al. 2007). Many predicted the potential for a "sustainable" niche for exports, linking the environment and livelihoods, such as ecotourism, organic farming, silk handicrafts, and medicinal plants and spices (Lazarus et al. 2006; Shaw et al. 2007). On the other hand, if not regulated, this could have detrimental effects on the environment. As development continues, a major challenge will be to enforce environmental regulations if natural resources are to be used sustainably for long-term growth (Sydara 2007).

The Government of Laos recognizes the opportunities that NTFPs hold for economic growth, as detailed in the 2012 Diagnostic Trade Integration Study produced by the Department of Planning and Cooperation under the Ministry of Industry and Commerce. Number eight of the National Export Strategy lists medicinal plants and spices. It also notes in the Sectoral Environmental Concerns that there is a need to "Develop and enforce laws and regulations related to the forestry sector as a whole, and especially to NTFPs, such as medicinal plants and spices." Further recommendations were to

> more scientific and community-oriented management of forest resources; improving statistics on harvesting and exporting; increased awareness of the potential impacts on rural communities…; ensuring that intellectual property considerations are adequately reflected in laws and regulations and improving the quality and quantity of NTFPs; and providing incentives for the private sector to sustainably manage these resources, such as through assurances of benefit sharing of any commercialized products.
>
> (MOIC 2012)

Conventional medicines from synthesis have high potential for treating various diseases, despite consumer concerns of possible side effects. Hence, conventional medicines still cannot solve all health problems. Many people continue to seek the answers to maintain their health using natural medicines.

At the beginning of 2019, the Food and Drug Department (FDD) and the ITM of Laos, in collaboration with Guangxi Botanical Garden of Medicinal Plants of PR China, published the *Lao Herbal Pharmacopoeia, Volume 1* (China Daily 2021). This pharmacopoeia consists of 160 medicinal plants and 40 formulations for the treatment of various diseases such as common cold, gastritis, dysentery-diarrhea, and chronic tonsilitis. This publication is one of a long list of the ITM's publication output on traditional medicine and medicinal plant use in Laos.

The Government of Laos promotes the agricultural sector to contribute to the economic development of the country. During the COVID-19 pandemic, the Ministry of Health promoted the use of three medicinal plants for the treatment of infected patients: (1) *Andrographis paniculata* (Burm.f.) Nees (Acanthaceae) (ວາຊານບ, ສາມພັນບ), (2) *Boesenbergia rotunda* (L.) Mansf. (Zingiberaceae) (ກະຊາຍ, ໂສມລາວ), and (3) *Houttuynia cordata* Thunb. (Saururaceae) (ຜກຄາວທອງ). The use of these medicinal plants contributed to the fight against COVID-19 and indirectly to the economic development of the country.

2.4   TRADITIONAL MEDICINE AND DRUG DISCOVERY

The traditional medicine system in Laos will continue to be a mainstay for the population so long as it continues to rely on medicinal plants and knowledge and is not lost due to deforestation, unsustainable agro-forestry practices, or the ascendancy of Western biomedicine at the expense of traditional medicinal beliefs and practices. While traditional medicine of Laos also encompasses the spiritual and psychological dimensions of illness, healing, and well-being, scientific research on the pharmacokinetics and efficacy of traditional medicine demonstrates that community-identified medicinal plants, in many cases, possess bioactive compounds that may act as leads in drug discovery and the development of new medicines (Gyllenhaal et al. 2011; Soejarto et al. 2012). The next chapter (Chapter 3) highlights research and identifications of new chemical constituents discovered in traditional medicine of Laos.

REFERENCES

Agouillal, F., Z.M. Taher, H. Moghrani, N. Nasrallah, and H. El Enshasy. 2017. "A review of genetic taxonomy, biomolecules chemistry and bioactivities of *Citrus hystrix* DC." *Biosciences, Biotechnology Research Asia* 14 (1): 285–305. doi:10.13005/BBRA/2446.

Almatar, M., H. Ekal, and Z. Rahmat. 2014. "A glance on medical applications of *Orthosiphon stamineus* and some of its oxidative compounds." *International Journal of Pharmaceutical Sciences Review and Research* 24 (2): 83–88.

Abirami, A., G. Nagarani, and P. Siddhuraju. 2014. "The medicinal and nutritional role of underutilized citrus fruit *Citrus hystrix* (Kaffir lime): A review." *Drug Invention Today* 6 (1): 1–5.

Ashraf, K., S. Sultan, and A. Adam. 2018. "*Orthosiphon stamineus* Benth. is an outstanding food medicine: Review of phytochemical and pharmacological activities." *Journal of Pharmacy And Bioallied Sciences* 10 (3): 109–118. doi:10.4103/JPBS.JPBS_253_17.

Bharali, P., A. Das, H. Tag, D. Kakati, and A. M. Baruah. 2015. "Ethnopharmacognosy and Nutritional Composition of *Stemona tuberosa* Lour.: A Potential Medicinal Plant From Arunachal Pradesh, India." *Journal of Bioresources* 2 (1): 23–32. Accessed on January 8, 2023. http://jbr.rgu.ac.in/img/pdf/23%20to%2032%202nd.pdf.

CABI 2022. "*Zingiber montanum* (Cassumunar Ginger)." *Invasive Species Compendium: Detailed Coverage of Invasive Species Threatening Livelihoods and the Environment Worldwide*. Accessed on January 8, 2023. https://www.cabi.org/isc/datasheet/57536.

Chan, E.W.C., P.Y. Lye, and S.K. Wong. 2016. "Phytochemistry, pharmacology, and clinical trials of *Morus alba*." *Chinese Journal of Natural Medicines* 14 (1): 17–30. doi:10.3724/SP.J.1009.2016.00017.

Chen, D., H. Li, W. Li, S. Feng, and D. Deng. 2018. "*Kaempferia parviflora* and its methoxy-flavones: Chemistry and biological activities." *Evidence-Based Complementary and Alternative Medicine*. Hindawi Limited. doi:10.1155/2018/4057456.

Chen, H., M. Ren, H. Li, Q. Xie, R. Ma, Y. Li, X. Guo, J. Wang, D. Gong, and T. Gao. 2020. "Neuroprotection of benzoinum in cerebral ischemia model rats via the ACE-AngI-VEGF pathway." *Life Sciences* 260: 118418. doi:10.1016/J.LFS.2020.118418.

China Daily 2021. "Laos to Take Center Stage at Expo." *ChinaDaily.Com.Cn*. Accessed on January 8, 2023. http://www.chinadaily.com.cn/regional/2021-09/08/content_37546699.htm.

Chouni, A. and S. Paul. 2018. "A review on phytochemical and pharmacological potential of *Alpinia galanga*." *Pharmacognosy Journal* 10 (1): 9–15. doi:10.5530/pj.2018.1.2.

Couto, C.L.L., D.F.C. Moraes, M. do S.S. Cartágenes, F.M.M. do Amaral, and R.N. Guerra. 2016. "*Eleutherine bulbous* (Mill.) Urb.: A review study." *Journal of Medicinal Plants Research* 10 (21): 286–97. doi:10.5897/JMPR2016.6106.

Devaraj, R.D., S.P.K. Jeepipalli, and B. Xu. 2020. "Phytochemistry and health promoting effects of Job's tears (*Coix lacryma*-Jobi) – A critical review." *Food Bioscience* 34: 100537. doi:10.1016/J.FBIO.2020.100537.

Dissanayake, K.G.C., W.A.L.C. Waliwita, and R.P. Liyanage. 2020. "A review on medicinal uses of *Zingiber officinale* (Ginger)." *International Journal of Health Sciences and Research* 10 (6): 142. Accessed on January 8, 2023. https://www.ijhsr.org/IJHSR_Vol.10_Issue.6_June2020/22.pdf.

Elkington, B.G., K. Sydara, J.F. Hartmann, B. Southavong, and D.D. Soejarto. 2013. "Folk epidemiology recorded in palm leaf manuscripts of Laos." *Journal of Lao Studies* 3 (1): 1–14.

Elliott, E.M. 2021. *Potent Plants, Cool Hearts: A Landscape of Healing in Laos* (PhD thesis, University College London). Accessed on January 8, 2023. https://discovery.ucl.ac.uk/id/eprint/10126896/.

Evans, G. 2002. A short history of Laos: The land in between. Edited by M. Osborne. *Short History of Asia Series*. Crows Nest: Allen & Unwin.

FDD of Lao PDR 2003. "List of Narcotic, Controlled Chemy and Precocure." *Narcotic Laws & Regulations*. Ministry of Health, Lao PDR. Food and Drug Department. Vientiane. Accessed on January 8, 2023. http://www.fdd.gov.la/content_en.php?contID=26.

GBIF 2022a. "*Zanthoxylum nitidum* (Roxb.) DC." *Global Biodiversity Information Facility*. Accessed on January 8, 2023. https://www.gbif.org/species/7269365.

——— 2022b. "*Amomum ovoideum* Pierre ex Gagnep." *Global Biodiversity Information Facility*. Accessed on January 8, 2023. https://www.gbif.org/species/5301776.

Gupta, B.M., K.K. Mueen Ahmed, S.M. Dhawan, and R. Gupta. 2018. "*Aloe vera* (medicinal plant) research: A scientometric assessment of global publications output during 2007–2016." *Pharmacognosy Journal* 10 (1): 1–8. doi:10.5530/PJ.2018.1.1.

Gupta, S. and A. Sharma. 2014. "Medicinal properties of *Zingiber officinale* Roscoe – A review." *IOSR Journal of Pharmacy and Biological Sciences (IOSR-JPBS)* 9 (5): 124–29. doi:10.9790/3008-0955124129.

Gyllenhaal, C., M.R. Kadushin, B. Southavong, K. Sydara, S. Bouamanivong, M. Xayveu, L.T. Xuan, et al. 2011. "Ethnobotanical approach versus random approach in the search for new bioactive compounds: Support of a hypothesis." *Pharmaceutical Biology* 50 (1). doi:10.3109/13880209.2011.634424.

Han, A.-R., H. Kim, D. Piao, C.-H. Jung, and E.K. Seo. 2021. "Phytochemicals and bioactivities of *Zingiber cassumunar* Roxb." *Molecules* 26 (8): 2377. doi:10.3390/MOLECULES26082377.

Higham, C. 2013. "Hunter-gatherers in southeast Asia: From prehistory to the present." *Human Biology* 85 (1/3): 21–43. doi:10.3378/027.085.0302.

Hua, S., Y. Zhang, J. Liu, L. Dong, J. Huang, D. Lin, and X. Fu. 2018. "Ethnomedicine, phytochemistry and pharmacology of *Smilax glabra*: An important traditional Chinese medicine." *The American Journal of Chinese Medicine* 46 (2): 261–97. doi:10.1142/S0192415X18500143.

Huang, S. 1987. "*Styrax tonkinensis*." *EFloras – Flora of China*. Volume 15. Page 256. Accessed on January 8, 2023. http://www.efloras.org/florataxon. aspx?flora_id=2&taxon_id=200017762.

Hughes, A.C. 2017. "Understanding the drivers of Southeast Asian biodiversity loss." *Ecosphere* 8 (1): e01624. doi:10.1002/ECS2.1624.

Inoguchi, A. and S. Bounthapandid. 2021. "What is the Forest Area of Lao People's Democratic Republic?" Vientiane. Accessed on January 8, 2023. https://www.fao. org/3/cb3000en/cb3000en.pdf.

Jantamat, P., N. Weerapreeyakul, and P. Puthongking. 2019. "Cytotoxicity and apoptosis induction of coumarins and carbazole alkaloids from *Clausena harmandiana*." *Molecules* 24 (18). doi:10.3390/MOLECULES24183385.

Kamarudin, A.A., N.H. Sayuti, N. Saad, N.A.A. Razak, and N.M. Esa. 2021. "*Eleutherine bulbosa* (Mill.) Urb. Bulb: Review of the pharmacological activities and its prospects for application." *International Journal of Molecular Sciences* 22 (13): 6747. doi:10.3390/IJMS22136747.

Kaur, A., R. Singh, C.S. Dey, S.S. Sharma, K.K. Bhutani, and I. P. Singh. 2010. "Antileishmanial phenylpropanoids from *Alpinia galanga* (Linn.) Willd." *Indian Journal of Experimental Biology* 48 (3): 314–17. Accessed on January 8, 2023. http://nopr.niscpr.res.in/handle/123456789/7407.

Kemp, S. 2021. "Digital Laos, 2021." *All the Statistics You Need in 2021 – DataReportal – Global Digital Insights*. Accessed on January 8, 2023. https://datareportal.com/reports/digital-2021-laos.

Kew 2022a. "*Andrographis paniculata* (Burm.f.) Nees." *Plants of the World Online. Facilitated by the Royal Botanic Gardens, Kew*. Accessed on January 8, 2023. https:// powo.science.kew.org/taxon/urn:lsid:ipni.org:names:45226-1.

———— 2022b. "*Centella asiatica* (L.) Urb." *Plants of the World Online. Facilitated by the Royal Botanic Gardens, Kew*. Accessed on January 8, 2023. https://powo.science. kew.org/taxon/urn:lsid:ipni.org:names:1197718-2.

———— 2022c. "*Aloe vera* (L.) Burm. F." *Plants of the World Online. Facilitated by the Royal Botanic Gardens, Kew*. Accessed on January 8, 2023. https://powo.science. kew.org/taxon/urn:lsid:ipni.org:names:530017-1.

———— 2022d. "*Coix lacryma*-Jobi L." *Plants of the World Online. Facilitated by the Royal Botanic Gardens, Kew*. Accessed on January 8, 2023. https://powo.science.kew.org/ taxon/urn:lsid:ipni.org:names:30100521-2.

———— 2022e. "*Eleutherine subaphylla* Gagnep." *Plants of the World Online. Facilitated by the Royal Botanic Gardens, Kew*. Accessed on January 8, 2023. https://powo. science.kew.org/taxon/urn:lsid:ipni.org:names:436897-1.

———— 2022f. *"Kaempferia parviflora* Wall. ex Baker." *Plants of the World Online. Facilitated by the Royal Botanic Gardens, Kew.* Accessed on January 8, 2023. https://powo.science.kew.org/taxon/urn:lsid:ipni.org:names:797188-1.

———— 2022g. *"Morus alba* L." *Plants of the World Online. Facilitated by the Royal Botanic Gardens, Kew.* Accessed on January 8, 2023. https://powo.science.kew.org/taxon/urn:lsid:ipni.org:names:30051955-2.

———— 2022h. *"Zingiber cassumunar* Roxb." *Plants of the World Online. Facilitated by the Royal Botanic Gardens, Kew.* Accessed on January 8, 2023. https://powo.science.kew.org/taxon/urn:lsid:ipni.org:names:798323-1.

———— 2022i. *"Zingiber officinale* Roscoe." *Plants of the World Online. Facilitated by the Royal Botanic Gardens, Kew.* Accessed on January 8, 2023. https://powo.science.kew.org/taxon/urn:lsid:ipni.org:names:798372-1.

Krishnaveni, M. and S. Mirunalini. 2010. "Therapeutic potential of *Phyllanthus emblica* (Amla): The Ayurvedic wonder." *Journal of Basic and Clinical Physiology and Pharmacology* 21 (1): 93–105. doi:10.1515/JBCPP.2010.21.1.93.

Kustina, E.Z. and S. Misfadhila. 2020. "Traditional uses, phytochemistry and pharmacology of *Curcuma xanthorriza* Roxb.: A review." *International Journal of Science and Healthcare Research* 5 (3): 494–500.

Lazarus, K., P. Dubeau, C. Bambaradeniya, R. Friend, and L. Sylavong. 2006. *"An Uncertain Future: Biodiversity and Livelihoods along the Mekong River in Northern Lao PDR."* Edited by IUCN: Bangkok.

Libman, A., H. Zhang, C. Ma, B. Southavong, K. Sydara, S. Bouamanivong, G.T. Tan, H.H.S. Fong, and D.D. Soejarto. 2008. "A first new antimalarial pregnane glycoside from *Gongronema napalense." Asian Journal of Traditional Medicines* 3 (6): 203–10.

Liu, L., F.-R. Xu, and Y.-Z. Wang. 2020. "Traditional uses, chemical diversity and biological activities of *Panax* L. (Araliaceae): A review." *Journal of Ethnopharmacology* 263: 112792. doi:10.1016/J.JEP.2020.112792.

Lorrillard, M. 2006. "Lao history revisited: Paradoxes and problems in current research." *South East Asia Research* 14 (3): 387–401. doi:10.5367/000000006778881582.

Lu, Q., R. Ma, Y. Yang, Z. Mo, X. Pu, and C. Li. 2020. *"Zanthoxylum nitidum* (Roxb.) DC: Traditional uses, phytochemistry, pharmacological activities and toxicology." *Journal of Ethnopharmacology* 260: 112946. doi:10.1016/J.JEP.2020.112946.

Lundberg, K.V. 2008. *Women Weaving Well-Being: The Social Reproduction of Health in Laos.* (PhD Dissertation) Lawrence, KS: University of Kansas. Accessed on January 8, 2023. https://www.proquest.com/docview/621762210/840EA11948E14FDFPQ/3?accountid=14552.

Ma, G., X. Chai, G. Hou, F. Zhao, and Q. Meng. 2022. "Phytochemistry, bioactivities and future prospects of mulberry leaves: A review." *Food Chemistry* 372: 131335. doi:10.1016/J.FOODCHEM.2021.131335.

MAF 2005. *National Forest Cover Assessment.* Vientiane: Ministry of Agriculture and Forestry.

———— 2016. "Lao PDR National Agro-Biodiversity Programme and Action Plan II (2015–2025)." Accessed on January 8, 2023. https://www.nafri.org.la/wp-content/uploads/2017/03/NABP-II_ENGLISH.pdf.

Mailina, J., M.A.N. Azah, Y,Y. Sam, S.L.L. Chua, and J. Ibrahim. 2007. "Chemical composition of the essential oil of *Amomum uliginosum." Journal of Tropical Forest Science* 19 (4): 240–42. Accessed on January 8, 2023. https://www.jstor.org/stable/43595393?seq=1.

Manivong, P. 2008. Environmental Impacts of Trade Liberalization in the Organic Agriculture Sector of the Lao PDR © 2007. International Institute for Sustainable Development (IISD) Published by the International Institute for Sustainable Development." *Rapid Trade and Environment Assessment (RTEA)*. Accessed on January 8, 2023. https://www.iisd.org/publications/report/ environmental-impacts-trade-liberalization-organic-agriculture-sector-lao-pdr.

MOIC 2012. *Diagnostic Trade Integration 2012: Trade and Private Sector Development Roadmap*. Vientiane. https://dangay.files.wordpress.com/2012/10/lao-dtis.pdf.

MoNRE and IUCN 2016. *Fifth National Report to the United Nations Convention on Biological Diversity*. Vientiane. Accessed on January 8, 2023. https://www.cbd.int/ doc/world/la/la-nr-05-en.pdf.

Okagu, I.U., J.C. Ndefo, E.C. Aham, and C.C. Udenigwe. 2021. "*Zanthoxylum* species: A comprehensive review of traditional uses, phytochemistry, pharmacological and nutraceutical applications." *Molecules* 26 (13): 4023. doi:10.3390/ MOLECULES26134023.

Okhuarobo, A., J.E. Falodun, O. Erharuyi, V. Imieje, A. Falodun, and P. Langer. 2014. "Harnessing the medicinal properties of *Andrographis paniculata* for diseases and beyond: A review of its phytochemistry and pharmacology." *Asian Pacific Journal of Tropical Disease* 4 (3): 213–22. doi:10.1016/S2222-1808(14)60509-0.

Peltzer, K., K. Sydara, and S. Pengpid. 2016. "Traditional, complementary and alternative medicine use in a community population in Lao PDR." *African Journal of Traditional, Complementary and Alternative Medicines* 13 (3): 95–100. doi:10.4314/ ajtcam.v13i3.12.

Pottier, R. 2007. *Yu Di Mi Heng; Etre Bien. Avoir de la Force; Essai Sur Les Pratiques Therapeutiques Lao*. Paris: Ecole Francaise D'extreme Orient.

Prasad, S. and A.K. Tyagi. 2015. "Ginger and its constituents: Role in prevention and treatment of gastrointestinal cancer." *Gastroenterology Research and Practice* 2015 (Article ID 142979). doi:10.1155/2015/142979.

President of the Lao PDR 1999. *Environmental Protection Law – Part I General Provisions. President's Decree No. 09/PO*. https://www.ajcsd.org/chrip_search/dt/pdf/Useful_ materials/Lao_PDR/Environmental_Protection_Law_Revised_Version.pdf.

——— 2000. *Law on Drugs and Medical Products. Decree No. 13/PO*. Amended, 2011. file:///C:/Users/dsoejarto/Downloads/%E3%80%90%E5%AF%AE%E5%9C%8B% E3%80%91Law_on_Drugs_and_Medical_Products(Amended).pdf.

Prime Minister of Laos 2003. "Decree on Natural Resources for Medicines." Vol. Decree 155.

Pulbutr, P., W. Caichompoo, P. Lertsatitthanakorn, M. Phadungkit, and S. Rattanakiat. 2012. "Antibacterial activity, antimutagenic activity and cytotoxic effect of an essential oil obtained from *Amomum uliginosum* K.D. Koenig." *Journal of Biological Sciences* 12 (6): 355–60. doi:10.3923/JBS.2012.355.360.

Ragasa, C., J.L. Caro, L.G. Lirio, and C.-C. Shen. 2014. "Chemical constituents of *Coix lacryma*-Jobi." *Research Journal of Pharmaceutical, Biological and Chemical Sciences* 5 (6): 344–48.

Rahman, N.N.N.A., T. Furuta, S. Kojima, K. Takane, and M.A. Mohd. 1999. "Antimalarial activity of extracts of Malaysian medicinal plants." *Journal of Ethnopharmacology* 64 (3): 249–54. doi:10.1016/S0378-8741(98)00135-4.

Rehman, S.U., K. Choe, and H.H. Yoo. 2016. "Review on a traditional herbal medicine, *Eurycoma longifolia* Jack (Tongkat Ali): Its traditional uses, chemistry, evidence-based pharmacology and toxicology." *Molecules* 21 (3): 331. doi:10.3390/ MOLECULES21030331.

Sánchez, M., E. González-Burgos, I. Iglesias, and M.P. Gómez-Serranillos. 2020. "Pharmacological update properties of *Aloe vera* and its major active constituents." *Molecules* 25 (6): 1324. doi:10.3390/MOLECULES25061324.

Shaw, S., A. Cosbey, H. Baumuller, T. Collander, L. Sylavong, and International Institute for Sustainable Development. 2007. *Rapid Trade and Environment Assessment (RTEA)*. Edited by IISD. *National Report for Lao PDR*: Winnipeg. Accessed January 5, 2023. Accessed on January 8, 2023. https://www.iisd.org/publications/report/rapid-trade-and-environment-assessment-rtea-national-report-lao-pdr.

She, T., L. Qu, L. Wang, X. Yang, S. Xu, J. Feng, Y. Gao, et al. 2015. "Sarsaparilla (*Smilax glabra* Rhizome) extract inhibits cancer cell growth by S phase arrest, apoptosis, and autophagy via redox-dependent ERK1/2 pathway." *Cancer Prevention Research* 8 (5): 464–74. doi:10.1158/1940-6207.

Shetty, G.R. and S.I. Monisha. 2015. "Pharmacology of an endangered medicinal plant *Alpinia galanga* – A review." *Semantic Scholar* 6: 499–511.

Singh, C.B., N. Manglembi, N. Swapana, and S.B. Chanu. 2015. "Ethnobotany, phytochemistry and pharmacology of *Zingiber cassumunar* Roxb. (Zingiberaceae)." *Journal of Pharmacognosy and Phytochemistry* 4 (1): 1–6. Accessed on January 8, 2023. https://www.phytojournal.com/archives/2015/vol4issue1/PartA/3.1-921.pdf.

Soejarto, D.D., C. Gyllenhaal, M.R. Kadushin, B. Southavong, K. Sydara, S. Bouamanivong, M. Xaiveu, et al. 2012. "An ethnobotanical survey of medicinal plants of Laos toward the discovery of bioactive compounds as potential candidates for pharmaceutical development." *Pharmaceutical Biology* 50 (1): 42–60.

Southavong, B., K. Sydara, and O. Souliya. 2014. *Medicinal Plants and Herbs in the Lao PDR*. Vientiane, Lao PDR. Accessed on January 10, 2023. https://catalogue.nla.gov.au/Record/6571316.

STEA & UNEP 2006. *National Environmental Performance Assessment (EPA) Report*. Vientiane.

Stuart-Fox, M. 1997. *A History of Laos*. Cambridge, UK: Cambridge University Press.

Sun, B., L. Wu, Y. Wu, C. Zhang, L. Qin, M. Hayashi, M. Kudo, M. Gao, and T. Liu. 2020. "Therapeutic potential of *Centella asiatica* and its triterpenes: A review." *Frontiers in Pharmacology* 11: 1373. Accessed on January 10, 2023. doi:10.3389/fphar.2020.568032.

Sydara, K. 2007. "Environmental impacts of trade liberalization in the medicinal plants & spices sector, Lao PDR." Edited by S. Shaw and T. Callander. *Rapid Trade and Environment Assesment for Lao PDR: IISD*. Accessed on January 10, 2023. https://www.iisd.org/publications/report/environmental-impacts-trade-liberalization-medicinal-plants-and-spices-sector.

Sydara, K., S. Gneunphonsavath, R. Wahlstrom, S. Freudenthal, K. Houamboun, G. Tomson, T. Falkenberg, et al. 2005. "Use of traditional medicine in Lao PDR." *Complementary Therapies in Medicine* 13 (3): 199–205. doi:10.1016/j.ctim.2005.05.004.

Tabana, Y.M., F.S.R. Al-Suede, M.B. Khadeer Ahamed, S.S. Dahham, L.E. Ahmed Hassan, S. Khalilpour, M. Taleb-Agha, D. Sandai, A.S. Abdul Majid, and A.M.S.A. Majid. 2016. "Cat's whiskers (*Orthosiphon stamineus*) tea modulates arthritis pathogenesis via the angiogenesis and inflammatory cascade." *BMC Complementary and Alternative Medicine* 16 (1): 1–11. doi:10.1186/s12906-016-1467-4.

UN Office on Drugs and Crime 1961. "*Single Convention on Narcotic Drugs*." Geneva. Accessed on January 8, 2023. https://www.unodc.org/unodc/en/treaties/single-convention.html.

UNDESA 2021. "Graduation of Bangladesh, Lao People's Democratic Republic and Nepal from the LDC Category – Department of Economic and Social Affairs." *United Nations Department of Economic and Social Affairs Economic Analysis*. Accessed on January 8, 2023. https://www.un.org/development/desa/dpad/2021/graduation-of-bangladesh-lao-peoples-democratic-republic-and-nepal-from-the-ldc-category/.

UNEP-WCMC 2011. "UNEP_WCMC Species Database: CITES-Listed Species." Accessed on January 8, 2023. http://www.cites.org/eng/resources/species.html.

United Nations Development Programme 2020. "Latest Human Development Index Ranking | Human Development Reports." *Human Development Reports*. Accessed on January 8, 2023. http://hdr.undp.org/en/countries/profiles/LAO.

Venkatramna Industries 2022. "Benzoin Resin (*Styrax tonkinensis*) Pure Essential Oil." Accessed on January 8, 2023. https://www.venkatramna-perfumers.com/ProductDetail.aspx?Category=Natural+Essential+Oils&&Title=Benzoin+Resin+(Styrax+tonkinensis)+Pure+Essential+Oil.

Viravong, M.S. 1964. *History of Laos*. New York: Paragon Book Reprint Corp. Accessed January 10, 2023. https://www.amazon.com/History-Laos-Martin-Stuart-Fox/dp/0521597463/ref=asc_df_0521597463/?tag=hyprod-20&linkCode=df0&hvadid=312154640153&hvpos=&hvnetw=g&hvrand=16693545043726962073&hvpone=&hvptwo=&hvqmt=&hvdev=c&hvdvcmdl=&hvlocint=&hvlocphy=9021710&hvtargid=pla-459707800018&psc=1&asin=0521597463&revisionId=&format=4&depth=1.

Wang, F., H. Hua, Y. Pei, D. Chen, and Y. Jing. 2006. "Triterpenoids from the resin of *Styrax tonkinensis* and their antiproliferative and differentiation effects in human leukemia HL-60 cells." *Journal of Natural Products* 69 (5): 807–10. doi:10.1021/np050371z.

WFO 2022a. "Alpinia Galanga (L.) Willd." *World Flora Online*. Accessed on January 8, 2023. http://www.worldfloraonline.org/taxon/wfo-0000338490.

——— 2022b. "Search." Accessed on January 8, 2023. http://www.worldfloraonline.org/.

WHO 2018. "WHO Expert Committee On Drug Dependence: Forty-First Report." *World Health Organization - Technical Report Series*. Vol. No. 729. Geneva, Switzerland. Accessed on January 8, 2023. https://www.who.int/publications/i/item/who-expert-committee-on-drug-dependence-forty-first-report.

——— 2019. "*WHO Global Report on Traditional and Complementary Medicine 2019*." Geneva. Accessed on January 8, 2023. https://www.who.int/publications/i/item/978924151536.

Wiart, C. 2021. "Antiparasitic asian medicinal plants in the Clade Malvids." *Medicinal Plants in Asia and Pacific for Parasitic Infections*. Academic Press, pp. 233–348. doi:10.1016/B978-0-12-816811-0.00008-1.

The World Bank 2011. "Lao PDR Now a Lower-Middle Income Economy." Press Release. Accessed on January 8, 2023. https://www.worldbank.org/en/news/press-release/2011/08/17/lao-pdr-now-lower-middle-income-economy.

WPR 2021. "Laos Population 2021 (Demographics, Maps, Graphs)." *World Population Review*. Accessed on January 8, 2023. https://worldpopulationreview.com/countries/laos-population.

WTO 2019. "Trade Policy Review WT/TPR/S/394 (19-6592)." Accessed on January 8, 2023. https://www.wto.org/english/tratop_e/tpr_e/s394_e.pdf.

Wu, Q., X. Zhao, C. Chen, Z. Zhang, and F. Yu. 2020. "Metabolite profiling and classification of developing *Styrax tonkinensis* kernels." *Metabolites* 10 (1): 21. doi:10.3390/METABO10010021.

Yan, S., C.-C. Wang, H.-X. Zheng, W. Wang, and Z.-D. Qin. 2014. "Y chromosomes of 40% Chinese descend from three neolithic super-grandfathers." *PLoS ONE* 9 (8): e105691. doi:10.1371/journal.pone.0105691.

Yoshino, S., R. Awa, Y. Miyake, I. Fukuhara, H. Sato, T. Ashino, S. Tomita, and H. Kuwahara. 2018. "Daily intake of *Kaempferia parviflora* extract decreases abdominal fat in overweight and preobese subjects: A randomized, double-blind, placebo-controlled clinical study." *Diabetes, Metabolic Syndrome and Obesity: Targets and Therapy* 11: 447–58. doi:10.2147/DMSO.S169925.

Yu, Q., G. Ye, Z. Lei, R. Yang, R. Chen, T. He, and S. Huang. 2021. "An isolated com-
pound from stems and leaves of *Coix lacryma*-Jobi L. and its anticancer effect." *Food
Bioscience* 42: 101047. doi:10.1016/J.FBIO.2021.101047.
Zhang, C., L. Fan, S. Fan, J. Wang, T. Luo, Y. Tang, Z. Chen, and L. Yu. 2019. "*Cinnamomum
cassia* Presl: A review of its traditional uses, phytochemistry, pharmacology and tox-
icology." *Molecules* 24 (19): 3473. doi:10.3390/MOLECULES24193473.
Zhang, Q. 2018. "Global situation and WHO strategy on traditional medicine." *Traditional
Medicine and Modern Medicine* 01 (01): 11–13. doi:10.1142/S257590001820001X.
Zhou, Y.-Q., H. Liu, M.-X. He, R. Wang, Q.-Q. Zeng, Y. Wang, W.-C. Ye, and Q.-W. Zhang.
2018. "A review of the botany, phytochemical, and pharmacological properties of
Galangal." *Natural and Artificial Flavoring Agents and Food Dyes*. Academic Press,
351–96. doi:10.1016/B978-0-12-811518-3.00011-9.

3 Discovery of New Bioactive Compounds from Medicinal Plants of Laos

Djaja Djendoel Soejarto
University of Illinois at Chicago, Chicago, IL, USA
Field Museum of Natural History, Chicago, IL, USA

Yulin Ren
The Ohio State University, Columbus, USA

Bethany Gwen Elkington
University of Illinois at Chicago, Chicago, IL, USA
Field Museum of Natural History, Chicago, IL, USA

Guido F. Pauli
University of Illinois at Chicago, Chicago, IL, USA

Joshua Matthew Henkin
University of Illinois at Chicago, Chicago, IL, USA
Field Museum of Natural History, Chicago, IL, USA

Kongmany Sydara
Institute of Traditional Medicine, Ministry of Health, Vientiane, Laos
University of Health Sciences, Vientiane, Laos

A. Douglas Kinghorn
The Ohio State University, Columbus, USA

DOI: 10.1201/9781003216636-3

CONTENTS

3.1   INTRODUCTION

Although the tropical flora of Laos is rich, the exact number of medicinal plants used in the country is still unknown. Laos belongs to the Indo-Burma biodiversity hotspot, with its flora being around 8,000 to 11,000 species of flowering plants (Schmid 1989; MAF and STEA 2003; National Herbarium of Laos I BRAHMS 2011). A recent paper (Zhu 2017) mentioned that there are at least 5,005 species of seed plants belonging to 1,373 genera and 188 families in Laos.

In the four-volume book series about medicinal plants of Cambodia, Laos, and Vietnam (Petelot 1952, 1954), 1,392 species of 169 families are listed, although the actual number of medicinal plants in Laos is not defined. In the book *Medicinal Plants of Vietnam, Cambodia and Laos* (Nguyen 1993), 674 species of medicinal plants are listed; however, no breakdown is given on the number of species found in Laos. Vidal listed 1,000 species of useful plants, with many of these having medicinal uses (Vidal 1959). According to Pottier, the people of Laos used about 4,000 plants medicinally in the 1960s (Pottier 1971), and Lao books on medicinal plants collectively contain about 3,000 medicinal plant species. The most experienced traditional practitioners may be familiar with more than 1,000 medicinal plant species, although the majority normally use less than 500 in their day-to-day practices (Sydara et al. 2014).

More recently, an inventory of plants collected in Laos for the period of 2007 to 2014 by the ITM, with the collaboration of Laos-based and foreign-based institutions, found that more than 7,000 collections have been made. These are stored in the ITM Herbarium, which include 2,000 to 3,000 plant species with records of medicinal uses (Sydara et al. 2014).

3.2   PLANTS AS POTENTIAL SOURCES OF NEW MEDICINES

Many articles have reviewed the importance of plants as sources of drugs (Fabricant and Farnsworth 2001; Heinrich 2003; Newman and Cragg 2020), and numerous effective plant-derived agents have been developed for the treatment of a number of diseases, including cancer and infection (Newman and Cragg 2020; Talib et al. 2020; Dehelean et al. 2021; Aldrich et al. 2022). As a consequence of the continuing search for antineoplastic and anti-infective agents from plants growing in Laos, substances targeting unusual biochemical mechanisms may be identified that offer new approaches to improving human health (Soejarto et al. 2006; Kinghorn et al. 2016). A reservoir of a large number of medicinal plants growing in Laos offers the opportunity for such a search (Henkin et al. 2017). Of these, eight species (listed below) have been investigated for the identification of bioactive components.

3.3   THE SEARCH FOR BIOACTIVE COMPOUNDS FROM PLANTS OF LAOS

During the past 20 years, the search for bioactive compounds from medicinal plants of Laos was undertaken, in two phases, through several collaborative efforts between the ITM and foreign institutions, as described below.

3.3.1   THE INTERNATIONAL COOPERATIVE BIODIVERSITY GROUPS (ICBG) PROGRAM

From 1998 to 2008, under the ICBG Program, a consortium of collaborating scientists was formed, which included UIC, the Vietnam Academy of Science and Technology (VAST) in Hanoi, the Cuc Phuong National Park in Ninh Binh province (near Hanoi), the ITM in Vientiane, and a pharmaceutical partner, Glaxo Research and Development in the United Kingdom (during its Phase I, 1998–2003), and Bristol-Myers Squibb (BMS), Princeton, NJ (during its Phase II, 2003–2008). This consortium was funded by the Fogarty International Center of the NIH, Bethesda, MD, USA, but the research as part of this consortium—performed at Glaxo, on the one hand, and at BMS on the other—was funded by each of these pharmaceutical partners.

A main purpose of the ICBG Program was to conduct research to discover bioactive compounds from medicinal plants of Laos as candidates for pharmaceutical development (Soejarto et al. 1999, 2002, 2006, 2007). However, the goals of the ICBG program were broader, namely drug discovery from plants of Vietnam and Laos, the conservation of biodiversity—specifically concerning medicinal plants, and economic development. Accordingly, biodiversity-based ("random") and ethnobotany-based collection methods were used. The biodiversity-based method, performed specifically in Cuc Phuong National Park, Vietnam, was intended to maximize the taxonomic diversity of the collections, while the ethnobotany-based method was used to select plants of Laos. The overall drug discovery accomplishments of the Vietnam-Laos ICBG Program were recently reviewed (Zhang et al. 2016; Ren et al. 2021), and the ethnobotanical research and accomplishments based in Laos were summarized separately (Soejarto et al. 2012).

3.3.2 A Research Collaboration between the Kunming Institute of Botany (KIB) of the People's Republic of China (PRC) and the ITM

This cooperation, under the name "Bioprospecting on Biological Materials of Laos" (2007–2015), functioned under a bilateral agreement between the KIB and the ITM and was renewed annually. The goals of this project included the collection of 100 wild plant species of Laos by the ITM and the establishment of a methanol extraction bank, followed by extract screening at KIB for the development of nutraceuticals or natural drug candidates. A total of 800 plant samples were collected, of which several samples were selected and evaluated biologically at KIB, but the results have not been published (K. Sydara, personal communication, June 7, 2022).

3.3.3 Research Cooperation between the ITM and the Korean Research Institute of Bioscience and Biotechnology of the Republic of Korea and with the Graduate School of Pharmaceutical Sciences, Kyoto University of Japan

No formal research agreements for these collaborations are recorded, but several papers have been published based on these efforts (Kim et al. 2018; Baek et al. 2019; Jung et al. 2019; Kim et al. 2020a, 2020b, 2020c), although no chemical isolation studies have been reported.

3.3.4 Cooperation between the UIC and the ITM

From 2015 to 2019, a team of scientists from OSU, UIC, and the ITM explored and collected plants from forests located in the Medicinal Plant and Medicinal Biodiversity Preserves (MPPs and MBPs; see Chapter 4) of Laos toward the discovery of potential anticancer agents under a program project with funding from the US NCI (P01) (Henkin et al. 2017; Henkin 2019).

3.4 DISCOVERIES OF NEW BIOACTIVE COMPOUNDS FROM PLANTS OF LAOS

Annonaceae

Marsypopetalum modestum (Pierre) B. Xue & R.M.K. Saunders

M. modestum is a large shrub or small tree native to the Andaman Islands, Myanmar, Thailand, Laos, Vietnam, and Cambodia (Kew 2022a). As part of the effort to discover new anti-tuberculosis (TB) agents from medicinal plants of

Laos, the literature was searched, including information recorded in old palm leaf manuscripts (see Chapter 2) (Elkington et al. 2012, 2014). Since palm leaf manuscripts were written in old scripts and languages, the original records were translated into the modern Lao language and script by expert translators. The use of a plant that might be indicative for treating symptoms of TB was recorded, and the taxonomy of the plant was identified.

These literature searches were followed by field interviews with healers in five different provinces: Bokeo, Bolikhamxay, Champasak, Luang Prabang, and Vientiane. The goal was to find possible connections between plants used to treat symptoms of TB that were recorded in the old palm leaf manuscripts and their utilization in contemporary practice. A total of 58 healers were interviewed, and 341 plant samples were collected and documented by a set of voucher herbarium specimens. After a comparative analysis, 77 species were submitted to various in vitro assays to determine growth inhibition against virulent *Mycobacterium tuberculosis* H37Rv (Mtb). Of these, an ethanol extract of *Marsypopetalum modestum* was found to be the most active. Dipyrithione (see Figure 3.1 below) was identified as a major active component, with a minimum inhibitory concentration (MIC) being less than 0.15 μM. Dipyrithione is a known compound, but its discovery from the genus *Marsypopetalum* was reported for the first time (Elkington et al. 2014). It is important to note that analogues of dipyrithione have been synthesized, which supports a better understanding of the structure-activity relationships (Liu et al. 2008; O'Donnell et al. 2009; Tailler et al. 2012; Krejčová et al. 2014; Srivastava et al. 2015).

Apocynaceae

Gongronema napalense (Wall.) Decne.

G. napalense is a vine with a distribution range spanning India east through the Himalayas, Nepal, Bangladesh, Myanmar, southern China, Laos, and Vietnam (Li et al. 1995; Kew 2022b). As part of an ICBG interview with a healer in Pontong District of Champasak Province (southern Laos), a sample (SL-7530) of the entire plant (Lao name "Kheua nguan mu") was collected by Amey Libman and documented by voucher herbarium specimen AL-148, in deposit at the John G. Searle Herbarium of the Field Museum (accession number 2252471). According to the healer, the plant is used locally in combination with one other species to treat polio. An equal amount of the vine is mixed with the stem wood of "Kok mak kham." One handful of each is boiled in 2 L of water until the volume reaches 500 ml. The decoction is then taken orally as needed for 7 to 15 days (Libman et al. 2008). A decoction of *G. napalense* has also been recorded in the literature to be used to treat leucorrhea, blennorrhea, and traumatic injuries (Li et al. 1995).

Following encouraging results from initial tests, a larger amount of the sample was recollected for isolation studies. From this recollected sample, a new steroidal glycoside, gongroneside A (Figure 3.1), was identified, which showed antimalarial activity against both the chloroquine-sensitive clone D6 and the chloroquine-resistant clone W2 of *Plasmodium falciparum* (Libman et al. 2008).

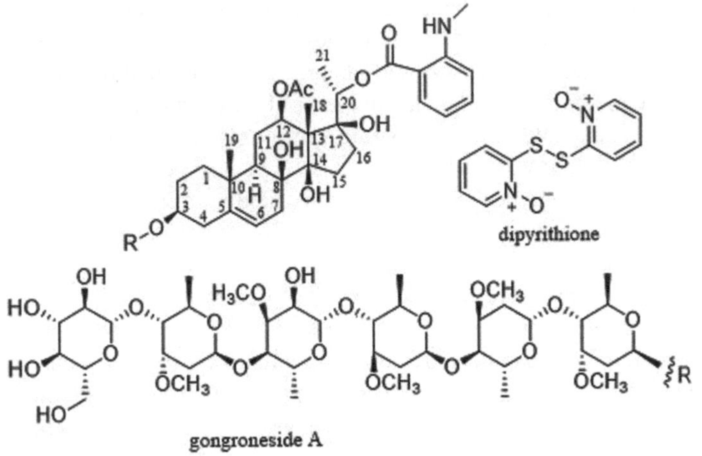

dipyrithione

gongroneside A

FIGURE 3.1 Structures of dipyrithione and gongroneside A isolated from the plants, *Marsypopetalum modestum* and *Gongronema napalense*, respectively (Ren et al. 2021). Due to the large size of the gongroneside A molecule, two structural units of this molecule are shown, namely the aglycone (upper left) and the glycose moiety (R, bottom level).

ASPARAGACEAE

Asparagus cochinchinensis (Lour.) Merr.

A. cochinchinensis, native to China, Korea, Japan, Laos, and Vietnam (Kew 2022c), is a delicate and tuberous root-bearing vine with white flowers and red fruits. Under the ICBG project, Kongmany Sydara carried out an interview with a healer in Salavan Province in 1999 and collected the root sample of this plant for biological testing. The interview, documented by voucher herbarium specimen Kongmany et al. KS-037, is in deposit at the FM Herbarium (accession number FM-2222116). According to the healer, the roots of this plant are mixed with other plants to treat chronic fever. Around 15–20 g of dried roots is mixed with 20 g of each of the other plants and 1.5 L of water, and the mixture is boiled for 15–20 min. The decoction is cooled, then drunk. Five to seven pieces of the roots may also be macerated in a glass of water for 3–4 hours, and the cooled decoction taken orally. A search of the literature indicated this plant has been used in mainland China to treat many health issues, including fever, cough, kidney diseases, and benign breast tumors (Jiangsu Medical College 1985).

Initial testing showed that the root extract exhibited activity against cancer cells, and seven compounds were isolated from the follow-up laboratory studies of the recollected sample (initial sample SL7037; recollected sample SLA7037). These include four norlignans, 3-hydroxy-4-methoxy-4-dehydroxynyasol, 3-methoxynyasol, nyasol, and 1,3-bis-di-p-hydroxyphenyl-4-penten-1-one, as well as three steroid derivatives, asparacoside and asparacosins A and B (Zhang et al. 2004). See Figure 3.2.

These compounds were evaluated for their cytotoxicity against Col-2 colon, KB oral epidermoid, LNCaP prostate, and Lu-1 lung human cancer cells, and 3-hydroxy-4-methoxy-4-dehydroxynyasol, 3-methoxynyasol, and asparacoside, which were deemed active, while other compounds isolated were inactive. These

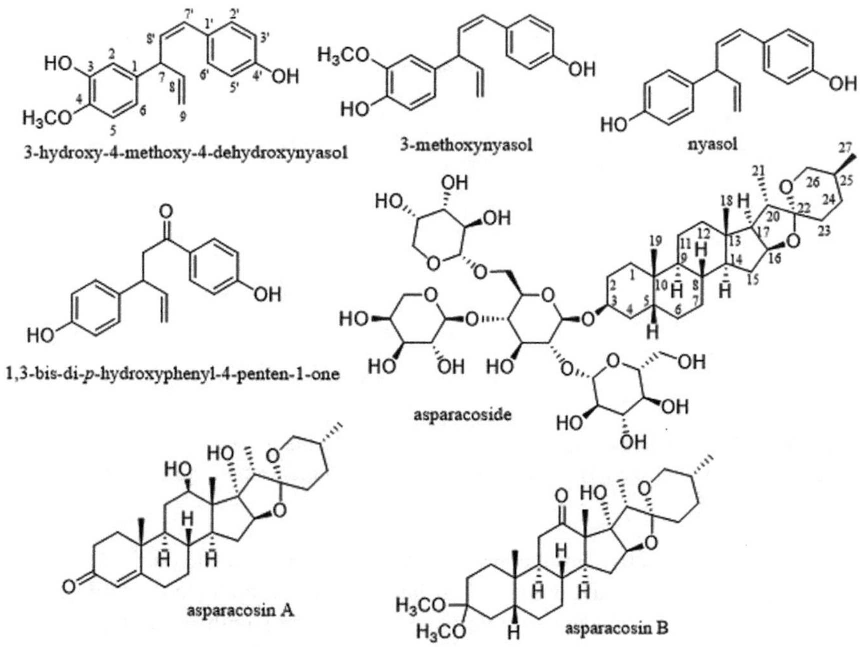

FIGURE 3.2 Structures of the norlignans, 3-hydroxy-4-methoxy-4-dehydroxynyasol, 3-methoxynyasol, nyasol, and 1,3-bis-di-*p*-hydroxyphenyl-4-penten-1-one, and the steroid derivatives, asparacoside and asparacosins A and B, isolated from the roots of *Asparagus cochinchinensis* collected in Laos (Zhang et al. 2004).

results demonstrated that the 3,4-*ortho*-hydroxy and methoxy groups are important for the mediation of cancer cell line cytotoxicity by 3-hydroxy-4-methoxy-4-dehydroxynyasol, but neither transposing the substituents between the C-3 and C-4 positions nor introducing a carbonyl group at the C-9′ position was found to contribute to the resultant cytotoxic potency (Zhang et al. 2004).

CONNARACEAE

Rourea minor (Gaertn.) Alston

R. minor is native to Tropical Africa, Madagascar, Tropical & Subtropical Asia to the SW Pacific (Kew 2022d). It is a liana (woody vine) found growing in Pakkadan Village, Paksan District of Bolikhamxay Province, Laos. As part of the ICBG Project, Somsanith Bouamanivong, a scientist at the ITM, interviewed a healer in Pakkadan Village, who claimed he used the dried stem bark to treat dengue fever. The remedy is prepared by boiling 30 g of the plant sample together with three other plants in two liters of water for 15–20 minutes. The solution is cooled and then taken orally by patients. Following this interview, a sample of the stems was collected, documented by herbarium voucher specimen Bouamanivong 006, in deposit at the FM (accession number 2222148). Initial testing of the stem bark extract indicated an activity against malaria parasites.

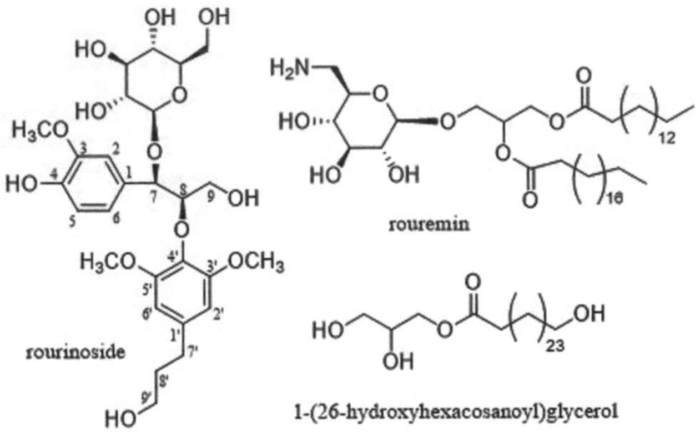

FIGURE 3.3 Structures of the lignan rourinoside and the glycerol esters, rouremin and 1-(26-hydroxyhexacosanoyl)-glycerol, isolated from *Rourea minor* collected in Laos (He et al. 2006).

A new in vitro-active antimalarial lignan glycoside, rourinoside, was isolated from the follow-up bioassay-guided isolation of the recollected sample (initial sample SL7006; recollected sample SLA7006), along with several glycerides, including rouremin and 1-(26-hydroxyhexacosanoyl)-glycerol (see Figure 3.3). All these compounds exhibited activity against both the chloroquine-sensitive clone D6 and the chloroquine-resistant clone W2 of *P. falciparum*, with IC50 values determined in the range 2–13 μM (He et al. 2006).

EBENACEAE

Diospyros cf. *quaesita* Thw.

D. quaesita is native to Sri Lanka (Kew 2022e), originally collected by Thwaites in April 1855 (Thwaites 1855) and later described as a new species (Thwaites and Hooker 1864). As part of an ICBG field interview, a stem bark sample of a *Diospyros* species was collected in the village of Tana (Sayabouli Province) by Somsanith Bouamanivong; voucher specimen: Bouamanivong 062 (Figure 3.4-A), in deposit at the FM (accession number FM-2245716). This plant was originally identified as *Diospyros quaesita* Thw. (Ma et al. 2008). After further review of the geographic distribution of this species and further examination of the voucher herbarium specimen of the first collection (Figure 3.4A) and that of the recollected sample (Figure 3.4B), the identity of this collection is given a provisional status in the present narrative.

According to the healer interviewed, the leaves and stems of *Diospyros* cf. *quaesita* are used to treat cases of alcohol and drug dependence, as well as to alleviate headaches. The remedy is prepared by boiling 30 g of the plant sample in 1½ L of water for 15–20 minutes. The cooled liquid is drunk by the patient.

Initial laboratory testing of the extract of the leafy branches of *Diospyros* cf. *quaesita* showed activity against malaria parasites, and pyracrenic acid was isolated (see Figure 3.5 below). Pyracrenic acid showed in vitro antimalarial activity

(a)   (b)

FIGURES 3.4 A and B Photographs of the plant from the first collection (Figure 3.4A) and from the second collection (Figure 3.4B) of Bouamanivong 62 (*Diospyros* cf. *quaesita*). Photo credit: S. Bouamanivong.

against *Plasmodium falciparum* clones D6 (chloroquine-sensitive) and W2 (chloroquine-resistant), with IC50 values of 1.40 and 0.98 µM, respectively. Evaluation of pyracrenic acid against the human oral epidermoid (KB) cancer cell line revealed cytotoxicity, with an ED50 value of 4.0 µM (Ma et al. 2008). Other studies showed pyracrenic acid activity against cell lines of human A549 and SK-Lu-1 lung, AGS gastric, COLO 205 and HCT-15 colon, HeLa cervical, HepG2 liver, MCF-7 breast, and SK-OV-3 ovarian cancer and SK-MEL-2 melanoma (Pan et al. 2008; Kim et al. 2010; Kim et al. 2017; Wang et al. 2018; Quang et al. 2020).

Rubiaceae

Nauclea orientalis (L.) L.

N. orientalis is a small tree in tropical Asia with a distribution spanning from Bangladesh east through Myanmar, Thailand, Indo-China, the Philippines, Indonesia, Papua New Guinea, and northern Australia (Kew 2022f). As part an ICBG Project field interview, a stem sample of this plant growing in Keng Sa Dok Village of Paksan District in Bolikhamxay Province, Laos, was collected and supported by voucher herbarium specimen Vanthanouvong 0711, in deposit at the Field Museum Herbarium (accession number FM-2283084). The healer claimed that the dried stems (i.e., pieces of stem wood and stem bark) are used to treat fatigue. The remedy is prepared by boiling 15–20 g of the dried stems of *N. orientalis* with 20 g each of two other plants in 1½ L of water for 15–20 minutes and then given orally to patients. Initial laboratory testing indicated that both the methanol extract and the chloroform partition were active against the chloroquine-sensitive clone D6 and the chloroquine-resistant clone W2 of *Plasmodium falciparum*. Follow-up experimental studies resulted in the isolation of two new and one known indole-like alkaloid glucosides, namely naucleaorine, epimethoxynaucleaorine, and strictosidine lactam, along with several known

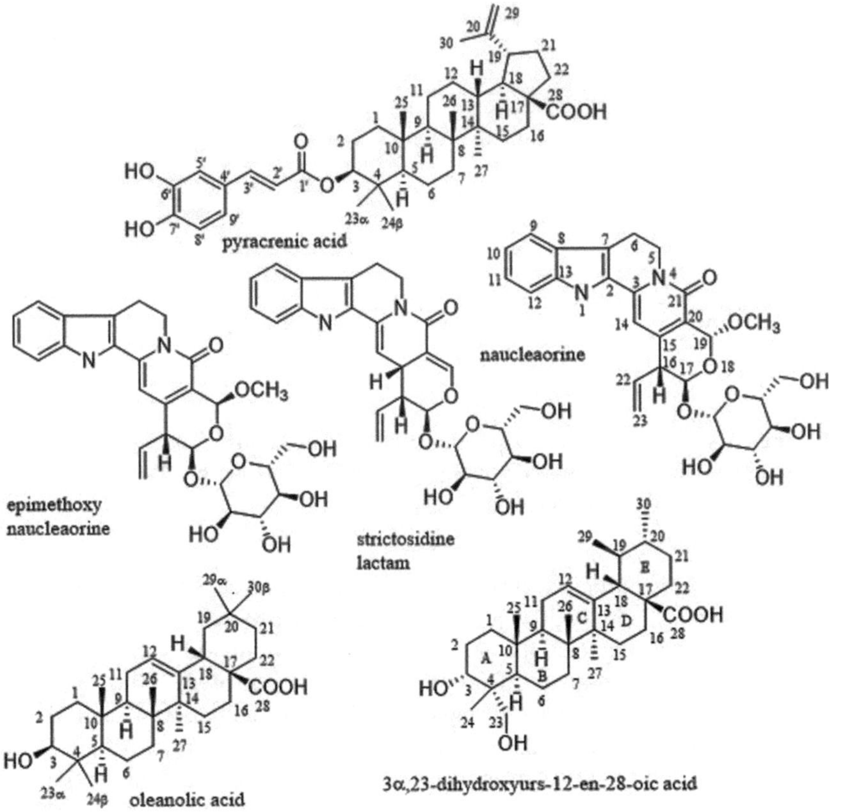

FIGURE 3.5 Structures of the triterpenoid pyracrenic acid isolated from the plant, *Diospyros* cf. *quaesita* (Lomchid et al. 2017; Ma et al. 2008; Shi et al. 2014), and the alkaloids, naucleaorine, epimethoxynaucleaorine, and strictosidine lactam, and the triterpenoids oleanolic acid and 3α,23-dihydroxyurs-12-en-28-oic acid isolated from *Nauclea orientalis* (Z.-D. He et al. 2005), collected in Laos.

triterpenoids, including oleanolic acid and 3α,23-dihydroxyurs-12-en-28-oic acid, and their analogues (see Figure 3.5). Evaluation of antimalarial activity showed that several of these compounds were active (Z.-D. He et al. 2005).

STEMONACEAE

Stemona pierrei Gagn. and *Stemona tuberosa* Lour.

Stemona is a genus of 27 species of lianas and creeping plants, with a distribution range primarily extending from India east to Myanmar, Southern China, Indo China, Thailand, Malaysia, the Philippines, Indonesia, and Papua New Guinea to Australia. Of these, *Stemona pierrei* is a vine native to Laos, Thailand, and Vietnam (Kew 2022g), while *Stemona tuberosa* is a climbing plant native to

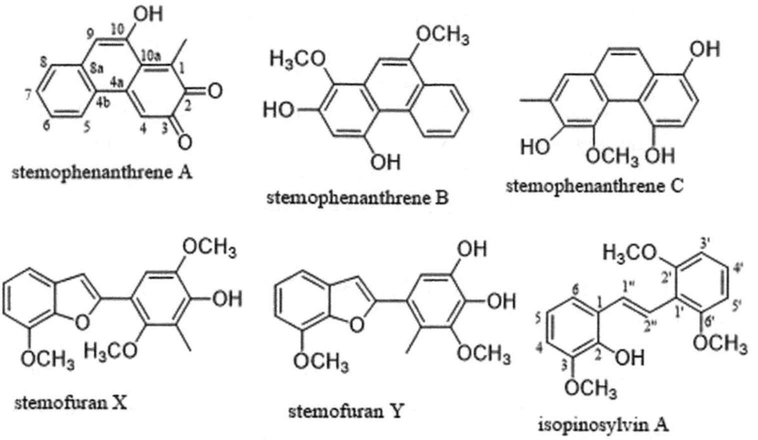

FIGURE 3.6 Structures of the phenanthrenes, stemophenanthrenes A, B, and C, and the stilbenoids, stemofuran Y and isopinosylvin A isolated from *Stemona tuberosa*, and the stilbenoid stemofuran X isolated from *Stemona pierrei*, collected in Laos (Khamko, Quang, and Dien 2013; Quang et al. 2014).

Bangladesh, Cambodia, China, India, Indonesia, Laos, Myanmar, New Guinea, the Philippines, Sri Lanka, Thailand, and Vietnam (Kew 2022h). In a search for potential antitumor agents from *Stemona* species, the roots of *S. pierrei* (KPNO2) were collected in Savannakhet Province, while the roots of *S. tuberosa* (KPS201101) were collected in Attapeu Province (Lao PDR), and phytochemical investigations were performed. Three new phenanthrenes, namely, stemophenanthrenes A, B, and C (see Figure 3.6), were identified from the roots of *S. tuberosa*, which exhibited cytotoxicity against human HepG2 liver, KB oral epidermal, MCF7 breast, and SK-Lu-1 lung cancer cells (Khamko et al. 2013). A new phenylbenzofuran-type stilbenoid, stemofuran X, was isolated from the roots of *S. pierrei*, and two new normal stilbenoids, stemofuran Y and isopinosylvin A, were characterized from the roots of *S. tuberosa* (Figure 3.6), but none of these compounds showed discernible activity against HepG2, KB, MCF7, and SK-Lu-1 cancer cells (Khamko et al. 2013; Quang et al. 2014).

3.5 ONGOING EFFORTS TOWARD NEW DISCOVERIES

During an expedition to the Xieng Khouang MBP and Bolikhamxay MPP in 2015, as part of a US NCI (P01) project (see 3.3.4, above) on the discovery of anticancer agents, 201 distinct samples from 96 species of flowering plants were collected along with appropriate herbarium documentation. Plant samples were extracted in azeotropic ethanol and evaporated to dryness for initial biological evaluation against HT-29 colon adenocarcinoma cells. Six species that exhibited notable cytotoxicity were the stem material of *Cryptolepis dubia* (Burm.f.) M.R.Almeida (A07194/ST; Apocynaceae); the aerial parts of *Rubia argyi* (H.Lév. & Vaniot)

Hara ex Lauener (A07196/PX; Rubiaceae); the fruits of *Reevesia pubescens* Mast. (A07214/FR; Malvaceae); the combined leaves, twigs, and fruits of *Maclura tricuspidata* Carrière (A07257/LF+TW+FR; Moraceae); the stem material of *Millettia pachyloba* Drake (A07338/ST; Fabaceae-Papilionoideae); and the leaves and twigs of *Gardenia annamensis* Pit. (A07365/LF+TW; Rubiaceae) (Henkin et al. 2017). Larger amounts of these plant samples have been recollected, and phytochemical work was ongoing at the time of the writing of this chapter.

3.6 RATIONALE FOR CONTINUING EFFORTS TOWARDS NEW DISCOVERIES

The traditional use of medicinal plants in Laos has proven useful for the discovery of new bioactive compounds, and ethnomedical collections can be supportive of plant-based drug discovery programs (Gyllenhaal et al. 2012). In fact, historically, ethnobotany has played an important role in the discovery of new drugs (McClatchey et al. 2009). Field-based interviews with healers in Laos aided in identifying several plant-derived lead compounds with activities toward cancer, human immunodeficiency virus (HIV), malaria, and tuberculosis (TB) infectious diseases (Soejarto et al. 2012). These outcomes provide a strong rationale to continue the search for therapeutic agents from traditionally used medicinal plants (Soejarto et al. 2002; Ren et al. 2021). This is further supported by the discovery of a potential anti-tuberculosis hit compound using field surveys that indicated the uses of the source plants by contemporary healers in Laos (Elkington et al. 2014).

Further rationales for the value of continuing drug discovery efforts from plants in general can be gleaned from recent developments in methodologies for mechanistic biological study and screening. While transferring the ongoing proteomic exploration of the biosynthetic potential of microbial species (Du and van Wezel 2018) to vascular plants remains a scientific challenge, proteomic and/or transcriptomic evaluation of plant metabolomes is already feasible. Recently described methodologies include arrayed transcriptomic platforms for the screening of such complex mixtures of natural products. Examples include using single-cell RNA sequencing with CRISPR methodology (Datlinger et al. 2017), as well as large-scale transcriptomics with panels of cell lines (Iwata et al. 2017). These transcriptomic approaches can unravel the complex network of signaling pathways affected by metabolomic mixtures. Such methodologies can yield important clues about clusters of biological mechanisms triggered by an ethnobotanical and its preparations and can be used to follow these bioactivity patterns throughout the course of fractionation, towards the identification of potential hit compounds.

One additional potential use of gene-expression assays is their ability to study favorable interactions (often referred to as "synergy") between natural products and synthetic drugs. In the recently demonstrated success in the study of interaction between the marine pigment, fucoxanthin, and an anti-cancer drug (Pruteanu et al. 2020), this approach can potentially be used to study how ethnobotanicals can support allopathic therapies.

REFERENCES

Aldrich, L.N., J.E. Burdette, E.C. de Blanco, C.C. Coss, A.S. Eustaquio, J.R. Fuchs, A.D. Kinghorn, et al. 2022. "Discovery of anticancer agents of diverse natural origin." *Journal of Natural Products* 85 (3): 702–19. doi:10.1021/acs.jnatprod.2c00036.

Baek, J., H. Jeong, Y. Ham, Y.H. Jo, M. Choi, M. Kang, B. Son, et al. 2019. "Improvement of spinal muscular atrophy via correction of the SMN2 splicing defect by Brucea javanica (L.) Merr. extract and Bruceine D." *Phytomedicine: International Journal of Phytotherapy and Phytopharmacology* 65. doi:10.1016/J.PHYMED.2019.153089.

Datlinger, P., A.F. Rendeiro, C. Schmidl, T. Krausgruber, P. Traxler, J. Klughammer, L.C. Schuster, A. Kuchler, D. Alpar, and C. Bock. 2017. "Pooled CRISPR screening with single-cell transcriptome readout." *Nature Methods* 14 (3): 297–301. doi:10.1038/nmeth.4177.

Dehelean, C.A., I. Marcovici, C. Soica, M. Mioc, D. Coricovac, S. Iurciuc, O.M. Cretu, and I. Pinzaru. 2021. "Plant-derived anticancer compounds as new perspectives in drug discovery and alternative therapy." *Molecules* 26 (4): 1109. doi:10.3390/MOLECULES26041109.

Du, C., and G.P. van Wezel. 2018. "Mining for microbial gems: Integrating proteomics in the postgenomic natural product discovery pipeline." *Proteomics* 18 (18). doi:10.1002/PMIC.201700332.

Elkington, B.G., K. Sydara, J.F. Hartmann, B. Southavong, and D.D. Soejarto. 2012. "Folk epidemiology recorded in palm leaf manuscripts of Laos." *Journal of Lao Studies* 4 (1): 1–14. Accessed on January 5, 2023. http://laostudies.org/system/files/subscription/JLS-v3-i1-Oct2012-elkington.pdf.

Elkington, B.G., K. Sydara, A. Newsome, C.H. Hwang, D.C. Lankin, C. Simmler, J.G. Napolitano, et al. 2014. "New finding of an anti-TB compound in the genus *Marsypopetalum* (Annonaceae) from a traditional herbal remedy of Laos." *Journal of Ethnopharmacology* 151 (2): 903–11. doi:10.1016/j.jep.2013.11.057.

Fabricant, D.S., and N.R. Farnsworth. 2001. "The value of plants used in traditional medicine for drug discovery." *Environmental Health Perspectives* 1: 69–75.

Gyllenhaal, C., M.R. Kadushin, B. Southavong, K. Sydara, S. Bouamanivong, M. Xayvue, L.T. Xuan, et al. 2012. "Ethnobotanical approach versus random approach in the search for new bioactive compounds: Support of a hypothesis." *Pharmaceutical Biology* 50 (1). doi:10.3109/13880209.2011.634424.

He, Z.-D., C.-Y. Ma, G.T. Tan, K. Sydara, P. Tamez, B. Southavong, S. Bouamanivong, et al. 2006. "Rourinoside and rouremin, antimalarial constituents from *Rourea minor*." *Phytochemistry* 67 (13). 1378–84. doi:10.1016/j.phytochem.2006.04.012.

He, Z.-D., C.-Y. Ma, H.-J. Zhang, G.T. Tan, P. Tamez, K. Sydara, S. Bouamanivong, et al. 2005. "Antimalarial constituents from *Nauclea orientalis* (L.) L." *Chemistry & Biodiversity* 2 (10): 1378–86. doi:10.1002/cbdv.200590110.

Heinrich, M. 2003. "Ethnobotany and natural products: The search for new molecules, new treatments of old diseases or a better understanding of indigenous cultures?" *Current Topics in Medicinal Chemistry* 3 (2): 141–54. doi:10.2174/1568026033392570.

Henkin, J.M. 2019. *Medical Ethnobotany and the Search for New Anticancer Agents from Plants of Laos and Vietnam*. Ph.D. Dissertation, University of Illinois at Chicago. Accessed on January 5, 2023. https://indigo.uic.edu/articles/thesis/Medical_Ethnobotany_And_The_Search_For_New_Anticancer_Agents_From_Plants_Of_Laos_And_Vietnam/12481115/1.

Henkin, J.M., K. Sydara, M. Xayvue, O. Souliya, A.D. Kinghorn, J.E. Burdette, W.-L. Chen, B.G. Elkington, and D.D. Soejarto. 2017. "Revisiting the linkage between ethnomedical use and development of new medicines: A novel plant collection strategy towards the discovery of anticancer agents." *Journal of Medicinal Plants Research Research* 11 (40): 621–34. doi:10.5897/jmpr2017.6485.

Iwata, M., R. Sawada, H. Iwata, M. Kotera, and Y. Yamanishi. 2017. "Elucidating the modes of action for bioactive compounds in a cell-specific manner by large-scale chemically-induced transcriptomics." *Scientific Reports* 7 (1): 1–15. doi:10.1038/srep40164.

Jiangsu Medical College. 1985. *Zhongyaodacidian (A Dictionary of Traditional Chinese Medicines)*. Shanghai.

Jung, S., J. Shin, J. Oh, G. Enkhtaivan, S.W. Lee, J. Gopal, K. Sydara, R.K. Saini, Y.-S. Keum, and J.-W. Oh. 2019. "Cytotoxic and apoptotic potential of *Phyllodium elegans* extracts on human cancer cell lines." *Bioengineered* 10 (1): 501–12. doi:10.1080/21655979.2019.1682110.

Kew 2022a. *Marsypopetalum modestum* (Pierre) B. Xue and R.M.K. Saunders. Plants of the World Online, Kew Science. Accessed on January 5, 2023. https://powo.science.kew.org/taxon/urn:lsid:ipni.org:names:77111132-1.

――― 2022b. *Gongronema napalense* (Wall.) Decne. Plants of the World Online, Kew Science. Accessed on January 5, 2023. https://powo.science.kew.org/taxon/urn:lsid:ipni.org:names:97744-1.

――― 2022c. *Asparagus cochinchinensis* (Lour.) Merr. Plants of the World Online, Kew Science. Accessed on January 5, 2023. https://powo.science.kew.org/taxon/urn:lsid:ipni.org:names:531041-1.

――― 2022d. *Rourea minor* (Gaertn.) Alston. Plants of the World Online, Kew Science. Accessed on January 5, 2023. https://powo.science.kew.org/taxon/urn:lsid:ipni.org:names:264810-1.

――― 2022e. *Diospyros quaesita* Thwaites. Plants of the World Online, Kew Science. Accessed on January 5, 2023. https://powo.science.kew.org/taxon/322922-1.

――― 2022f. *Nauclea orientalis* (L.) L. Plants of the World Online, Kew Science. Accessed on January 5, 2023. https://powo.science.kew.org/taxon/urn:lsid:ipni.org:names:757181-1.

――― 2022g. *Stemona pierrei* Gagnep. Plants of the World Online, Kew Science. Accessed on January 5, 2023. https://powo.science.kew.org/taxon/urn:lsid:ipni.org:names:821994-1.

――― 2022h. *Stemona tuberosa* Gagnep. Plants of the World Online, Kew Science. Accessed on January 5, 2023. https://powo.science.kew.org/taxon/urn:lsid:ipni.org:names:742815-1.

Khamko, V.A., D.N. Quang, and P.H. Dien. 2013. "Three new phenanthrenes, a new stilbenoid isolated from the roots of *Stemona tuberosa* Lour. and their cytotoxicity." *Natural Product Research* 27 (24): 2328–32. doi:10.1080/14786419.2013.832677.

Kim, C.S., L. Subedi, J. Oh, S.Y. Kim, S.U. Choi, and K.R. Lee. 2017. "Bioactive triterpenoids from the twigs of *Chaenomeles sinensis*." *Journal of Natural Products* 80 (4): 1134–40. doi:10.1021/acs.jnatprod.7b00111.

Kim, H.S., H.S. Kwon, C.H. Kim, S.W. Lee, K. Sydara, and S.J. Cho. 2018. "Effects of methanol extracts from *Diospyros malabarica* stems on growth and biofilm formation of oral bacteria 서 론." *Journal of Life Science* 28 (1): 110–15. doi:10.5352/JLS.2018.28.1.110.

Kim, J.G., M.J. Kim, J.S. Lee, K. Sydara, S. Lee, S. Byun, and S.K. Jung. 2020b. "*Smilax guianensis* vitman extract prevents LPS-induced inflammation by inhibiting the NF-κB pathway in RAW 264.7 cells." *Journal of Microbiology and Biotechnology* 30 (6): 822–29. doi:10.4014/JMB.1911.11042.

Kim, K.H., S.U. Choi, and K.R. Lee. 2010. "Bioactivity-guided isolation of cytotoxic triterpenoids from the trunk of *Berberis koreana*." *Bioorganic & Medicinal Chemistry Letters* 20 (6): 1944–47. doi:10.1016/J.BMCL.2010.01.156.

Kim, M.J., J.G. Kim, K.M. Sydara, S.W. Lee, and S.K. Jung. 2020a. "*Croton hirtus* L'Hér. *extract prevents inflammation in RAW264.7 macrophages via inhibition of NF-KB signaling pathway.*" *Journal of Microbiology and Biotechnology* 30 (4): 490–96. doi:10.4014/JMB.1908.08045.

Kim, M.-O., S.U. Lee, H.J. Yuk, H.-J. Jang, J.-W. Lee, E.-B. Kwon, J.-H. Paik, et al. 2020c. "Metabolomics approach to identify the active substances influencing the antidiabetic activity of *Lagerstroemia* species." *Journal of Functional Foods* 64. doi:10.1016/J.JFF.2019.103684.

Kinghorn, A.D., E.J. Carcache de Blanco, D.M. Lucas, H.L. Rakotondraibe, J. Orjala, D.D. Soejarto, N.H. Oberlies, et al. 2016. "Discovery of anticancer agents of diverse natural origin." *Anticancer Research* 36 (11): 5623–37. doi:10.21873/ANTICANRES.11146.

Krejčová, P., P. Kučerová, G.I. Stafford, A.K. Jäger, and R. Kubec. 2014. "Antiinflammatory and neurological activity of pyrithione and related sulfur-containing pyridine *N*-oxides from Persian shallot (*Allium stipitatum*)." *Journal of Ethnopharmacology* 154 (1): 176–82. doi:10.1016/J.JEP.2014.03.066.

Li, P.-T., M.G. Gilbert, and W.D. Stevens. 1995. "Asclepiadaceae 萝沸科 Luo Mo Ke." *Flora of China*, 16: 189–270.

Libman, A., H. Zhang, C. Ma, B. Southavong, K. Sydara, S. Bouamanivong, G.T. Tan, H.H.S. Fong, and D.D. Soejarto. 2008. "A first new antimalarial pregnane glycoside from *Gongronema napalense*." *Asian Journal of Traditional Medicines* 3 (6): 203–10.

Liu, Z., Y. Fan, Y. Wang, C. Han, Y. Pan, H. Huang, Y. Ye, L. Luo, and Z. Yin. 2008. "Dipyrithione inhibits lipopolysaccharide-induced INOS and COX-2 up-regulation in macrophages and protects against endotoxic shock in mice." *FEBS Letters* 582 (12): 1643–50. doi:10.1016/j.febslet.2008.04.016.

Lomchid, P., P. Nasomjai, S. Kanokmedhakul, J. Boonmak, S. Youngme, and K. Kanokmedhakul. 2017. "Bioactive lupane and hopane triterpenes from *Lepisanthes senegalensis*." *Planta Medica* 83(3/4): 334–40. doi:10.1055/s-0042-116438.

Ma, C.-Y., S.F. Musoke, G.T. Tan, K. Sydara, S. Bouamanivong, B. Southavong, D.D. Soejarto, H.H.S. Fong, and H.-J. Zhang. 2008. "Study of antimalarial activity of chemical constituents from *Diospyros quaesita*." *Chemistry and Biodiversity* 5 (11): 2442–48. doi:10.1002/cbdv.200890209.

MAF, and STEA. 2003. "Biodiversity Country Report Lao People's Democratic Republic Peace Independence Democracy Unity Prosperity." Vientiane. Accessed on January 8, 2023. http://lad.nafri.org.la/fulltext/780-0.pdf.

McClatchey, W.C., G.B. Mahady, B.C. Bennett, L. Shiels, and V. Savo. 2009. "Ethnobotany as a pharmacological research tool and recent developments in CNS-active natural products from ethnobotanical sources." *Pharmacology & Therapeutics* 123 (2): 239–54. doi:10.1016/J.PHARMTHERA.2009.04.002.

National Herbarium of Laos | BRAHMS. 2011. "Flore Du Laos." Accessed on January 8, 2023. https://herbaria.plants.ox.ac.uk/bol/HNL/Home/Index.

Newman, D.J., and G.M. Cragg. 2020. "Natural products as sources of new drugs over the nearly four decades from 01/1981 to 09/2019." *Journal of Natural Products* 83 (3): 770–803. doi:10.1021/acs.jnatprod.9b01285.

Nguyen, V.D. 1993. *Medicinal Plants of Vietnam, Cambodia, and Laos*. Santa Monica, CA: Nguyen Van Duong.

O'Donnell, G., R. Poeschl, O. Zimhony, M. Gunaratnam, J.B.C. Moreira, S. Neidle, D. Evangelopoulos, et al. 2009. "Bioactive pyridine-*N*-oxide disulfides from *Allium stipitatum*." *Journal of Natural Products* 72 (3): 360–65. doi:10.1021/np800572r.

Pan, M.-H., C.-M. Chen, S.-W. Lee, and Z.-T. Chen. 2008. "Cytotoxic triterpenoids from the root bark of *Helicteres angustifolia.*" *Chemistry & Biodiversity* 5 (4): 565–74. doi:10.1002/CBDV.200890053.

Petelot, A. 1952. *Les Plantes Medicinales Du Cambodge, Du Laos et Du Vietnam - Tome II - Caprifolaciees a Plantaginacees.* Edited by Centre National de Recherches Scientifiques et Techniques. Vol. 3.

——— 1954. *Les Plantes Médicinales Du Cambodge, Du Laos et Du Vietnam - Tome III.* Edited by Centre National de Recherches Scientifiques et Techniques. Vol. 3.

Pottier, R. 1971. "Le végétal dans la pharmacopée traditionnelle Lao." *Bulletin de La Société Botanique de France* 118 (3–4): 263–73. doi:10.1080/00378941.1971.108 38897.

Pruteanu, L.-L., L. Kopanitsa, D. Módos, E. Kletnieks, E. Samarova, A. Bender, L.D. Gomez, and D.S. Bailey. 2020. "Transcriptomics predicts compound synergy in drug and natural product treated glioblastoma cells." *PLOS ONE* 15 (9). doi:10.1371/ JOURNAL.PONE.0239551.

Quang, D.N., V.A. Khamko, N.T. Trang, L.T.H. Yen, and P.H. Dien. 2014. "Stemofurans X-Y from the roots of *Stemona* species from Laos." *Natural Product Communications* 9 (12): 1741–42. Accessed on January 8, 2023. https://journals.sagepub.com/doi/pdf/ 10.1177/1934578X1400901220.

Quang, D.N., C.T. Pham, L.T.K. Le, Q.N. Ta, N.K. Dang, N.T. Hoang, and D.H. Pham. 2020. "Cytotoxic constituents from *Helicteres hirsuta* collected in Vietnam." *Natural Product Research* 34 (4): 585–89. doi:10.1080/14786419.2018.1490907.

Ren, Y., B.G. Elkington, J.M. Henkin, K. Sydara, A.D. Kinghorn, and D.D. Soejarto. 2021. "Bioactive small-molecule constituents of Lao plants." *Journal of Medicinal Plants Research* 15 (12): 540–59. doi:10.5897/JMPR2021.7137.

Schmid, M. 1989. "Vietnam, Kampuchea and Laos." In *Floristic Inventory of Tropical Countries: The Status of Plant Systematics, Collections, and Vegetation, plus Recommendations for the Future*, edited by D.G. Campbell and H.D. Hammond, pp. 83–91. New York: The New York Botanical Garden. Accessed on January 8, 2023. https://www.gbv.de/dms/bs/toc/025351877.pdf.

Soejarto, D.D., C. Gyllenhaal, J.A.T. Sørensen, H.S. Fong, L.T. Xuan, L.T. Binh, et al. 2007. "Bioprospecting arrangements: Cooperation between the north and the south." In *Proceedings: The 2nd International Conference on Medicinal Mushroom and the International Conference on Biodiversity and Bioactive Compounds*, edited by Biotec, NSTDA, and Ministry of Science and Technology, pp. 5–22. Pattaya. Accessed on January 8, 2023. https://ipmall.law.unh.edu/sites/default/files/hosted_ resources/IP_handbook/ch16/ipHandbook-Ch%2016%2005%20Soejarto%20et%20 al%20Bioprospecting%20Arrangements%20rev.pdf.

Soejarto, D.D., C. Gyllenhaal, M.R. Kadushin, B. Southavong, K. Sydara, S. Bouamanivong, M. Xaiveu, et al. 2012. "An ethnobotanical survey of medicinal plants of Laos toward the discovery of bioactive compounds as potential candidates for pharmaceutical development." *Pharmaceutical Biology* 50 (1) 42–60. doi:10.3109/13880209.2011. 619700.

Soejarto, D.D., C. Gyllenhaal, J.C. Regalado, J.M. Pezzuto, H.S. Fong, G.T. Tan, N.T. Hiep, et al. 2002. "An international collaborative program to discover new drugs from tropical biodiversity of Vietnam and Laos." *Natural Product Sciences* 8 (1): 1–15.

Soejarto, D.D., C. Gyllenhaal, J. Regalado, J. Pezzuto, H. Fong, G.T. Tan, N.T. Hiep, et al. 1999. "Studies on biodiversity of Vietnam and Laos: The UIC-based ICBG program." *Pharmaceutical Biology* 37 (4): 100–13.

Soejarto, D.D., H.J. Zhang, H.H. Fong, G.T. Tan, C.Y. Ma, C. Gyllenhaal, M.C. Riley, et al. 2006. "'Studies on biodiversity of Vietnam and Laos' 1998–2005: Examining the impact." *Journal of Natural Products* 69 (3): 473–81. doi:10.1021/np058107t.

Srivastava, G., A. Matta, G. Fu, R.T. Somasundaram, A. Datti, P.G. Walfish, and R. Ralhan. 2015. "Anticancer activity of pyrithione zinc in oral cancer cells identified in small molecule screens and xenograft model: Implications for oral cancer therapy." *Molecular Oncology* 9 (8): 1720–35. doi:10.1016/J.MOLONC.2015.05.005.

Sydara, K., M. Xayvue, O. Souliya, B.G. Elkington, and D.D. Soejarto. 2014. "Inventory of medicinal plants of the Lao People's Democratic Republic: A mini review." *Journal of Medicinal Plant Research* 8 (43): 1262–74. doi:10.5897/JMPR2014.5534.

Tailler, M., L. Senovilla, E. Lainey, S. Thépot, D. Métivier, M. Sébert, V. Baud, et al. 2012. "Antineoplastic activity of ouabain and pyrithione zinc in acute myeloid leukemia." *Oncogene* 31 (30): 3536–46. doi:10.1038/ONC.2011.521.

Talib, W.H., I. Alsalahat, S. Daoud, R.F. Abutayeh, and A.I. Mahmod. 2020. "Plant-derived natural products in cancer research: Extraction, mechanism of action, and drug formulation." *Molecules* 25 (22): 5319. doi:10.3390/MOLECULES25225319.

Thwaites, G.H.K. 1855. "*Diospyros quaesita.*" *JSTOR Global Plants.* Accessed on January 8, 2023. https://plants.jstor.org/stable/10.5555/al.ap.specimen.k000792466.

Thwaites, G.H.K., and J.D. Hooker. 1864. *Enumeratio Plantarum Zeylaniae: An Enumeration of Ceylon Plants, with Descriptions of the New and Little-Known Genera and Species, Observations on Their Habitats, Uses, Native Names, etc.* London: Dulau & Co. Accessed on January 8, 2023. https://archive.org/details/mobot31753000821030/mode/2up.

Vidal, J. 1959. "Noms Vernaculaires de Plantes En Usage Au Laos." In *Extrait Du Bulletin de l'Ecole Francaise d'Extrême-Orient, Tome XLIX, Fascicule 2.* Paris: l'Ecole Francaise d'Extrême-Orient (EFEO). pp. 435–608. Accessed on January 8, 2023. https://www.persee.fr/doc/befeo_0336-1519_1959_num_49_2_1493.

Wang, L., H. Tang, K. Chen, L.-L. Xue, H.-Y. Ye, L.-F. Ma, and Z.-Y. Li. 2018. "Two new triterpenoids from the stems of *Celastrus orbiculatus* Thunb." *Phytochemistry Letters* 27: 90–93. doi:10.1016/J.PHYTOL.2018.07.001.

Zhang, H.-J., W.-F. Li, H.H.S. Fong, and D.D. Soejarto. 2016. "Discovery of bioactive compounds by the UIC-ICBG drug discovery program in the 18 years since 1998." *Molecules (Basel, Switzerland)* 21 (11): 1448. doi:10.3390/molecules21111448.

Zhang, H.-J., K. Sydara, G.T. Tan, C. Ma, B. Southavong, D.D. Soejarto, J.M. Pezzuto, and H.H.S. Fong. 2004. "Bioactive constituents from *Asparagus cochinchinensis.*" *Journal of Natural Products* 67 (2): 194–200. doi:10.1021/np030370b.

Zhu, H. 2017. "Floristic characteristics and affinities in Lao PDR, with a reference to the biogeography of the Indochina Peninsula." *PLOS ONE* 12 (9): e0184716. doi:10.1371/JOURNAL.PONE.0179966.

4 Conservation of Medicinal Plants of Laos

Djaja Djendoel Soejarto
University of Illinois at Chicago, Chicago, IL, USA
Field Museum of Natural History, Chicago, IL, USA

Kongmany Sydara
Institute of Traditional Medicine, Ministry of Health,
Vientiane, Laos
University of Health Sciences, Vientiane, Laos

Bethany Gwen Elkington
University of Illinois at Chicago, Chicago, IL, USA
Field Museum of Natural History, Chicago, IL, USA

*Bounleuane Douangdeuane, Onevilay Souliya,
and Mouachanh Xayvue*
Institute of Traditional Medicine, Ministry of Health,
Vientiane, Laos

CONTENTS

DOI: 10.1201/9781003216636-4

4.1 INTRODUCTION: THE NEED FOR MEDICINAL PLANT PROTECTION

Laos is the only landlocked country in southeast Asia. It is bordered by Thailand to the west, Myanmar (Burma) to the northwest, China to the north, Vietnam to the east, and Cambodia to the south. The capital and largest city of the country is Vientiane. This area of Southeast Asia is shown on the map in Figure 4.1, which also depicts a network of Medicinal Plant Preserves (MPPs) and Medicinal

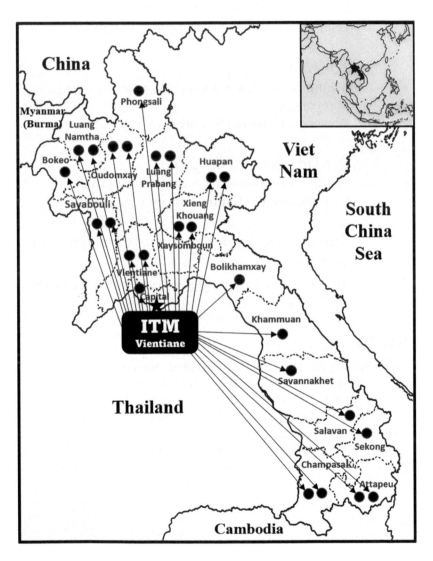

FIGURE 4.1 The Network of MPPs-MBPs in Laos. Credit: O. Souliya (ITM) and D. Soejarto (UIC). The first and only ex situ MPP established in Vientiane Capital, Xaythany District, Sam Sa At Village is marked with a star symbol.

Biodiversity Preserves (MBPs) (indicated by arrows) of the Lao PDR (see also Table 4.1 and Section 4.5). Further discussion about the MPPs and MBPs is presented in detail below.

The Government of Laos has taken measures toward the protection of the rich biodiversity of the country by establishing the National Protected Area (NPA) system (ປ່າສະຫງວນແຫ່ງຊາດ), also referred to as National Biodiversity Conservation Areas (NBCA) (ອຸທິຍານແຫ່ງຊາດ) (ICEM 2003; Chanthakoummane 2008; Gilmour and Tsechalicha 2000; de Koning et al. 2017, WCS Lao PDR 2021). These are forest-covered lands ranging in size from 400 to more than 3,700 sq km, with the majority around 1,000 to 2,000 sq km. The largest of these NPAs is the Nakai Nam Theun (ນາກາຍນ້ຳເທິນ) NPA, with an extension of 3,710 sq km (Clarke n.d.).

Medicinal plants are, without a doubt, protected through this measure. However, these NPAs are not accessible to the population due to their remote locations, physical distance, and protected status. On the other hand, the population has easy access to the more degraded unprotected forest areas, which are still home to many medicinal plants and where healers and members of the community freely enter and collect medicinal plants. Over-collection in such forests may lead to scarcity of many plant species. Unless these forest areas are protected, the supply of medicinal plants could eventually disappear. It is necessary to protect these forest locations that are strategically situated near human populations and to regulate access to them so the medicinal plant species found there can continue to serve the day-to-day health maintenance needs of the population in a more sustainable manner. Such protection will also allow time for researchers to continue to study these medicinal plants to identify their medicinal properties.

4.2   RARE, THREATENED, AND HEALTH-PROMOTING VALUE OF SPECIES OF MEDICINAL PLANTS OF LAOS

Many species of the country's plants are considered at-risk, either threatened or endangered. There were 96 at-risk species of medicinal plants listed as worthy of protection in Prime Minister Decree 155 of 2003 (Prime Minister of Laos 2003). In a later document, the list was reduced to 30 species (Prime Minister of Laos 2007). In both cases, the plants listed face a significant risk, hence their protection is a serious matter. In our fieldwork to study medicinal plants, we had a firsthand opportunity to observe, document, and collect some of these and other rare plants.

The following factors appear to be some of the causes that led to the risks and concerned status of these species:

- Lack of awareness of sustainable use of these natural resources.
- Weakness of implementation of the rules, regulations, and laws of biodiversity conservation.
- Lack of cooperation among regulatory authorities.
- Inappropriate techniques in the collection of plant materials.
- Increased demand for raw materials in international markets.

TABLE 4.1

MPPs (ສວນອະນຸລັກພືດເປັນຢາ) and MBPs (ສວນອະນຸລັກຊີວະນາໆພັນຫຼົດເປັນຢາ) of Lao PDR as of December 31, 2021

Name of Preserve; Village/District/ Province Location	Geographic Coordinates; Altitude	Size (hectares)	Year Established	Supported by*
1. Attapeu Province MBP (in-situ) (ສວນອັດຕະປື); Phouvong (ເມືອງພູວົງ) District	14°42' N 106°43' E; 105 m asl	36	2014	GoL
2. Attapeu Province MBP (in-situ); Sanxai District (ເມືອງຊານໄຊ)	14°40' N 106°34' E	40	2016	GoL
3. Bokeo Province MBP (In-Situ); Namepouk, Huaysai District (ບ້ານນ້ຳປຸກ, ເມືອງຫ້ວຍຊາຍ)	20°13' N 100°34' E; 372 m asl	100	2013	LBF
4. Bolikhamxay Province MPP (in-situ); Somsavath Village, Paksan District (ບ້ານສົມສະຫວາດ, ເມືອງປາກຊັນ)	18°27' N 103°48' E; 163 m asl	15	2004	GoL, ICBG
5. Champasak Province MBP (in-situ); Champasak District (ເມືອງຈຳປາສັກ)	14°52' N 105°50' E; 107 m asl	40	2018	GoL
6. Champasak Province MPP (in-situ); Dongkalong Village, Pakse District (ເມືອງປາກເຊ)	15°09' N 102°34' E; 180 m asl	4	2008	GoL, LBF
7. Huapan Province (ສວນຫົວພັນ) Xamtai District (ເມືອງຊຳໄຕ້)	19°59' N 104°37' E; 680 m asl	200	2018	GoL
8. Huapan Province MBP (in-situ); Hiem District (ເມືອງຫ້ຽມ MBP in-situ)	20°05' N 103°22' E	30	2019	GoL
9. Khammuan Province MBP (in-situ); Khounkham District (ເມືອງຄູນຄຳ)	17°57' N 104°45' E; 142 m asl	300	2019	GoL, KRIBB
10. Luang Namtha Province MBP (in-situ); Chaleunsouk Village, Namtha District (ເມືອງນ້ຳທາ)	20°52' N 101°20' E; 798 m asl	10	2010	GoL, LBF
11. Luang Namtha Province MBP (in-situ); Lak Khan Mai Village, Namtha District (ເມືອງນ້ຳທາ)	21°08' N 101°16' E; 798 m asl	300	2021	GoL
12. Luang Prabang Province MBP (in-situ); Huaykhing Village, Phonxay District (ເມືອງໂພນໄຊ)	19°56' N 102°45' E; 1300 m asl	2.352	2013	TABI
13. Luang Prabang Province MBP (in-situ); Pak Ou District (ເມືອງປາກອູ)	20°02' N 102°12' E; 477 m asl	245	2017	GoL
14. Oudomxay Province MBP (in-situ); Donxay Village, Xay District (ບ້ານດອນໄຊ, ເມືອງໄຊ)	20°40' N 101°57' E; 680 m asl	15	2009	GoL, LBF

Site	Coordinates	No.	Year	Organization
15 Oudomxay Province MBP (in-situ); Nga District (ເມືອງງາ)	20°34' N 102°06' E	40	2018	GoL
16. Phongsali Province MBP (in-situ); 24 Phoufa Village, Phongsaly District (ເມືອງໜອງຄ້າວ)	21°41' N 102°06' E	300	2019	GoL
17. Salavan Province MBP (in-situ); Salavan District (ເມືອງສາລະວັນ)	15°43' N 106°24' E	40	2019	GoL
18. Savannakhet Province MPP (in-situ); Donghang Village, Kaysone District (ເມືອງໄກສອນ)	16°39' N 104°51' E; 159 m asl	15	2008	GoL, LBF
19. Sekong Province MBP (in-situ); Palengtai Village, Thateng District (ເມືອງທ່າແຕງ)	15°28' N 106°20' E; 798 m asl	30	2013	GoL, LBF
20. Vientiane Capital MBP (in-situ); Sangthong District (ເມືອງສັງທອງ)	18°13' N 102°20' E	10	2018	GoL
21. Vientiane Capital MPP (ex-situ); Sam Sa At Village, Xaythany District (ເມືອງໄຊທານີ)	18°10' N 105°46' E; 130 m asl	6	1998	GoL, ICBG
22. Vientiane MBP (in-situ); Vientiane Province, Feuang District (ເມືອງເຟືອງ)	18°50 N 101°04' E; 257 m asl	22	2014	GoL, LBF
23. Vientiane MBP (in-situ); Vientiane Province, Met District (ເມືອງແມດ)	18°41 N 102°55' E; 320 m asl	49	2020	GoL
24. Sayabouli Province MBP (in-situ); Pha Xang Sayabouli District (ພາຊ້າງ, ເມືອງໄຊຍະບູລີ)	19°04' N 101°45' E; 816 m asl	30	2014	ITM
25. Sayabouli Province MBP (in-situ); Paklay District (ເມືອງປາກລາຍ)	19°04' N 101°45' E; 816 m asl	30	2021	ITM
26. Xieng Khouang Province MBP (in-situ); Tha Village, Kham District (ເມືອງຄຳ)	19°42' N 103°35' E; 1129 m asl	500	2010	GoL, LBF
27. Xieng Khouang Province MBP (in-situ); Phoukout District (ເມືອງພູກູດ)	19°32' N 103°00' E; 1148 m asl	323	2015	ABP

* Acronyms. *ABP* - Agro-biodiversity Project of the Food and Agriculture Organization of the United Nations (ໂຄງການຊີວະນາໆພັນກະສິກຳຂອງອົງການອາຫານ ແລະ ການກະເສດ ອົງການສະຫະປະຊາຊາດ); *GoL* - Government of the Lao PDR (Laos) (ລັດຖະບານແຫ່ງ ສປປ ລາວ ຫຼື ລາວ); *ICBG* - International Cooperative Biodiversity Groups (ກຸ່ມປະສານງານສາກົນດ້ານຊີວະນາໆພັນ); *ITM* - Institute of Traditional Medicine (ສະຖາບັນການແພດແຜນບູຮານ ແລະ ການແພດພື້ນເມືອງ); *KRIBB* - Korea Research Institute of Bioscience & BioTechnology (ສະຖາບັນຄົ້ນຄວ້າວິທະຍາສາດ ແລະ ຊີວະເຕັກໂນໂລຊີຂອງເກົາຫຼີ); *LBF* - Lao Biodiversity Fund (ກອງທຶນຊີວະນາໆພັນຂອງລາວ); *TABI* - The Agro-Biodiversity Initiative (ໂຄງການລິເລີ່ມຄວາມຫຼາກຫຼາຍຂອງຊີວະນາໆພັນ).

The collection of medicinal plants is either a day-to-day chore in response to the health care needs of the population, or, on a larger scale, an effort to reduce poverty in the local communities. Inefficient collection methods and lack of properly coordinated management of collection practices have had a serious negative impact on the richness of the biodiversity (Schippmann et al. 2002). Medicinal plants are just one component of the biodiversity that is lost. Other living organisms are affected as well. Rare medicinal plant species may enter into a risk domain as a result of uncontrolled, over-collection practices to generate income, in particular to satisfy demands due to the increased volume of exportation. Some of these species are listed in Figure 4.2 (Souliya and Sydara 2015) and Table 4.2 (Sydara et al. 2014).

A contemporary example of plants or plant group facing endangerment is provided by the medicinal plant genus *Dalbergia* (Fabaceae-Papilionoideae) (also see Chapter 5). On May 12, 2022, the Convention on International Trade and Endangered Species of Wild Fauna and Flora (CITES) Secretariat issued a "Notification to the Parties" No. 2022/030 "To continue to suspend commercial trade of specimens of the genus *Dalbergia* spp., including finished products, such as carvings and furniture from Lao PDR, until Lao PDR makes scientifically based non-detriment findings for trade in the relevant species, including *D. cochinchinensis* and *D. oliveri* ..." (CITES 2022). Another example is the species *Afzelia xylocarpa* (Kurz) Craib (Fabaceae-Caesalpinioideae) (see Chapter 5), which is listed in the International Union for Conservation of Nature (IUCN) Red List as an endangered species (IUCN Red List 2022).

The following provides brief summaries as examples to highlight the medicinal importance of the plant species listed in Table 4.2.

(Apocynaceae) *Rauvolfia serpentina* (Kha Yom Phou). Known as Rauwolfia or snakeroot, this plant is native to India and other Southeast Asian countries. It has been used to treat high blood pressure and many other diseases in Ayurvedic medicine for centuries. Its medicinal value comes from the presence of many indole alkaloids in the root that include reserpine, yohimbine, serpentine, deserpidine, ajmalicine, and ajmaline (Kumari et al. 2013). Reserpine is used as a first line treatment as an oral antihypertensive medication and was one of the first antihypertensive agents introduced into clinical practice (Shamon and Perez 2016; Cheung and Parmar 2022; PubChem 2021a). Another alkaloid, ajmalicine, a monoterpenoid indole alkaloid, is a drug also used to treat hypertension (PubChem 2021b). The world's needs for dried Rauwolfia roots (20,000 tons annually) for drug manufacturing is beyond the capacity of India, the largest supplier, to meet (ca. 650 tons). This situation is the result of continuing root collections from natural areas, which led to the endangerment of the species (Babu 2019). Only large-scale cultivation in suitable areas under good agricultural practices can supply the world's needs of Rauwolfia roots (Dutta, Chopra, and Kapoor 1963; Sharma 2011).

(Araliaceae) *Panax vietnamensis* (Pom Bi Kha Ting). The history of this species is rather recent, dating back to 1973, when the plant was discovered in Vietnam as a "secret medicine" of the Sedang ethnic group in the high mountains of the Truong

FIGURE 4.2 A poster titled "Some Threatened and Rare Medicinal Plant Species of Lao PDR." (Souliya and Sydara 2015).

TABLE 4.2
Rare and Threatened Medicinal Plant Species of Laos[a]

Local (Lao) Name	Scientific Name (Family) Species	Traditional Use and Disease Treated	Part Used
Kha Yom Phou ຂະຍອມຜູ້ນ້ອຍ	(Apocynaceae) *Rauvolfia serpentina* (L.) Benth. ex Kurz	Treat high blood pressure	Root
Pom Bi Kha Ting ປອມບິກະທິງ	(Araliaceae) *Panax vietnamensis* Ha et Grushv.	As a tonic	Rhizome
Lep Meu Nang ເລັບມືນາງ	(Araliaceae) *Schefflera elliptica* Harms	Treat rheumatism; as a tonic	Stem bark
Wan Hua Toh ຫວ້ານຫົວໂຕ້	(Asparagaceae) *Disporopsis longifolia* Craib	As a tonic	Rhizome
Chan Dai Deng ຈັນໄດແດງ	(Asparagaceae) *Dracaena cambodiana* Pierre ex Gagn.	As a blood tonic, purifier	Stem wood
Khing Pa ຂີງປ່າ	(Asparagaceae) *Polygonatum kingianum* Collett & Hemsl.	As a tonic; treat diabetes	Root/ rhizome
Man Kha Kai ມັນຂາໄກ່	(Campanulaceae) *Codonopsis pilosula* (Franch.) Nannf.	Used as a tonic	Rhizome
Leu Lang Lai ລືຫຼັງລາຍ	(Gesneriaceae) *Aeschynanthus longicaulis* Wall. ex R. Br.	Treatment of nervous disorders	Whole plant
Khao Kai ເຂົ້າໄກ່	(Hypoxidaceae) *Curculigo orchioides* Gaertn.	As a tonic	Root/ rhizome
Ii Tu Ton ອີຕູ່ຕົ້ນ	(Lauraceae) *Cinnamomum camphora* (L.) J.Presl	Treatment of vertigo/ dizziness	Essential oil
Mai Tha Lo ໄມ້ທະໂລ້	(Lauraceae) *Cinnamomum parthenoxylon* (Jack) Nees	Treatment of vertigo/ dizziness	Essential oil
Yang Bong ຍາງບົງ	(Lauraceae) *Litsea monopetala* (Roxb.) Pers.	Treat flatulence; shooting pain	Stem
Seng Reua ແສງເບືອ	(Loganiaceae) *Strychnos nux-vomica* L.	Treat rheumatism, joint pain, paralysis and as an antispasmodic	Seed
Tin Houng ຕີນຮຸ້ງ	(Melanthiaceae) *Paris marmorata* Stearn	As a tonic; treat stomach pain	Rhizome
Kheua Hem ເຄືອເຮັມ	(Menispermaceae) *Coscinium fenestratum* (Gaertn.) Colebr.	Treat dysentery, diarrhea, diabetes	Liana
Hoa Tom Ngeun ຫົວຕ່ອມເງິນ	(Menispermaceae) *Stephania rotunda* Lour.	Treat insomnia	Tuber
Hat Mee ຫາດມີ່	(Moraceae) *Artocarpus lakoocha* Roxb.	Treatment of tape worms	Stem wood
Koud Tin Houng ກູດຕີນຮຸ້ງ	(Ophioglossaceae) *Helminthostachys zeylanica* (L.) Hook.	Treat fever; as a tonic	Root
Man On Ling ມັນອ້ອນລີງ	(Polygonaceae) *Fallopia multiflora* (Thunb.) Czerep.	As a tonic; treat back pain, heat inside	Fruit
Ian Don ອ່ຽນດ່ອນ	(Simaroubaceae) *Eurycoma harmandiana* Pierre	As a tonic; to treat fever	Root
Mak Chong Ban ໝາກຈອງບານ	(Sterculiaceae) *Sterculia lychnophora* Hance	As a cold medicine; treat heat inside the body	Fruit

[a] Source: Updated from Sydara et al. 2014; ITM document.

Son Range. It was later given the scientific name *Panax vietnamensis* (Yamasaki 2000, Nguyen and Phuong 2019). The plant was regarded as a life-saving medicinal plant used to treat many serious diseases and to enhance health and physical strength that includes CNS stimulant, antifatigue, antibacterial and expectorant effect, hypoglycemic effect, improving heart contractility, decreasing blood cholesterol, and having an adaptogenic effect. The chemical constituents found in the rhizome of this plant that are responsible for the above effects are triterpene glycosides that include dammarane saponins named vina-ginsenosides-R-$_{1-24}$ (Nguyen and Nguyen 1998). Overall, the chemistry and pharmacology of *Panax vietnamensis* are similar to those of *Panax ginseng* and other cultivated *Panax* species. A recent review paper on the traditional uses, chemistry, and pharmacology of the genus *Panax* provides further details on the chemistry and pharmacology of *Panax vietnamensis* (Liu et al. 2020).

(Araliaceae) *Schefflera elliptica* (Lep Meu Nang). The resinous sap from this plant is used as a vulnerary, while the bark is used to treat coughs, and the leaf decoction is used as a bath to treat cases of scurvy (Fern et al. 2014). A recent paper (Wang et al. 2021) mentions that 14 species of the genus *Schefflera* have been documented in traditional medicines in China, India, Vietnam, Thailand, and Indonesia, specifically to manage rheumatism, pain, and trauma. Chemically, the main phytochemical constituents that have been identified include triterpenoids and saponins, with sesquiterpenes, phenylpropanoids, and lignans, while pharmacologically, the extracts and pure isolated compounds from the genus *Schefflera* have the properties of analgesic, anti-inflammatory, anticancer, hypoglycemic, antimicrobial, hepatoprotective, neuroprotective, antimalarial, and antiallergic effects (Wang et al. 2021).

(Asparagaceae) *Disporopsis longifolia* (Wan Hua Toh). In Vietnam this plant species is protected under the "endangered" status (Nguyen et al. 2016) and is medicinally used to "nourish yin," hence "to lessen fire," and to invigorate the liver and the kidney (Thu Ha et al. 2021). The phytochemistry of members of *Disporopsis* is mainly characterized by the presence of steroidal saponins and phenolic compounds, which have showed antitumor, anti-inflammatory, anti-diabetes, and anti-microbial activities (Thu Ha et al. 2021). A recent study of the rhizome of *D. longifolia* found two new and five known spirostanol glycosides that inhibit nitric oxide (NO) production in lipopolysaccaharide (LPS) stimulated RAW 264.7 macrophage cells (Thu Ha et al. 2021).

(Asparagaceae) *Dracaena cambodiana* (Chan Dai Deng). This plant is one of more than 100 species of the genus *Dracaena* belonging to the family Asparagaceae, distributed mainly in tropical Africa, Asia, and Australia (Kew 2022f). Many species of this genus exude a medicinally important red sap or resin (called "dragon's blood") when the stem is cut. In the Pharmacopoeia of the People's Republic of China, only *Dracaena cochinchinensis* is listed as the source of dragon's blood (Z. D. He et al. 2006; Zhonglian Zhang et al. 2019). According to a recent review (Y. Liu et al. 2021), eleven *Dracaena* species are producers of the red resin (dragon's blood) containing 300 compounds, mainly flavonoids, steroids and phenolics, while the pharmacological activities include anti-inflammatory, analgesic, antithrombotic, antioxidant, antimicrobial, antidiabetic, and anticancer. These activities are responsible for the dragon's blood's wound healing effects, preventive effects on cardiovascular and cerebrovascular diseases, dual-directional regulation of blood flow, neuroprotection, and radioprotective effects (Y. Liu et al. 2021). It must be noted, however, that several other species of angiosperms belonging to different genera also exude resinous red sap often called "dragon blood," which is not to be confused with the true dragon's blood derived from species of *Dracaena*. Of special interest is "dragon's blood" ("sangre de drago" in Spanish) derived from *Croton lechleri* Muell.- Arg., native to S. America from Colombia south to Bolivia (Kew 2022g), also locally used as a medicine, and from which the drug Crofelemer was developed (Mader 2013).

(Asparagaceae) *Polygonatum kingianum* (Khing Pa). This species is one of approximately 60 species of the genus *Polygonatum* distributed in the temperate regions of East Asia, including Laos. Various species of this genus have found medicinal uses among traditional healers to treat many human ailments, some of which have been validated experimentally, such as antihyperglycemic potential, anticancer, analgesic, antipyretic, diuretic, antimalarial, antioxidant, and antimicrobial effects. The phytochemicals responsible for these activities include saponins, phytohormones, glycosides, flavonoids, and alkaloids (Khan, Saeed, and Muhammad 2012a). Another study (Khan et al. 2012b) showed that the crude extract and fractions of the rhizome of another species, *P. verticillatum*, showed broad-spectrum antibacterial activity against gram negative bacteria. A study in rats fed a high fat diet showed that *P. kingianum* regulated the abundance and diversity of the intestinal microbial community through increasing the relative abundance of gut bacteria that produce short chain fatty acids (Gu et al. 2020). Consequently, this promotes the recovery of the intestinal permeability barrier, inhibits long chain polysaccharides' entry into the circulation, alleviates inflammation, and prevents glucose and lipid metabolism disorders.

(Campanulaceae) *Codonopsis pilosula* (Man Kha Kai). This plant is one of 42 species belonging to the genus *Codonopsis*, mainly distributed in Central, East, and Southeast Asia (J.-Y. He et al. 2015). The root of a number of species, including *C. pilosula*, known as Radix Codonopsis or Dangshen, has been used in traditional Chinese medicine as early as the Qing Dynasty and was claimed to strengthen the spleen, tonify the lung, nourish the blood, replenish energy deficiency, strengthen the immune system, lower blood pressure, and improve appetite (Li et al. 2009; Gao et al. 2018a). Chemical compounds found in the plant include polyacetylenes, polyenes, flavonoids, lignans, alkaloids, coumarins, terpenoids, steroids, organic acids, and saccharides, and are responsible for diverse pharmacological activities that regulate the immune system, hematopoiesis, cardiovascular protection, neuroprotection, gastrointestinal function, endocrine function regulation, cytotoxic and antibacterial effects, anti-aging, and anti-oxidation (Gao et al. 2018a).

(Gesneriaceae) *Aeschynanthus longicaulis* (Leu Lang Lai). This plant is one of the 150 species belonging to the genus *Aeschynanthus* with a distribution range spanning Nepal, India, and Sri Lanka, through Indo China, southern China,

Thailand, Malaysia, and Indonesia east to the Solomon Islands (Nasution 2003, 2016). The native range of *A. longicaulis*, however, is from Andaman & Nicobar Islands, China (S. Yunnan), Indo China to NW. Peninsula Malaysia (Kew 2022h). The species is a popular ornamental plant, and the leaves are used as a poultice to treat boils in Malaysia (Nasution 2003, 2016). It has not been documented from Laos (Newman et al. 2007), but there are records indicating the plant has been used in the country to treat cases of nervous disorders (Table 4.2 above). Chemically, the species contains caffeoyl esters, comoside, verbacoside, and iso-verbacoside (Jensen 1996).

(Hypoxidaceae) *Curculigo orchioides* (Khao Kai, Xi Din). *C. orchioides* has a wide distribution range in tropical and subtropical Asia from India to New Guinea, including China, Japan, Indo China, Malaysia, the Philippines, and Indonesia (Kew 2022i), and is considered an endangered species in India (Chauhan et al. 2010). In traditional Chinese medicine, the plant has been used for the treatment of various diseases that include impotence, limb limpness, arthritis of the lumbar and knee joints, and watery diarrhea. It is also used as a potent immunomodulator and aphrodisiac in the Ayurvedic medical system (Chauhan et al. 2010; Nie et al. 2013). The rhizome contains curculigoside (more than 0.1% as determined by HPLC analysis). *C. orchioides* and other species have been shown to have a wide spectrum of pharmacological activities, including adaptive, immunostimulatory, taste-modifying and sweet-tasting, antioxidant, mast cell stabilization, antihista-minic and anti-asthmatic, hepatoprotective, and neuroprotective activity. Toxicological testing indicated that *Curculigo orchioides* at the dose of 120 g/kg after administration on rats for 180 days may cause injury of liver and kidney (Nie et al. 2013).

(Lauraceac) *Cinnamomum camphora* (Ii Tu Ton). *C. camphora* is one of the approximately 250 species of the genus *Cinnamomum* distributed mainly in the tropics and subtropics of Asia, Australia, and the Pacific Islands (Li et al. 2008, 102). Many species are extensively used as local and traditional medicines for the treatment of various disorders. *C. camphora*, distributed in Japan, China, Korea, and northern Vietnam and cultivated in many countries (Li et al. 2008), is a source of camphor and camphor oil. It is used to treat rheumatism, sprains, bronchitis, asthma, indigestion, muscle pain, diarrhea, menstrual disorders, colds, and chills (Wang et al. 2020). Diverse chemical compounds that include essential oil chemo-types, phenylpropanoids, lignans, flavonoids, and aliphatic compounds are found in the aerial part of the plant, specifically the leaves, and also in the bark. D-camphor, 3-methyl-2-butenoic acid, eucalyptol, and 1,6-octadien-3-ol are responsible for the antimicrobial property of *C. camphora* (Kumar and Kumari 2019; Wang et al. 2020).

(Lauraceae) *Cinnamomum parthenoxylon* (Mai Tha Lo). *C. parthenoxylon* is a synonym of *Cinnamomum porrectum* (Roxb.) Kosterm. (WFO 2021) and is widely distributed in tropical Asia from India east to Myanmar, China, Vietnam, Cambodia, south to Thailand, Malaysia, the Philippines, and Indonesia (R.P.J. de Kok 2019), where it is used as a spice, flavoring, vegetable, carminative, tonic, febrifuge, and in postpartum care (Uphof 2001; Pukdeekumjorn et al. 2016).

Chemically, the plant is rich in essential oils containing monoterpenes and sesqui-terpenes (linalool, elemol, eucalyptol, β-caryophyllene, camphor, α-eudesmol, citral, L-α-terpineol), also flavonoid rutinosides (rutin, nicotiflorin, isorhoifolin) and phenolic oligomer, with diverse pharmacological activities that include hypo-glycemic, hepatoprotctive, antioxidant, anti-inflammatory, and antileukemic activities (Jia et al. 2009; Adfa et al. 2016; Pukdeekumjorn et al. 2016; Sukcharoen et al. 2017; Pardede et al. 2017; Qiu et al. 2019). *C. parthenoxylon* is listed in the IUCN Red List (de Kok 2020b).

(Lauraceae) *Litsea monopetala* (Yang Bong). *L. monopetala* is a synonym of *L. polyantha* Juss. It is an Asian member of the approximately 400 species of the genus *Litsea*, which is widely distributed in the tropics and subtropics of Asia, North America, and South America (R.P.J. de Kok 2021; Kew 2022c; Wang et al. 2016). In Ayurvedic medicine, the bark of the plant is used as a stimulant, astrin-gent, spasmolytic, and antidiarrheal, while the root is used externally to treat pains, bruises, and contusions (H. Hasan et al. 2014). Research papers on the biol-ogy and chemistry of *L. monopetala* have been published under both Latin bino-mials. Chemically, the leaf, stembark, and root contain essential oils, reducing sugar, tannins, flavonoids, alkaloids, terpenoids, and phenol, aldehydes, alcohols, and acids. Among the components of the essential oils are α-caryophyllene alco-hol, pentacosane, caryophyllene oxide, humulene oxide and tricosane (Choudhury et al. 1997; Biswas et al. 2017; Kumari et al. 2019), while biologically, the plant has been shown to have antioxidant, antibacterial, antifungal, antihyperglycemic, neuropharmacological, and cytotoxic activities (Hasan et al. 2014; Fakhrul Hasan et al. 2016; Biswas et al. 2017). *L. monopetala* is listed in the IUCN Red List as a plant of Least Concern (de Kok 2020b).

(Loganiaceae) *Strychnos nux-vomica* (Seng Beua). This plant is native to India east to Thailand, Indo-China, and Malay Peninsula, and is an introduced species in many tropical countries (Kew 2022e). In Asia, the dried seeds are traditionally used for the treatment of neural disorders, arthritis, diabetes, asthma, aphrodisiac, and to improve appetite (Bhati et al. 2012; Patel et al. 2017). Chemically, 84 com-pounds have been isolated from the seed, comprising, among others, two primary alkaloids, strychnine and brucine, plus other minor alkaloids protostrychnine, vomicine, *n*-oxystrychnine, pseudostrychnine, isostrychnine, together with chlo-rogenic acid and glycosidic alkaloids, iridoid glycosides, flavonoid glycosides, triterpenoids, steroids, and organic acids (Bhati et al. 2012; Guo et al. 2018). Pharmacological activities that include effects on nervous system disorders, pain, inflammation, microbial infection, gastrointestinal problems, nervous system dis-orders, problems related to cardiovascular systems, cancer, and diabetes, as well as antioxidant activity and antifeedant activity, have been validated (Bhati et al. 2012; Patel et al. 2017; Guo et al. 2018).

(Malvaceae) *Sterculia lychnophora* (Mak Chong Ban). The name *S. lych-nophora* Hance is a synonym of *Scaphium affine* (Mast.) Pierre, whose geographic distribution spans Indochina, Thailand, the Malay Peninsula, and western Indonesia (Wilkie 2009). The mature, ripened dried seeds of this plant (*Pangdahai* in Chinese as recorded in the 2015 edition of Chinese Pharmacopoeia) have been used widely in traditional Chinese medicine, Japanese folk medicine, Vietnamese traditional medicine, traditional Thai medicine, and Indian traditional medicine to treat constipation, pharyngitis, and pain (Li 2015; Oppong et al. 2018). These uses have the support of pharmacological studies, such as anticancer effects (Kawk et al. 2021), antibacterial effects against *Streptococcus mutans* (Yang et al. 2016), and a neuroprotective effect (Wang, Wu, and Geng 2013). Chemically, the seed of the plant is rich in polysaccharides, alkaloids, phenolics, glycosides, and peptides and fatty acids, with poorer presence of steroids, terpenoids, flavonoids, and organic acids (Wang et al. 2003; Zeng and Liu 2011; Yang et al. 2021).

(Melanthiaceae) *Paris marmorata* (Tin Houng). *P. marmorata* Stearn is a close relative of *P. polyphylla* Sm. and has been synonymized under the name of *Paris polyphylla* ssp. *marmorata* (Stearn) H. Hara (WFO 2022c). It has a geographic distribution that spans Nepal and Bhutan east to Southwest China (i.e., Yunnan, Sichuan, and Tibet). Taxonomically, *P. marmorata*, together with *P. luquanensis* H. Li, belong to *Paris* Section *Marmorata*, which contains *Rhizoma Paridis* saponins, including diosgenyl and pennogenyl saponins, active ingredients in the treatment of tumors, hemostasis, and inflammation (X. Gao et al. 2018b). Members of *Paris* Sect. *Marmorata* have long been used as a traditional oriental medicine in the regions of its native range, but unfortunately the natural populations of both *P. marmorata* and *P. luquanensis* have declined significantly due to overexploitation for their economic value (X. Gao et al. 2018b). Although the conservation status of *Paris* Sect. *Marmorata* is not known with certainty, its close relative, *P. polyphylla*, is listed as "vulnerable" in the IUCN Red List (Chauhan 2020).

(Menispermaceae) *Coscinium fenestratum* (Kheua Hem). This medicinal plant is indigenous to tropical Asia, spanning India east to Cambodia, Laos, Vietnam, Thailand, the Malay Peninsula, and western Indonesia (Kew 2022j). It is a critically endangered species and highly traded medicinal plant (Tushar et al. 2008; Roopashree et al. 2021), although its conservation status is ranked in the category "Data Deficient" in the IUCN Red List (Ved et al. 2015). *C. fenestratum*, which is medicinally used to treat inflammations, wounds, ulcers, jaundice, burns, skin diseases, abdominal disorders, diabetes, fever, and general debility, contains the isoquinoline alkaloid berberine as the active compound (Tushar et al. 2008). Berberine from *C. fenestratum* also has antioxidant activity (Roopashree et al. 2021), while the plant extract has been shown to have antiproliferative activity (Ueda et al. 2002).

(Menispermaceae) *Stephania rotunda* (Hoa Tom Ngeun). *S. rotunda* is native to the Indian subcontinent east to Nepal, Tibet, Bangladesh, Myanmar, Laos, Vietnam, Cambodia, and Thailand (Kew 2022d). The stem, leaves, and tubers have been used in the traditional medicine of these countries to treat about twenty health disorders, including asthma, headache, fever, and diarrhea (Desgrouas et al. 2014). The roots primarily contain l-tetrahydropalmatine (l-THP), whereas the tubers contain cepharanthine and xylopinine. These alkaloids exert antiplasmodial, anticancer, and immunomodulatory effects, while sinomenine, catharanthine, and l-stepholidine are the most promising and have been tested in humans (Desgrouas et al. 2014).

(Moraceae) *Artocarpus lakoocha* (Hat Mee). *A. lakoocha* Roxb., a synonym of *A. lacucha* Buch.-Ham. and *A. dadah* (WFO 2022a), is distributed from India east to S. Central China, and south to Indo-China and Thailand (Kew 2022a). Most published reviews and research papers on this plant have used the name *A. lakoocha*, rarely *A. lacucha*. It is listed as endangered in Singapore under the synonym of *A. dadah* (Singapore Government Agency 2022). The ripe fruit is an Ayurvedic medicine to strengthen digestion, to improve taste, and is used as an aphrodisiac, while traditionally the fresh pulp is used as a liver tonic, with the seeds and milky latex used as a purgative (D'Souza 2019), the bark to heal wounds and ulcers (D'Souza 2019) and to treat stomachache, headache, and boils (Islam et al. 2019). The fruit extract contains quercetin, gallic acid, vanillic acid, cinnamic acid, ferulic acid and kaempferol, and has been shown to be hepatoprotective (Saleem et al. 2018). It has also been shown to have antimicrobial activity against oral pathogens (Teanpaisan et al. 2014), while the bark has been shown to have antinociceptive effects (Islam et al. 2019). In Laos, the Pharmaceutical Development Center (Pharmaceutical Factory No. 3) used the heartwood of this plant to extract the active ingredients and marketed the product (tablets) under the name "Artocarpine" for the treatment of tapeworm (Sydara, personal communication, July 17, 2022).

(Ophioglossaceae) *Helminthostachys zeylanica* (Koud Tin Houng). *H. zeylanica* is a pteridophyte (fern) species with geographic distribution spanning from India and Sri Lanka east to Bangladesh, Myanmar, southern China, Japan, Indo-China, Malaysia, Philippines, Indonesia, New Guinea, Australia, and western Polynesia (Kew 2022b; GBIF 2021). It is a rare and endangered plant in India, Nepal, and China (Fellowes et al. 2003; Chandra et al. 2008; Bharali et al. 2017; Wu et al. 2017a; Ojha et al. 2021), with many medicinal uses, such as treatment of dysentery, catarrh, sciatica, malaria, fever, pneumonia, inflammation, and burns (Bharali et al. 2017; Wu et al. 2017a; Wu et al. 2017b). Laboratory studies have shown anti-inflammatory (Y.C. Huang et al. 2009; Y.L. Huang et al. 2017; Su et al. 2016; Hsieh et al. 2016), antimicrobial (Yenn et al. 2018), anti-osteoporotic (Y.L. Huang et al. 2017), antioxidant (Y.L. Huang et al. 2003), and melanogenic activities (Yamauchi et al. 2015). Chemically, *H. zeylanica* is rich in various types of flavonoids responsible for the biological activities of this plant (luteolin, quercetin, apigenin, chrysin, ugonins) (Y.L. Huang et al. 2003; Yamauchi et al. 2015; Su et al. 2016; Y.C. Huang et al. 2009; Y.L. Huang et al. 2017; Wu et al. 2017a; Wu et al. 2017b).

(Polygonaceae) *Fallopia multiflora* (Manh Onh Ling). *Fallopia multiflora* (Thunb.) Czerep, a synonym of *Reynoutria multiflora* (Thunb.) Moldenke, the legitimate name (WFO 2022b), has a geographic range from China south to Indochina, Taiwan, and Thailand. The nomenclature of this plant is confusing since various synonymous Latin binomials, including *Polygonum multiflorum*, *Fagopyrum multiflorum*, *Pleuropterus multiflorus*, *Tiniaria multiflora*, *Reynoutria multiflora*, and *Fallopia multiflora*, have been published to refer to this taxon. Research papers (Hao et al. 2015; Lin et al. 2015; T.T. Nguyen et al. 2018; Zhai et al. 2021) and public posting (DrugBank Online 2022) on the medicinal uses, chemistry, and pharmacology have used the binomial *Reynoutria multiflora*, *Fallopia multiflora*, and *Polygonum multiflorum*. This plant is listed in the Chinese

Pharmacopoeia under the name of *Polygonum multiflorum*. It is one of the most popular Chinese traditional medicines called "He shou wu" in China and Asia and "Fo-ti" in North America, and is used as a tonic to treat liver injury, cancer, diabetes, alopecia, atherosclerosis, and neurodegenerative diseases (Bounda and Feng 2015). Chemically, a series of chromones, lignans, anthraquinones, isobenzofurans, phenolic acid, and pyran are found in the plant (Z. Zhang et al. 2021).

(Simaroubaceae) *Eurycoma harmandiana* (Ian Don). *E. harmandiana* has a distribution range throughout Indo China (Kew 2022k) and is a rare plant in Laos, found in the southern part of the country (Savannakhet, Champasak). This plant is well known for its ability to enhance male sexual performance, as an antioxidant, and to treat cases of inflammation. The chemicals primarily responsible for these activities are a series of quasinoids, canthin-6-one alkaloids, and scopoletin (Kanchanapoom et al. 2001; Chaingam et al. 2021, 2022; Choonong et al. 2022).

4.3 THE SOMSAVATH MEDICINAL PLANT PRESERVE (MPP)

The Somsavath Medicinal Plant Preserve (MPP) (ສວນອະນຸລັກພືດຕົບຢາໆ ສົມສະຫວາດ), a tract of secondary forest (Figure 4.3) located about 1 km from, and within the jurisdiction of Somsavath Village (ບ້ານສົມສະຫວາດ), Paksan District, Bolikhamxay Province, was known to harbor many medicinal plant

FIGURE 4.3 A profile of Somsavath forest in 2004. Photo credit: D. Soejarto.

FIGURE 4.4 Somsavath forest in the background was under threat of elimination in 2004. Photo credit: D. Soejarto.

species at the time that the International Cooperative Biodiversity Group (ICBG) (ກຸ່ມປະສານງານຊີວະນາງພັນສາກົນ) Project operated in Laos in 2004 (Soejarto et al. 2009). Somsavath is approximately 20 km from Paksan, capital city of Bolikhamxay province, by road. Healers and members of the communities of Somsavath and other villages freely enter this tract of forest to collect medicinal plants. It is widely known that unregulated collection practices may lead to the exhaustion of the population of certain sought-after species. Such species may become rare and "at risk" or threatened (Table 4.2).

In the case of this Somsavath forest, aside from the prospect of declining populations of desired species because of over-collection, at that time there was also *real danger* that the entire tract of Somsavath forested land would be wiped out to be converted into rice fields. As witnessed, forest clearing was already ongoing with the speed of the fire that consumed it, starting from the far end of the Somsavath forest (Figure 4.4). Thus, there was a clear need to protect these medicinal plant resources for the health of present and future generations.

Somsavath forest (15 ha), today called the Somsavath Medicinal Plant Preserve (MPP), also referred to as the Bolikhamxay MPP or Somsavath MPP is the forest that lines the horizon beyond the devastated area in Figure 4.4. Together with the devastated area shown, it was part of a much larger tract of forest. The action taken to protect the Somsavath forest saved it from elimination.

An initial and somewhat improvised inventory of medicinal plants in the remaining tract of forest (the "Somsavath forest") was conducted by the Head of the Paksan "Traditional Medicine Station" (TMS - ສະຖານີຢາພື້ນເມືອງ), with guidance from ITM staff and the ICBG Project team. The results were presented to the Head of the Somsavath Village and the healers (ໝໍຢາພື້ນເມືອງ) of Somsavath (Figure 4.5) with the goal of impressing upon them the importance of protecting the Somsavath forest.

FIGURE 4.5 One of several meetings with the Head (ນາຍບ້ານ) of the Somsavath Village. From left foreground to the right in an arch sequence: Kongmany Sydara, Deputy Director of the ITM at that time, Ministry of Health, Vientiane; Gregg Dietzman, a Geographic Information System expert from White Point Systems, Inc., Friday Harbor, Washington State, USA; Djaja Djendoel Soejarto, Principal Investigator of the Vietnam-Laos ICBG Project, UIC; Healer of Somsavath; Member of the Village Council (ອຳນາດການປົກຄອງບ້ານ), Somsavath; Bounkong, Head of the TMS (ສະຖານີຢາພື້ນເມືອງ), Paksan District, Bolikhamxay Province; Head of the Somsavath Village, who has jurisdiction over Somsavath forest. Photo credit: M. Xayvue.

Following several meetings and discussions, the Head of the TMS of Paksan District, the Director of the ITM, and the ICBG project Principal Investigator jointly proposed to declare the remaining 15-hectare portion of this tract of land a protected forest. The Head of the Somsavath Village was receptive to the idea and collaborated in the effort. The implementation of this proposal was supported by funds from the ICBG Project.

The proposal was then submitted to the district authority in Paksan and to the Bolikhamxay provincial government office. Once the full documentation on the ownership (title) of the forested tract of land was completed and submitted, the provincial government made a declaration on the protection of that tract of secondary forest, naming it the "Medicinal Plant Preserve" (MPP) (ສວນອະນຸລັກພືດເບັນຢາ) of Somsavath. A road sign for the preserve was erected (Figure 4.7) and a shelter for visitors was built at the entrance of the preserve (see Chapter 5, Figure 5.7).

For the sake of record, the documents for the establishment of the preserve consisted of the following:

- Document #1. Origination Document (from the Paksan TMS and The Bolikhamxay Health Department)
- Document #2. Proposal for the use of the forested land (as a preserve for research and education)
- Document #3. Memorandum of Agreement (MOA) on the offer by the Head of the Village (between the village and the TMS and the Provincial Health Department)
- Document #4. Report of the field trip (land survey by the Forestry Department of Bolikhamxay Province)
- Document #5. Decree of the Director of the Provincial Health Department (designating the use of the land as a MPP and nominating a governing committee)
- Document #6. Decree of the District Governor (releasing the land to the Traditional Medicine Station and the Provincial Health Department)
- Document #7. Contract of the implementation (as a preserve, as well as a research and education site)

The originals of the above documents are deposited at the Paksan Traditional Medicine Station, and copies are at ITM, at the Provincial Health Department, and at the Village of Somsavath.

On April 11, 2006, the Deputy Director of the Cabinet of the Ministry of Health (ຮອງຫົວໜ້າຫ້ອງການກະຊວງສາທາລະນະສຸກ) inaugurated the preserve with much celebration and attendance of government officers, members of the Somsavath Village, and surrounding communities in the presence of the ITM staff and members of the ICBG project team (Figure 4.6). The unveiling of the street sign (Figure 4.7) accompanied the inauguration celebration.

Since then, the Somsavath MPP has been managed and regulated by the Village Board (ຄະນະບໍລິຫານງານບ້ານ) of Somsavath, under the auspices of the Bolikhamxay provincial government, specifically, the provincial Food and Drug Division (ຂະແໜງອາຫານ ແລະ ຢາແຂວງ). The ITM, with the support of the ICBG team, has served as the technical coordinator and was instrumental in promoting a good understanding among the Somsavath healers and community members on the importance of the preserve as a source of medicinal plant species to be used in a sustainable manner.

The ITM staff conduct periodic meetings with village members, especially with school children and younger members of the community, to educate them about growing medicinal plants in the preserve and the value of the forest as a continuing source of medicines for the future. The ITM staff placed name labels on trees around the visitor shelter to identify these plants. The ITM staff also conducts field classes and training for students at the faculty of pharmacy in Vientiane to learn about the identity of important medicinal plants found in the preserve.

FIGURE 4.6 Inauguration and cutting of the ribbon by the Deputy Director of the Cabinet of the Ministry of Health (ຮອງຫົວໜ້າຫ້ອງການກະຊວງສາທາລະນະສຸກ), of the Somsavath Medicinal Plants Preserve (MPP) (referred to in the chapter as Bolikhamxay Medicinal Plants Preserve), the first such Preserve in the Lao PDR. The inauguration took place on April 11, 2006, by the Deputy of the Cabinet of the Minister of Health (ຮອງຫົວໜ້າຫ້ອງການກະຊວງສາທາລະນະສຸກ), with the attendance of the Ministerial and Provincial Health officers, the ICBG Program officers and researchers from the USA, and members of the Somsavath and surrounding communities. Photo credit: O. Souliya.

The preserve has also served as a venue for training healers from other provinces and as a demonstration site for visiting scientists.

In summary, the original idea for a Medicinal Plant Preserve (MPP) was the belief that protection of such a degraded forest, and an attempt to protect the forest from further degradation, will prevent the complete loss of the medicinal plant resources to the local communities. The long-term goal is to protect medicinal plant genetic resources for the health of present and future generations.

4.4 THE LAO BIODIVERSITY FUND

Through funds provided by the Vietnam-Laos ICBG industrial partner, GlaxoSmithKline, a semi-private Trust Fund, named the "Lao Biodiversity Fund" (LBF) (ກອງທຶນຊີວະນາງພັນລາວ), was established through Ministerial Decree No. 2832/MOH, dated December 30, 2003 (Ministry of Health 2003). The LBF

FIGURE 4.7   Roadside sign at the entrance of the Somsavath Medicinal Plants Preserve. Photo credit: D. Soejarto.

is based at the ITM. The three main objectives of this fund are defined in the LBF by-laws as follows:

- To promote environmental education and to increase environmental awareness among communities of Laos.
- To promote economic development among communities of Laos.
- To promote research to value and protect traditional medicine of Laos.

To fulfill the objectives of the fund, the LBF provided funding to support poverty eradication projects and to support efforts to conserve medicinal plants of Laos. The cycle and process of funding starts with a request for support through a formal proposal from a particular community or from a provincial Traditional Medicine Station (TMS/TMU). Following an evaluation of the proposal by the LBF Board, site visits are conducted and the requested funds released if the proposal is approved. The Somsavath MPP serves as a model for the establishment of a medicinal plant preserve for others to follow. With the desire to broaden the scope of biodiversity protection, several of the newly established preserves were named "Medicinal Biodiversity Preserves" (MBPs) (ສວນອະນຸລັກຊີວະນາງພັນ). However, the function is the same, namely, protection of medicinal plants and other biodiversity components found growing in the MBP.

4.5   MEDICINAL PLANT CONSERVATION NETWORK OF LAOS

To date, the LBF has supported the establishment of eight MPPs and MBPs. In every award presentation, an inauguration ceremony was held in the presence of the LBF representatives, members of the local communities, and the district and provincial government authorities. All these preserves, including the Bolikhamxay MPP, are administered by the Food and Drug Division (ຂະແໜງອາຫານ ແລະ ຢາແຂວງ) of the Provincial Health Department (ພະແນກສາທາລະນະສຸກແຂວງ) through its Traditional Medicine Station/Unit (referred to here as TMS). The TMS works in close contact with the corresponding Head of the Village where the preserve falls. For example, in the Bolikhamxay preserve located in Somsavath Village, the Village Council and the Women's Association (ສະຫະພັນແມ່ຍິງ) of Somsavath lead the communities in the monitoring and managing of their preserve.

In 2013, one additional MBP was established through the support provided by The Agro-Biodiversity Initiative (TABI) of Laos, in the vicinity of the Huay Khing Village, Phonxay District, Luang Prabang Province. Beyond 2013, additional MBPs were established with funding from the Ministry of Health of Lao PDR. Today, Laos has a Medicinal Plant Conservation Network (ເຄືອຂ່າຍ/ ການບ່າງອະນຸລັກພຶດເປັນຢາ) made up of 27 MPPs/MBPs (Figure 4.1; Table 4.1). The ITM serves as the technical coordinating body of this network.

REFERENCES

Adfa, M., R. Rahmad, M. Ninomiya, S. Yudha, K. Tanaka, and M. Koketsu. 2016. "Antileukemic activity of lignans and phenylpropanoids of *Cinnamomum parthenox-ylon*." *Bioorganic & Medicinal Chemistry Letters* 26 (3): 761–64. doi:10.1016/J. BMCL.2015.12.096.

Babu, G. 2019. "Medicinal plant 'Sarpagandha' listed as endangered." *The New Indian Express*. Accessed on January 8, 2023. https://www.newindianexpress.com/cities/ chennai/2019/nov/06/medicinal-plant-sarpagandha-listed-as-endangered-2057650. html.

Bharali, P., H. Tag, A. Das, and M.C. Kalita. 2017. "Notes on *Helminthostachys zeylanica* (L.) Hook. - an endangered less known fern-ally from Arunachal Pradesh in NE India." *Pleione* 11 (1): 25–28.

Bhati, R., A. Singh, V. Saharan, V.A. Ram, and A. Bhandari. 2012. "*Strychnos nux-vomica* seeds: Pharmacognostical standardization, extraction, and antidiabetic activity." *Journal of Ayurveda and Integrative Medicine* 3 (2): 80–84. Accessed on January 8, 2023. https://www.ncbi.nlm.nih.gov/pmc/articles/PMC3371563/pdf/JAIM-3-80.pdf.

Biswas, N.N., A.K. Acharzo, S. Anamika, S. Khushi, and B. Bokshi. 2017. "Screening of natural bioactive metabolites and investigation of antioxidant, antimicrobial, antihy-perglycemic, neuropharmacological, and cytotoxicity potentials of *Litsea polyantha* Juss. Ethanolic root extract." *Evidence-Based Complementary and Alternative Medicine* 2017: 1–11. doi:10.1155/2017/3701349.

Bounda, G.-A., and Y.U. Feng. 2015. "Review of clinical studies of *Polygonum multiflorum* Thunb. and its isolated bioactive compounds." *Pharmacognosy Research* 7 (3): 225–36. doi:10.4103/0974-8490.157957.

Chaingam, J., R. Choonong, T. Juengwatanatrakul, T. Kanchanapoom, W. Putalun, and
     G. Yusakul. 2022. "Evaluation of anti-inflammatory properties of *Eurycoma longifolia*
     Jack and *Eurycoma harmandiana* Pierre *in vitro* cultures and their constituents." *Food
     and Agricultural Immunology* 33 (1): 530–45. doi:10.1080/09540105.2022.2100324.

Chaingam, J., T. Juengwatanatrakul, G. Yusakul, T. Kanchanapoom, and W. Putalun.
     2021. HPLC-UV-based simultaneous determination of Canthin-6-One alkaloids,
     quassinoids, and scopoletin: The active ingredients in *Eurycoma longifolia Jack* and
     Eurycoma harmandiana Pierre, and their anti-inflammatory activities. *Journal of
     AOAC International* 104 (3): 802–10. doi:10.1093/jaoacint/qsaa141.

Chandra, S., C.R. Fraser-Jenkins, A. Kumari, and A. Srivastava. 2008. "A summary of the
     status of threatened pteridophytes of India." *Taiwania* 53 (2): 170–209. Accessed on
     January 8, 2023. https://taiwania.ntu.edu.tw/pdf/tai.2008.53.170.pdf.

Chanthakoummane, S. 2008. "Finding Protected Area Funding Solutions." *IUCN*.
     Vientiane. Accessed on January 8, 2023. https://www.iucn.org/sites/default/files/
     import/downloads/powpa___Crochure_1.pdf.

Chauhan, H.K. 2020. "*Paris Polyphylla* (Love Apple)." *The IUCN Red List of Threatened
     Species* 2020. Accessed on January 11, 2023. https://www.researchgate.net/
     publication/350171807_Paris_polyphylla_Love_Apple_THE_IUCN_RED_LIST_
     OF_THREATENED_SPECIES.

Chauhan, N.S., V. Sharma, M. Thakur, and V.K. Dixit. 2010. *Curculigo orchioides*: The
     black gold with numerous health benefits. *Journal of Integrative Medicine*. 8 (7):
     613–23. doi:10.3736/jcim20100703.

Cheung, M., and M. Parmar. 2022. "Reserpine." In: *StatPearls* [Internet]. Treasure Island
     (FL): StatPearls Publishing. 133–38. Accessed on January 8, 2023. https://www.
     statpearls.com/ArticleLibrary/viewarticle/28402.

Choonong, R., J. Chaingam, R. Chantakul, S. Mukda, P. Temkitthawon, K. Ingkaninan, T.
     Juengwatanatrakul, G. Yusakul, T. Kanchanapoom, and W. Putalun. 2022.
     "Phosphodiesterase-5 inhibitory activity of canthin-6-one alkaloids and the roots of
     Eurycoma longifolia and *Eurycoma harmandiana*." *Chemistry and Biodiversity* 19
     (7): e202200121. doi:10.1002/cbdv.202200121.

Choudhury, S.N., A.C. Ghosh, M. Choudhury, and P.A. Leclercq. 1997. "Essential oils of
     Litsea monopetala (Roxb.) Pers. A new report from India." *Journal of Essential Oil
     Research* 9 (6): 635–39. doi:10.1080/10412905.1997.9700802.

CITES. 2022. "Recommendation to suspend commercial trade in live Asian elephants
     (*Elephas maximus*) and *Dalbergia* spp. and other recommendations." *Notification to
     the Parties Concerning Application of Article XIII in the Lao People's Democratic
     Republic*. No. 2022/030. Geneva. Accessed on January 8, 2023. https://cites.org/
     sites/default/files/notifications/E-Notif-2022-030_0.pdf.

Clarke, J.E. n.d. "Biodiversity and Protected Areas of Lao PDR." Accessed on January 8,
     2023. https://data.opendevelopmentmekong.net/dataset/0bbcda64-c9eb-4325-94a8-
     f4e27ca04f7d/resource/2607ebf9-649e-4743-a9f2-57f58f2ded51/download/0002547-
     environment-biodiversity-and-protected-areas-lao-p-d-r.pdf.

de Kok, R. 2020a. "*Cinnamomum parthenoxylon* (Selasian Wood)." *IUCN Red List of
     Threatened Species 2020*. doi:10.2305/IUCN.UK.2020-1.RLTS.T33198A2834736.en.

———. 2020b. "*Litsea monopetala*." *The IUCN Red List of Threatened Species*.
     doi:10.2305/IUCN.UK.2020-1.RLTS.T150218025A150219969.en.

de Kok, R.P.J. 2019. A revision of *Cinnamomum* Schaeff. (Lauraceae) for Peninsular
     Malaysia and Singapore. *Gardens' Bulletin Singapore* 71 (1): 89–139. doi:10.26492/
     gbs71(1).2019-07.

———. 2021. "A revision of *Litsea* (Lauraceae) in Peninsular Malaysia and Singapore."
     Gardens' Bulletin Singapore 73 (1): 81–178. doi:10.26492/gbs73(1).2021-07.

de Koning, M., T. Nguyen, M. Lockwood, S. Sengchanthavong, and S. Phommasane. 2017. "Collaborative goverance of protected areas: Success factors and prospects for Hin Nam No national protected area, Central Laos." *Conservation and Society* 15 (1): 87–99. Accessed on January 8, 2023. http://www.jstor.org/stable/26393273.

Desgrouas, C., N. Taudon, S.-S. Bun, B. Baghdikian, S. Bory, D. Parzy, and E. Ollivier. 2014. "Ethnobotany, phytochemistry and pharmacology of *Stephania rotunda* Lour." *Journal of Ethnopharmacology* 154 (3): 537–63. doi:10.1016/J.JEP.2014.04.024.

DrugBank Online. 2022. "*Reynoutria multiflora* root: Uses, interactions, mechanism of action | DrugBank Online." Accessed on January 8, 2023. https://go.drugbank.com/drugs/DB14295.

D'Souza, R. 2019. "Monkey Jack Fruit (Lakucha) – Uses, remedies, research." *Easy Ayurveda | Ayurvedic Herbs.* Accessed on January 8, 2023. https://www.easyayurveda.com/2019/08/02/monkey-jack-fruit-lakucha/amp/.

Dutta, P.K., I.C. Chopra, and L.D. Kapoor. 1963. "Cultivation of *Rauvolfia serpentina* in India." *Economic Botany* 17 (4): 243–51. doi:10.1007/BF02860133.

Fakhrul Hasan, M., M. Ashraful Iqbal, and M. Sahab Uddin. 2016. "Antibacterial and antifungal activity of *Litsea monopetala* leaves on selected pathogenic strains." *European Journal of Medicinal Plants* 12 (4): 1–8. doi:10.9734/EJMP/2016/23658.

Fellowes, J.R., B.P.L. Chan, M.W.N. Lau, S.-C. Ng, and G.L.P. Siu. 2003. "Report of rapid biodiversity assessments at Wuzhishan Nature Reserve, Central Hainan, China, 1999 and 2001, Kadoorie Farm and Botanic Garden in Collaboration with South China Forest Biodiversity Survey Report Series: No. 24" (Online Simplified Version) *Report of Rapid Biodiversity Assessments at Wuzhishan Nature Reserve, Central Hainan, China, 1999 and 2001.* Central Hainan, China. Accessed on January 8, 2023. https://www.kfbg.org/upload/Documents/Free-Resources-Download/Report-and-Document/E24_Wuzhishan_report_w.pdf.

Fern, K., A. Fern, and R. Morris. 2014. "Schefflera elliptica." *Useful Tropical Plants Database.* Accessed on May 27, 2022. https://tropical.theferns.info/viewtropical.php?id=Schefflera+elliptica.

Gao, S.-M., J.-S. Liu, M. Wang, T.-T. Cao, Y.-D. Qi, B.-G. Zhang, X.-B. Sun, H.-T. Liu, and P.-G. Xiao. 2018a. "Traditional uses, phytochemistry, pharmacology and toxicology of *Codonopsis*: A review." *Journal of Ethnopharmacology* 219: 50–70. doi:10.1016/j.jep.2018.02.039.

Gao, X., X. Zhang, H. Meng, J. Li, D. Zhang, and C. Liu. 2018b. "Comparative chloroplast genomes of *Paris* Sect. *Marmorata*: Insights into repeat regions and evolutionary implications." *BMC Genomics* 19 (10): 133–44. doi:10.1186/S12864-018-5281-X/FIGURES/7.

GBIF. 2021. "*Helminthostachys zeylanica* (L.) Hook." *GBIF Backbone Taxonomy.* Accessed on January 8, 2023. https://www.gbif.org/species/2650123.

Gilmour, D.A., and X. Tsechalicha. 2000. "Forest rehabilitation in Lao PDR: Issues and constraints." *IUCN Library System.* Accessed on January 8, 2023. https://portals.iucn.org/library/node/7866.

Gu, W., Y. Wang, L. Zeng, J. Dong, Q. Bi, X. Yang, Y. Che, S. He, and J. Yu. 2020. "Polysaccharides from *Polygonatum kingianum* improve glucose and lipid metabolism in rats fed a high fat diet." *Biomedicine & Pharmacotherapy* 125 (May). doi:10.1016/J.BIOPHA.2020.109910.

Guo, R., T. Wang, G. Zhou, M. Xu, X. Yu, X. Zhang, F. Sui, C. Li, L. Tang, and Z. Wang. 2018. "Botany, phytochemistry, pharmacology and toxicity of *Strychnos nux-vomica* L.: A review." *The American Journal of Chinese Medicine* 46 (1): 1–23. doi:10.1142/S0192415X18500015.

Hao, D.C., X.-J. Gu, and P.G. Xiao. 2015. "Phytochemical and biological research of Polygoneae medicinal resources." In *Medicinal Plants: Chemistry, Biology and Omics*, 465–529. Woodhead Publishing.

Hasan, H., S. Al Azad, Z. Islam, and S.K.M. Rahman. 2014. "(PDF) antihyperglycemic activity of methanolic extract of *Litsea monopetala* (Roxb.) Pers. leaves." *Advances in Natural and Applied Sciences* 8 (1): 51–55. Accessed on January 8, 2023. http://www.aensiweb.com/old/anas/2014/51-55.pdf.

He, J.-Y., N. Ma, S. Zhu, K. Komatsu, Z.-Y. Li, and W.-M. Fu. 2015. "The genus *Codonopsis* (Campanulaceae): A review of phytochemistry, bioactivity and quality control." *Journal of Natural Medicines* 69 (1): 1–21. doi:10.1007/s11418-014-0861-9.

He, Z.-D., C.-Y. Ma, G.T. Tan, K. Sydara, P. Tamez, B. Southavong, S. Bouamanivong, et al. 2006. "Rourinoside and rouremin, antimalarial constituents from *Rourea minor*." *Phytochemistry* 67 (13): 1378–84. doi:10.1016/j.phytochem.2006.04.012.

Hsieh, H.-L., S.-H. Yang, T.-H. Lee, J.-Y. Fang, and C.-F. Lin. 2016. "Evaluation of anti-inflammatory effects of *Helminthostachys zeylanica* extracts via inhibiting Bradykinin-induced MMP-9 expression in brain astrocytes." *Molecular Neurobiology* 53 (9): 5995–6005. doi:10.1007/S12035-015-9511-9.

Huang, Y.-C., T.-L. Hwang, C.-S. Chang, Y.-L. Yang, C.-N. Shen, W.-Y. Liao, S.-C. Chen, and C.C. Liaw. 2009. "Anti-inflammatory flavonoids from the rhizomes of *Helminthostachys zeylanica*." *Journal of Natural Products* 72 (7): 1273–78. doi:10.1021/np900148a.

Huang, Y.-L., C.-C. Shen, Y.-C. Shen, W.-F. Chiou, and C.-C. Chen. 2017. "Anti-Inflammatory and antiosteoporosis flavonoids from the rhizomes of *Helminthostachys zeylanica*." *Journal of Natural Products* 80 (2): 246–53. doi:10.1021/acs.jnatprod.5b01164.

Huang, Y.-L., P.-Y. Yeh, C.-C. Shen, and C.-C. Chen. 2003. "Antioxidant flavonoids from the rhizomes of *Helminthostachys zeylanica*." *Phytochemistry* 64 (7): 1277–83. doi:10.1016/j.phytochem.2003.09.009.

ICEM. 2003. "Lao People's Democratic Republic National Report on Protected Areas and Development." Queensland, Australia. Accessed on January 8, 2023. https://portals.iucn.org/library/sites/library/files/documents/2003-106-2.pdf.

Islam, S., S. Shajib, R.B. Rashid, M.F. Khan, A. Al-Mansur, B.K. Datta, and M.A. Rashid. 2019. "Antinociceptive activities of *Artocarpus lacucha* Buch-ham (Moraceae) and its isolated phenolic compound, catechin, in mice." *BMC Complementary and Alternative Medicine* 19 (1): 1–13. doi:10.1186/s12906-019-2565-x.

IUCN Red List. 2022. "IUCN Red List of Threatened Species." *International Union for the Conservation of Nature*. Accessed on January 8, 2023. https://www.iucnredlist.org/.

Jensen, S.R. 1996. Caffeoyl phenylethanoid glycosides in *Sanango racemosum* and in the Gesneriaceae. *Phytochemistry* 43 (4): 777–83. Accessed on January 8, 2023. https://gesneriads.info/wp-content/uploads/2020/02/Jensen_S_1996.pdf.

Jia, Q., X. Liu, X. Wu, R. Wang, X. Hu, Y. Li, and C. Huang. 2009. "Hypoglycemic activity of a polyphenolic oligomer-rich extract of *Cinnamomum parthenoxylon* bark in normal and streptozotocin-induced diabetic rats." *Phytomedicine* 16 (8): 744–50. doi:10.1016/J.PHYMED.2008.12.012.

Kanchanapoom, T., R. Kasai, P. Chumsri, and K. Yamasaki. 2001. Quassinoids from *Eurycoma harmandiana*. *Phytochemistry* 57 (8): 1205–08. doi:10.1016/s0031-9422(01)00235-7.

Kawk, H.W., G.-H. Nam, M.J. Kim, S.-Y. Kim, and Y.-M. Kim. 2021. "*Scaphium affine* ethanol extract induces anoikis by regulating the EGFR/Akt pathway in HCT116 colorectal cancer cells." *Frontiers in Oncology* 11 (May): 1317. doi:10.3389/fonc.2021.621346.

Kew. 2022a. "*Artocarpus lacucha* Buch.-Ham." *Plants of the World Online. Facilitated by the Royal Botanic Gardens, Kew*. Accessed on January 8, 2023. https://powo.science. kew.org/taxon/urn:lsid:ipni.org:names:927744-1.

———. 2022b. "*Helminthostachys zeylanica* (L.) Hook." *Plants of the World Online. Facilitated by the Royal Botanic Gardens, Kew*. Accessed on January 8, 2023. https:// powo.science.kew.org/taxon/urn:lsid:ipni.org:names:17116690-1.

———. 2022c. "*Litsea monopetala* (Roxb.) Pers." *Plants of the World Online. Facilitated by the Royal Botanic Gardens, Kew*. Accessed on January 8, 2023. https://powo. science.kew.org/taxon/urn:lsid:ipni.org:names:465855-1.

———. 2022d. "*Stephania rotunda* Lour." *Plants of the World Online. Facilitated by the Royal Botanic Gardens, Kew*. Accessed on January 8, 2023. https://powo.science. kew.org/taxon/urn:lsid:ipni.org:names:581482-1.

———. 2022e. "*Strychnos nux-vomica* L." *Plants of the World Online. Facilitated by the Royal Botanic Gardens, Kew*. Accessed on January 8, 2023. https://powo.science. kew.org/taxon/urn:lsid:ipni.org:names:547371-1.

———. 2022f. "*Dracaena cambodiana* Pierre & Gagn." *Plants of the World Online. Facilitated by the Royal Botanic Gardens, Kew*. Accessed on January 8, 2023. https:// powo.science.kew.org/taxon/urn:lsid:ipni.org:names:534134-1.

———. 2022g. "*Croton lechleri* Mull. Arg." *Plants of the World Online. Facilitated by the Royal Botanic Gardens, Kew*. Accessed on January 8, 2023. https://powo.science. kew.org/taxon/urn:lsid:ipni.org:names:342842-1.

———. 2022h. "*Aeschynanthus longicaulis* Wall. ex R.Br." *Plants of the World Online. Facilitated by the Royal Botanic Gardens, Kew*. Accessed on January 8, 2023. https:// powo.science.kew.org/taxon/urn:lsid:ipni.org:names:77130618-1.

———. 2022i. "*Curculigo orchioides* Gaertn." *Plants of the World Online. Facilitated by the Royal Botanic Gardens, Kew*. Accessed on January 8, 2023. https://powo.science. kew.org/taxon/urn:lsid:ipni.org:names:64163-1.

———. 2022j. "*Coscinium fenestratum* (Gaertn.) Colebr." *Plants of the World Online. Facilitated by the Royal Botanic Gardens, Kew*. Accessed on January 8, 2023. https:// powo.science.kew.org/taxon/urn:lsid:ipni.org:names:580661-1.

———. 2022k. "*Eurycoma harmandiana* Pierre" Plants of the World Online. Facilitated by the Royal Botanic Gardens, Kew. Accessed on January 8, 2023. https://powo. science.kew.org/taxon/urn:lsid:ipni.org:names:813713-1.

Khan, H., M. Saeed, and N. Muhammad. 2012a. "Pharmacological and phytochemical updates of genus *Polygonatum*." *Phytopharmacology* 3 (2): 286–308. Accessed on January 8, 2023. http://inforesights.com/phytopharmacology/files/pp3v2i6.pdf.

Khan, H., M. Saeed, N. Muhammad, R. Ghaffar, S. Khan, and S. Hassan. 2012b. "Antimicrobial activities of rhizomes of *Polygonatum verticillatum*: Attributed to its total flavonoidal and phenolic contents." *Pakistan Journal of Pharmaceutical Sciences* 25 (2): 463–67.

Kumar, S., and R. Kumari. 2019. "*Cinnamomum*: Review article of essential oil com- pounds, ethnobotany, antifungal and antibacterial effects." *Open Access Journal of Science* 3 (1): 13–16. doi:10.15406/OAJS.2019.03.00121.

Kumari, R., B. Rathi, A. Rani, and S. Bhatnagar. 2013. "*Rauvolfia serpentina* L. Benth. ex Kurz: Phytochemical, pharmacological and therapeutic aspects." *International Journal of Pharmaceutical Sciences Review and Research* 23 (January): 348–55. Accessed on January 8, 2023. https://globalresearchonline.net/journalcontents/v23-2/56.pdf.

Li, C.-Y., H.-X. Xu, Q.-B. Han, and T.-S. Wu. 2009. "Quality assessment of Radix Codonopsis by quantitative nuclear magnetic resonance." *Journal of Chromatography A* 1216 (11): 2124–29. doi:10.1016/J.CHROMA.2008.10.080.

Li, C. 2015. *Sterculia lychnophora* Hance 胖大海 (Pangdahai, Malva Nut Tree). *Dietary Chinese Herbs*. Springer, Vienna. doi:10.1007/978-3-211-99448-1_61.

Li, S., X.-W. Li, J. Li, P. Huang, F.-N. Wei, H. Cui, and H. van der Werff. 2008. *Efloras. Org: Flora of China: "Lauraceae"*. 7. Accessed on January 8, 2023. http://www.efloras.org/florataxon.aspx?flora_id=2&taxon_id=10479.

Lin, L., B. Ni, H. Lin, M. Zhang, X. Li, X. Yin, C. Qu, and J. Ni. 2015. "Traditional usages, botany, phytochemistry, pharmacology and toxicology of *Polygonum multiflorum* Thunb.: A review." *Journal of Ethnopharmacology* 159 (January): 158–183. doi:10.1016/J.JEP.2014.11.009.

Liu, L., F.-R. Xu, and Y.-Z. Wang. 2020. "Traditional uses, chemical diversity and biological activities of *Panax* L. (Araliaceae): A review." *Journal of Ethnopharmacology* 263 (December). doi:10.1016/J.JEP.2020.112792.

Liu, Y., X. Zhao, R. Yao, C. Li, Z. Zhang, Y. Xu, and J.-H. Wei. 2021. "Dragon's blood from *Dracaena* worldwide: Species, traditional uses, phytochemistry and pharmacology." *The American Journal of Chinese Medicine* 49 (6): 1315–67. doi:10.1142/S0192415X21500634.

Mader, L.S. 2013. Update: FDA approves Crofelemer as first oral botanical drug. *HerbalGram* 97:68–69. Accessed on January 8, 2023. https://www.herbalgram.org/resources/herbalgram/issues/97/table-of-contents/hg97-legreg-crofelemer/.

Ministry of Health. 2003. "Ministerial Decree on the Establishment of the Lao Biodiversity Fund (LBF). Ref. No. 2832/MoH." Vientiane.

Nasution, R.E. 2003. "Aeschynanthus." In *PROSEA: Plant Resources of Southeast Asia*. 12(3), edited by R.H.M.J. Lemmens and N. Bunyapraphatsara, 41–42. Backhuys Publishers, Leiden. Medicinal Plants 3. Online edition March 13, 2016. Accessed on January 8, 2023. https://uses.plantnet-project.org/en/Aeschynanthus_(PROSEA).

Newman, M., S. Ketphanh, B. Svengsuksa, P. Thomas, K. Sengdala, V. Lamxay, and K. Armstrong. 2007. *A checklist of the vascular plants of Lao PDR*. Royal Botanic Garden Edinburgh. Accessed on January 8, 2023. https://portals.iucn.org/library/efiles/documents/2007-014.pdf.

Nguyen, M.D., and T.N. Nguyen. 1998. "Chemical composition and pharmacological activities of Vietnamese Ginseng, *Panax vetnamensis*." In *Proceedings of the Ginseng Society Conference*, edited by The Korean Society of Ginseng, 127–37. Ho Chi Minh City: The Korean Society of Ginseng. Accessed on January 8, 2023. https://www.koreascience.or.kr/article/CFKO199811919377940.pdf.

Nguyen, T.H., T.T.H. Chu, The Cuong Nguyen, A.T. Nguyen, and H.T. Tran. 2016. "Study on knowledge of medicinal plants used of Tay ethnic minority in Na Hang special-use forest, Tuyen Quang Province." *Journal of Vietnamese Environment* 8 (5): 277–83. doi:10.13141/jve.vol8.no5.pp277-283.

Nguyen, T.H., and T.T. Phuong. 2019. Vietnamese Ginseng (*Panax vietnamensis* Ha and Grushv.): Phylogenetic, phytochemical, and pharmacological profiles. *Pharmacognosy Reviews* 13 (26): 59–62. Accessed on January 8, 2023. https://www.phcogrev.com/sites/default/files/PharmacognRev-13-26-59.pdf.

Nguyen, T.T., T.B. Pham, N.P. Thao, N.H. Dang, V.H. Nguyen, V.C. Pham, C.V. Minh, Q.H. Tran, and N.T. Dat. 2018. "Phenolic constituents from *Fallopia multiflora* (Thunberg) Haraldson." *Journal of Chemistry* 2018. doi:10.1155/2018/4851439.

Nie, Y., X. Dong, Y. He, T. Yuan, T. Han, K. Rahman, L. Qin, Q. Zhang 2013. Medicinal plants of genus *Curculigo*: Traditional uses and a phytochemical and ethnopharmacological review. *Journal of Ethnopharmacology* 147 (3): 547–63.

Ojha, R., H. Prasad, D. Correspondence, and H.P. Devkota. 2021. "Edible and medicinal pteridophytes of Nepal: A review." *Ethnobotany Research and Applications* 22 (August): 1–16. doi:10.32859/era.22.16.1-16.

Oppong, M.B., Y. Li, P.O. Banahene, S.-M. Fang, and F. Qiu. 2018. "Ethnopharmacology, phytochemistry, and pharmacology of *Sterculia lychnophora* Hance (Pangdahai)." *Chinese Journal of Natural Medicines* 16 (10): 721–31. doi:10.1016/S1875-5364(18)30112-2.

Pardede, A., M. Adfa, A.J. Kusnanda, M. Ninomiya, and M. Koketsu. 2017. "Flavonoid rutinosides from *Cinnamomum parthenoxylon* leaves and their hepatoprotective and antioxidant activity." *Medicinal Chemistry Research* 26 (9). 2074–79. doi:10.1007/S00044-017-1916-8.

Patel, K., D. Laloo, G.K. Singh, M. Gadewar, and D.K. Patel. 2017. "A review on medicinal uses, analytical techniques and pharmacological activities of *Strychnos nuxvomica* Linn.: A concise report." *Chinese Journal of Integrative Medicine*, 1–13. doi:10.1007/S11655-016-2514-1.

Prime Minister of Laos. 2003. "Decree on Natural Resource for Medicines." Vol. Decree 155. Accessed on January 8, 2023. http://www.laotradeportal.gov.la/kcfinder/upload/files/Decree_No.155_-_Eng.pdf.

———. 2007. "Ordinance/Guidance No. 252/MOH on the Implementation of Prime-Minister Decree on Pharmaceutical Natural Resource No. 155/PM." Vientiane, Lao PDR. Accessed on January 8, 2023. http://www.laotradeportal.gov.la/kcfinder/upload/files/Additional_Notice_on_Implementation_of_Decree_155.pdf. [in Lao language]

PubChem. 2021a. "Compound Summary: Ajmalicine." *National Libary of Medicine; National Center for Biotechnology Information.* Accessed on January 8, 2023. https://pubchem.ncbi.nlm.nih.gov/compound/Ajmalicine.

———. 2021b. "Compound summary: Reserpine." *National Library of Medicine; National Center for Biotechnology Information.* Accessed on January 8, 2023. https://pubchem.ncbi.nlm.nih.gov/compound/Reserpine.

Pukdeekumjorn, P., S. Ruangnoo, and A. Itharat. 2016. "Anti-inflammatory activities of extracts of *Cinnamomum porrectum* (Roxb.) Kosterm. Wood (Thep-tha-ro) - PubMed." *Journal of the Medical Association of Thailand = Chotmaihet Thangphaet* 99 (suppl 4): S138–43.

Qiu, F., H. Yang, T. Zhang, X. Wang, S. Wen, and X. Su. 2019. "Chemical composition of leaf essential oil of *Cinnamomum porrectum* (Roxb.) Kosterm." *Journal of Essential Oil Bearing Plants* 22 (5): 1313–21. doi:10.1080/0972060X.2019.1689178.

Roopashree, T.S., D. Kuntal, N. Prashanth, and K.R. Kumar. 2021. "Pharmacognostic and chromatographic evaluation of male and female flowers of *Coscinium fenestratum* for berberine content and its effect on antioxidant activity." *Thai Journal of Pharmaceutical Sciences (TJPS)* 45 (2): 155–64.

Saleem, M., A. Asif, M.F. Akhtar, and A. Saleem. 2018. "Hepatoprotective potential and chemical characterization of *Artocarpus lakoocha* fruit extract." *Bangladesh Journal of Pharmacology* 13 (1): 90–97. doi:10.3329/BJP.V13I1.34117.

Schippmann, U., D.J. Leaman, and A.B. Cunningham. 2002. "Impact of Cultivation and Gathering of Medicinal Plants on Biodiversity: Global Trends and Issues." Published in FAO. 2002. *Biodiversity and the Ecosystem Approach in Agriculture, Forestry and Fisheries. Satellite event on the occasion of the Ninth Regular Session of the Commission on Genetic Resources for Food and Agriculture.* Rome, 12–13 October 2002. Inter-Departmental Working Group on Biological Diversity for Food and Agriculture. Rome. Accessed on January 8, 2023. https://www.fao.org/3/aa010e/AA010E00.pdf.

Shamon, S.D., and M.I. Perez. 2016. "Blood pressure-lowering efficacy of reserpine for primary hypertension." *The Cochrane Database of Systematic Reviews* 12 (12): CD007655. doi:10.1002/14651858.CD007655.pub3.

Sharma, B.K. 2011. "Rauwolfia: Cultivation and collection." *Biotech Articles*, May. Accessed on January 8, 2023. https://www.biotecharticles.com/Agriculture-Article/ Rauwolfia-Cultivation-and-Collection-892.html.

Singapore Government Agency. 2022. "*Artocarpus dadah* Miq." *National Parks Flora & Fauna Web*. Accessed January 8, 2023. https://www.nparks.gov.sg/florafaunaweb/ flora/3/5/3571.

Soejarto, D.D., B. Southavong, K. Sydara, S. Bouamanivong, M.C. Riley, A.S. Libman, M.R. Kadushin, and C. Gyllenhaal. 2009. "Studies on medicinal plants of Laos – A collaborative program between the University of Illinois at Chicago (UIC) and the Traditional Medicine Research Center (TMRC), Lao PDR: Accomplishments 1998–2005." In *Contemporary Lao Studies: Research on Development, Culture, Language, and Traditional Medicine*, edited by C.J. Compton, J.F. Hartmann, and V. Sysamouth, pp. 307–23. San Fransisco, De Kalb: Center for Lao Studies and the Center for Southeast Asian Studies, Northern Illinois University.

Souliya, O., and K. Sydara. 2015. "Some Threatened and Rare Medicinal Plant Species of Lao PDR." Poster. Vientiane, Lao PDR.

Su, L.-H., Y.-P. Li, H.-M. Li, W.-F. Dai, D. Liu, L. Cao, and R.-T. Li. 2016. "Anti-inflammatory prenylated flavonoids from *Helminthostachys zeylanica*." *Chemical and Pharmaceutical Bulletin* 64 (5): 497–501. doi:10.1248/CPB.C15-00661.

Sukcharoen, O., P. Sirirote, and D. Thanaboripat. 2017. "Control of aflatoxigenic strains by *Cinnamomum porrectum* essential oil." *Journal of Food Science and Technology* 54 (9): 2929–35. doi:10.1007/S13197-017-2731-4.

Sydara, K., M. Xayvue, O. Souliya, B.G. Elkington, and D.D. Soejarto. 2014. "Inventory of medicinal plants of the Lao People's Democratic Republic: A mini review." *Journal of Medicinal Plants Research* 8 (43): 1262–74. Accessed on January 11, 2023. https:// web-prod-2.pharm.uic.edu/webdav/belkin3/intellcont/JMPR_Sydara_2014-1.pdf.

Teanpaisan, R., S. Senapong, and J. Puripattanavong. 2014. "*In vitro* antimicrobial and anti-biofilm activity of *Artocarpus lakoocha* (Moraceae) extract against some oral pathogens." *Tropical Journal of Pharmaceutical Research* 13 (7): 1149–55. doi:10.4314/ TJPR.V13I7.20.

Thu Ha, T.T., D.K. Vui, H.H. Nguyen, B.H. Tai, and P. Van Kiem. 2021. "Dispolongiosides A and B, two new fucose containing spirostanol glycosides from the rhizomes of *Disporopsis longifolia* Craib., and their nitric oxide production inhibitory activities." *Natural Product Communications* 16 (10). doi:10.1177/1934578X211055013.

Tushar, K.V., S. George, A.B. Remashree, and I. Balachandran. 2008. "*Coscinium fenestratum* (Gaertn.) Colebr. – A review on this rare, critically endangered and highly-traded medicinal species." *Journal of Plant Sciences* 3 (2): 133–45. doi:10.3923/ JPS.2008.133.145.

Ueda, J.-Y., Y. Tezuka, A.H. Banskota, Q. Le Tran, Q.K. Tran, Y. Harimaya, I. Saiki, and S. Kadota. 2002. "Antiproliferative activity of Vietnamese medicinal plants." *Biological and Pharmaceutical Bulletin* 25 (6): 753–60. doi:10.1248/BPB.25.753.

Uphof, J.C.T. 2001. *Dictionary of Economic Plants*. Dehra Dun, India; Ruggell, Germany; Köenigstein, Germany: Bishen Singh Mahendra Pal Singh; A.R.G. Gantner; Koeltz Scientific Books [distributor].

Ved, D., D. Saha, K. Ravikumar, and K. Haridasan. 2015. "*Coscinium fenestratum*." *The IUCN Red List of Threatened Species* 2015. doi:10.2305/IUCN.UK.2015-4.RLTS. T50126585A50131325.en.

Wang, Y., F.-A. Khan, M. Siddiqui, M. Aamer, C. Lu, Atta-ur-Rahman, Atia-tul-Wahab, and M.I. Choudhary. 2021. "The genus *Schefflera*: A review of traditional uses, phytochemistry and pharmacology." *Journal of Ethnopharmacology* 279 (28). doi:10.1016/J.JEP.2020.113675.

Wang, J., B. Su, H. Jiang, N. Cui, Z. Yu, Y. Yang, and Y. Sun. 2020. "Traditional uses, phytochemistry and pharmacological activities of the genus *Cinnamomum* (Lauraceae): A review." *Fitoterapia* 146 (October). doi:10.1016/J.FITOTE.2020.104675.

Wang, Y.-S., Z.-Q. Wen, B.-T. Li, H.-B. Zhang, and J.-H. Yang. 2016. "Ethnobotany, phytochemistry, and pharmacology of the genus *Litsea*: An update." *Journal of Ethnopharmacology* 181 (April). 66–107. doi:10.1016/J.JEP.2016.01.032.

Wang, R.-F., X.-W. Wu, and D. Geng. 2013. "Two cerebrosides isolated from the seeds of *Sterculia lychnophora* and their neuroprotective effect." *Molecules* 18 (1): 1181–87. doi:10.3390/MOLECULES18011181.

Wang, R., X. Yang, C. Ma, M. Shang, S. Yang, M. Wang, and S. Cai. 2003. "[Analysis of fatty acids in the seeds of *Sterculia lychnophora* by GC-MS] - PubMed." *Zhongguo Zhong Yao Za Zhi - Chinese Journal of Materia Medica* 28 (6): 533–35.

WCS Lao PDR. 2021. "WCS Lao PDR program position statement regarding the recent discovery of SARS-CoV-2-like viruses in bats in Northern Laos." *Latest News.* November 1. Accessed on January 8, 2023. https://laos.wcs.org/About-Us/Latest-News.aspx.

WFO. 2021. "Cinnamomum Parthenoxylon (Jack) Meisn." *World Flora Online.* Accessed on January 8, 2023. http://www.worldfloraonline.org/search?query=Cinnamomum+parthenoxylon.

———. 2022a. "*Artocarpus lakoocha* Roxb." *The World Flora Online.* Accessed on January 8, 2023. 2023. http://www.worldfloraonline.org/taxon/wfo-0000550518.

———. 2022b. "*Fallopia multiflora* (Thunb. Ex Murray) Czerep." *The World Flora Online.* Accessed on January 8, 2023. http://www.worldfloraonline.org/taxon/wfo-0001101485.

———. 2022c. "*Paris marmorata* Stearn." *The World Flora Online.* Accessed on January 8, 2023. http://www.worldfloraonline.org/taxon/wfo-0000715194.

Wilkie, P. 2009. "A revision of *Scaphium (Sterculioideae, Malvaceae/Sterculiaceae)*." *Edinburgh Journal of Botany* 66 (2): 283–328. doi:10.1017/S0960428609005411.

Wu, K.-C., S.-S. Huang, Y.-H. Kuo, Y.-L. Ho, C.-S. Yang, Y.-S. Chang, and G.-J. Huang. 2017a. "Ugonin M, a *Helminthostachys zeylanica* constituent, prevents LPS-induced acute lung injury through TLR4-mediated MAPK and NF-κB signaling pathways." *Molecules: A Journal of Synthetic Chemistry and Natural Product Chemistry* 22 (4): 573. doi:10.3390/MOLECULES22040573.

Wu, K.-C., C.-P. Kao, Y.-L. Ho, and Y.-S. Chang. 2017b. "Quality control of the root and rhizome of *Helminthostachys zeylanica* (*Daodi-Ugon*) by HPLC using Quercetin and Ugonins as markers." *Molecules: A Journal of Synthetic Chemistry and Natural Product Chemistry* 22 (7): 1115. doi:10.3390/MOLECULES22071115.

Yamasaki, K. 2000. "Bioactive Saponins in Vietnamese Ginseng, *Panax vietnamensis*." *Pharmaceutical Biology* 38 (supplement): 16–24. doi:10.1076/phbi.38.6.16.5956.

Yamauchi, K., T. Mitsunaga, Y. Itakura, and I. Batubara. 2015. "Extracellular melanogenesis inhibitory activity and the structure-activity relationships of ugonins from *Helminthostachys zeylanica* roots." *Fitoterapia* 104 (July). 69–74. doi:10.1016/J.FITOTE.2015.05.006.

Yang, W.-Q., W. Tang, X.-J. Huang, J.-G. Song, Y.-Y. Li, Y. Xiong, C.-L. Fan, Z.-L. Wu, Y. Wang, and W.-C. Ye. 2021. "Quassinoids from the roots of *Eurycoma longifolia* and their anti-proliferation activities." *Molecules* 26 (19): 5939. doi:10.3390/MOLECULES26195939.

Yang, Y. B.-I. Park, E.-H. Hwang, and Y.-O. You. 2016. "Composition analysis and inhibitory effect of *Sterculia lychnophora* against biofilm formation by *Streptococcus mutans*." *Evidence-Based Complementary and Alternative Medicine* 2016. doi:10.1155/2016/8163150.

Yenn, T.W., L.C. Ring, K.A. Zahan, M.S.A. Rahman, W.-N. Tan, B. Jauza, and S. Alaudin. 2018. "Chemical composition and antimicrobial efficacy of *Helminthostachys zeylanica* against foodborne *Bacillus cereus*." *Natural Product Sciences* 24 (1): 66–70. doi:10.20307/nps.2018.24.1.66.

Zeng, C., and Z. Liu. 2011. "Microwave extraction of polysaccharides from the traditional Chinese medicinal plant, *Sterculia lychnophora*." *2011 International Conference on Remote Sensing, Environment and Transportation Engineering, RSETE 2011 - Proceedings*, pp. 6931–34. doi:10.1109/RSETE.2011.5965958.

Zhai, X.R., Z.S. Zou, Jia Bo Wang, and Xiao He Xiao. 2021. "Herb-induced liver injury related to *Reynoutria multiflora* (Thunb.) Moldenke: Risk factors, molecular and mechanistic specifics." *Frontiers in Pharmacology* 12 (September). doi:10.3389/FPHAR.2021.738577.

Zhang, Z., Y. Zhang, M. Song, Y. Guan, and X. Ma. 2019. "Species identification of *Dracaena* using the complete chloroplast genome as a super-barcode." *Frontiers in Pharmacology* 10: 1441. doi:10.3389/fphar.2019.01441.

Zhang, Z., Y. Li, Q. Cheng, G. Wu, D. Zhang, W. Huang, C. Deng, Z. Wang, and X. Song. 2021. "Chemical constituents from the roots of *Fallopia multiflora* var. Ciliinerve." *Biochemical Systematics and Ecology* 99. doi:10.1016/J.BSE.2021.104340.

5 A Rapid Ethnobotanical Inventory of Medicinal Plants

A Model for Future Conservation Efforts

Djaja Djendoel Soejarto
University of Illinois at Chicago, Chicago, IL, USA
Field Museum of Natural History, Chicago, IL, USA

Kongmany Sydara
Institute of Traditional Medicine, Ministry of Health,
Vientiane, Laos
University of Health Sciences, Vientiane, Laos

Bethany Gwen Elkington and Joshua Matthew Henkin
University of Illinois at Chicago, Chicago, IL, USA
Field Museum of Natural History, Chicago, IL, USA

Mouachanh Xayvue and Onevilay Souliya
Institute of Traditional Medicine, Ministry of Health,
Vientiane, Laos

CONTENTS

DOI: 10.1201/9781003216636-5

5.1 INTRODUCTION AND RATIONALE

A first step toward assuring continuing availability of medicinal plant resources to the community, for the present and for the future, is the establishment of a forest preserve for the purpose of protecting the medicinal plants found therein. We have established such preserves in Laos, and presently there is a network of 27 plant preserves, each named as either a Medicinal Plant Preserve (MPP) or a Medicinal Biodiversity Preserve (MBP). (See Chapter 4 for more details).

What is the next step?

The logical next step is to demonstrate the value and benefits of these preserves to members of the communities that use them. The short-term and immediate benefits are the continuing availability of medicinal plant species for the healer(s) and members of the community. In this context, healers (male or female) are defined as individuals in a community one consults with to seek medical treatment for health problems. The long-term benefits are the availability of the medicinal plants for future generations and the opportunity to perform biochemical studies of all the plants inside these preserves for their potential to provide new medicines. Chapters 3 and 4 have laid out the platform toward these ends. A review of the literature intended to provide scientific support for use of a plant by a healer in disease treatment is provided in Chapter 6 for selected medicinal plant species.

5.2 SELECTION OF THE PRESERVES AND THE INTERVIEW PROCESS

The first step in the valuation attempts of medicinal plant species found in these preserves was to inventory them. Six medicinal plant preserves and one forest preserve were selected. Each was inventoried using a rapid ethnobotanical method through interviews with healers living near the preserves. Through trial and error, information passed down by word of mouth and/or written documents (Elkington et al. 2012), along with other ways, healers discovered plants with medicinal virtues and then prescribed them to treat various diseases. A healer's deep knowledge of the medicinal plant species found within a defined preserve is their medicinal armamentarium.

With funding support from the California Community Foundation (CCF), we were able to undertake rapid ethnobotanical interviews in the following MBPs/MPPs (geographically from west to east and north to south): Oudomxay MBP, Luang Prabang MBP, Xieng Khouang MBP, Bolikhamxay MPP, Savannakhet MBP, Dong Natat Protected Forest, and Sekong MBP (see Chapter 1, Figure 1.1; Chapter 4, Figure 4.1). A variety of considerations led to the selection of these medicinal plant preserves and the protected forest.

It should be emphasized that fieldwork in these preserves was intended to perform a rapid survey that would provide an initial understanding on the richness and diversity of medicinal plants found in each of the six preserves and the protected forest. Depending on the results of these preliminary surveys, a more detailed study of the medicinal plant richness and diversity of a particular preserve may be undertaken in the future.

In addition to surveying medicinal plants, each fieldwork trip was used to train one or two member(s) of the provincial Traditional Medicine Stations (TMS) physically located at the Provincial Health Department in each province. This helped to ensure that medicinal plant inventories continue in the future.

In each case, the process consisted of interviewing a well-reputed healer within a community (village) located in the vicinity of the preserve, recording and documenting the healer's knowledge, collecting, and taxonomically identifying the voucher herbarium specimens. Interviews were conducted in the Lao and English languages with interpretation/translation by the ITM investigators.

5.3 BRIEF DESCRIPTION OF THE INVENTORIED PRESERVES

5.3.1 OUDOMXAY MBP: SEPTEMBER 2014. FIGURE 5.1

The 15-hectare Oudomxay MBP is located approximately 6 km to the south of Oudomxay City (also called Muang Xay). It is a forested hill that stands approximately 730 m above sea level (asl) at its base and 1,000 m at its peak. The forest cover is a dense, mixed submontane, broad-leaved tropical rain forest with a closed canopy, though uneven, due to the hilly terrain (Figure 5.1). Concrete posts in various locations mark the boundaries. The preserve sign and a guardhouse are located on the north side of the MBP hill.

FIGURE 5.1 Oudomxay MBP. A view of the interior of the forest on a slope. Photo credit: D. Soejarto.

Permission to carry out interviews and collect voucher herbarium specimens to document the interviews was granted by the Food and Drug Division (FDD) of the Provincial Health Department. The Head and staff of the Oudomxay TMS provided support to the research team and assisted with the recruitment of a healer. Healer #59 was recruited from the outskirts of Oudomxay city. This healer had frequented the forest preserve before its declaration as a preserve, as well as other forests located in the vicinity of the preserve area. The interview team consisted of two staff scientists of the ITM (Sydara and Xayvue) and one from UIC (Soejarto), together with the Head of the Oudomxay TMS and a local worker. Interviews took place on September 25 to 27, 2014.

5.3.2 LUANG PRABANG MBP: FEBRUARY 2020. FIGURE 5.2

Luang Prabang MBP covers a 245-hectare limestone mountain range that rises from an altitude of 302 m at the base at Somsanouk Village, Pak-Ou District, to more than 477 m asl. A monsoonal tropical forest (Figure 5.2) covers its lower and middle elevations. Bare rock, with patches of shrubby vegetation mixed with trees, is on the upper elevations all the way to the peak. Barbed wire on the lower end delimits the boundary of the preserve from the settled area (Somsanouk village). During the time of our field work (dry season, February to March, 2020) the forest cover was practically leafless with piles of dry leaves on the forest floor.

FIGURE 5.2 Luang Prabang MBP (a): covers 245 hectares of a limestone mountain range that climbs from 320 m to more than 477 m asl. (b): At middle elevations, bare and semi-bare forest trees are standing, with the forest floor littered with dry leaves. (c): A section of the forest on the lower elevation, where small streams flow, was lush green at the time of fieldwork. Photo credit: D. Soejarto.

Dense green vegetation was only found in areas at lower elevations, with some humidity and small streams flowing (Figure 5.2c).

Permission to carry out interviews and collect voucher herbarium specimens to document the interviews was granted by the Health District Office of Pak-Ou District, which monitors the preserve. The Head and staff of this office also provided generous support with field laborers and assisted in the recruitment of the healer to be interviewed. A renowned male healer (#62) who frequented the forests in the area was recruited from Somsanouk Village. The interview team consisted of

(a)   (b)

FIGURE 5.3 A panoramic view of Xieng Khouang MBP (a): The forests in the fore-ground and on the mountain slopes just beyond (not the one on the distant mountain that lines the horizon) are part of this MBP. (b): View of the interior of the MBP forest. Photo credit: D. Soejarto.

two staff scientists of the ITM (Sydara and Xayvue) and one from UIC (Soejarto). Interviews were conducted every day from February 10 to 13, 2020.

5.3.3   XIENG KHOUANG MBP: MAY 2014. FIGURE 5.3

This MBP is located 15 km north of Kham District and about 51 km from the pro-vincial capital city of Phonsavan. The preserve is about 1,100 m asl. It is the larg-est MBP established with funds from the LBF (see Chapter 4), having a total land surface area of 500 hectares. The forest cover is a mixed forest with somewhat open secondary forest on the southern end, intergrading into a denser, closed-canopy, submontane, broad-leaved evergreen forest to the north. There are signs of past con-version toward the northern end and along the slopes on the east side of the highway.

Permission to carry out interviews and collect voucher herbarium specimens to document the interviews was granted by the provincial FDD. A male healer (#56) from the Tha village was recruited by the Head of the TMS of Xieng Khouang City/Province. The interview team consisted of two staff scientists of the ITM (Sydara and Xayvue) and one from UIC (Soejarto), one driver-assistant, the Head of the Xieng Khouang TMS, and with the support of local workers. Interviews were conducted on May 16 and 17, 2014.

5.3.4   BOLIKHAMXAY MPP: MAY 2014. FIGURE 5.4

Bolikhamxay MPP is the first medicinal plant preserve established in Laos through the initiative and funding support from the ICBG Program and from the Government of Laos (Ministry of Health) in 2004 (see Chapter 4 for details). The

FIGURE 5.4 Visitor shelter just inside the entrance to the Bolikhamxay (Somsavath) MPP. Photo credit: D. Soejarto.

preserve is within walking distance from Somsavath Village in Paksan District, and is located approximately 20 km from Paksan city, the capital of Bolikhamxay Province. Paksan is a 2½ hour drive from Vientiane.

For the present project, fieldwork took place inside this 15 ha. preserve, which is delimited by a barbed wire fence anchored to concrete posts. The forest is a secondary, semi-deciduous, broad-leaved tropical rain forest with remnants of large forest tree species scattered within.

Permission to carry out interviews and collect voucher herbarium specimens was granted by the provincial FDD and by the Head of the village of Somsavath. A male healer (#57) from Somsavath Village was recruited by the Head of the village. The interview team consisted of one staff scientist of the ITM (Souliya) and one from UIC (Soejarto), a driver-assistant, and three local workers. The Village Head joined the field team during the first day of fieldwork. Interviews were conducted on May 24 to 26, 2014.

5.3.5 SAVANNAKHET MBP: DECEMBER 2013. FIGURE 5.5

This MBP was formerly a part of the Dong Natat and Nong Lom Lake Protected Area (see 5.3.6 below), intended to promote ecotourism. An unpaved forest road separates the MBP and the Dong Natat protected area. In 2008, following legal

FIGURE 5.5 Visitor center at the entrance to the Savannakhet MBP. Photo credit: D. Soejarto.

proceedings, a 15-ha area that lines the highway, on one side, and the Dong Natat Forest, on the other, was separated and delineated by barbed wire to become what is now known as the Savannakhet MBP, under the jurisdiction of Kaysone District. The preserve is approximately 19 km from the Savannakhet City Center. The forest is a lowland, mixed broad-leaved evergreen and somewhat degraded dipterocarp forest. The administrative responsibility for this MBP lies in the hands of the Savannakhet Province FDD.

Fieldwork interviews were conducted on December 17 and 18, 2013. The interview team consisted of two staff scientists of the ITM (Sydara and Xayvue), a UIC scientist (Soejarto), a driver-assistant, a healer (#58) who lives in a village located south of the Savannakhet MBP, and support provided by locally recruited workers. An observer from the Savannakhet FDD accompanied the field team.

5.3.6 DONG NATAT PROTECTED FOREST: MARCH 2014. FIGURE 5.6

This protected forest (Figure 5.6a) includes the Nong Lom Lake on the eastern side of the forest. It is under the administrative authority of the Kaysone Phomvihan District, Savannakhet Province. The area covers about 8,300 hectares of forest. It was established as a provincial preserve in 1962 during the reign of King Sisavang Vatthana and is a zone for ecotourism development where activities like trekking, cycling, field study, or simply relaxing and observing nature can be enjoyed. A large section of this protected area comprises original, semi-dry, deciduous diptero-carp forest. A section of this forest that borders the Savannakhet MBP still harbors

(a)   (b)

FIGURE 5.6 Profile of the Dong Natat Protected Area (a): showing dipterocarp forest across the Nong Lom Lake. (b): A majestic tree of *Vatica odorata* (Dipterocarpaceae). Photo credit: D. Soejarto.

original dipterocarp tree species. One such example is *Vatica odorata* (Griff.) Symington (Dipterocarpaceae), which majestically rises to 30–45 m in height with a bole diameter of up to 2 m (Figure 5.6b). Extraction of useful resin (nyang oil) from another dipterocarp, commonly called the Nyang tree, (*Dipterocarpus alatus* Roxb. ex G. Don) currently threatens the survival of this species.

Although this forest preserve is not part of the MBP-MPP inventory study, it was explored due to its adjacent location to the Savannakhet MBP and to enrich the information on medicinal plants of Laos. A permit to enter the forest to undertake ethnobotanical field interviews with a healer was secured from the District Office of Savannakhet Province.

An ethnobotanical survey was performed on December 17–18, 2014. The interview team consisted of two staff scientists of the ITM (Sydara and Xayvue), one driver-assistant, a healer (#58), and two locally recruited workers. A district representative accompanied the field team.

5.3.7 SEKONG MBP: NOVEMBER 7–13, 2019. FIGURE 5.7

While Sekong Province is one of the smallest, least populated, and monetarily poorest provinces (Lao Statistics Bureau and World Bank 2020), it is also one of the most ethnically diverse area of Laos (Tagwerker 2009). The province is well-forested (Wittmer and Gundimeda 2012) and represents an area rich in traditional herbal medicine knowledge. Sekong Province is situated on the Bolaven Plateau, between Vietnam on the east, Attapeu Province to the south, Salavan Province to the north, and Champasak Province to the west.

The MBP of Sekong Province (ເຊກອງ) falls under the jurisdiction of the Palengtai Village (ບ້ານພະແລງໃຕ້) of Thateng District (ເມືອງທ່າແຕງ) and is

FIGURE 5.7 Secondary forest at the Sekong MBP showing the interior of the forest. Photo credit: B. Elkington.

approximately 30 hectares in size. It was established in 2013 with funds from the Government of Laos and the LBF (See chapter 4; Section 4.4). The vegetation of the preserve is a combination of mixed deciduous and semi-evergreen forest (Figure 5.7). It is approximately 500–700 m asl. At the time of this expedition, the sign at the entry to the preserve had been destroyed, and there was no apparent indication of the preserve boundaries.

The interviews took place from November 17–23, 2019. Before beginning the fieldwork, the team met with the Provincial Health Department representative, the Deputy Head of the Sekong FDD, the Deputy of the TMS, the Director of the local hospital, and several other provincial health department representatives. Following these meetings, the research team (one ITM scientist—Mouachanh Xayvue; one UIC scientist—Bethany Elkington) met at the home of a traditional healer (#60). The Village Head joined for the second day of fieldwork.

5.4 METHODS

5.4.1 TRANSPORTATION AND ACCOMMODATIONS

For the sake of simplicity and clarity, the term "medicinal plant preserve" is used here to refer to any of the preserves listed in Table 4.2 in Chapter 4, irrespective of whether the preserve is designated as a "Medicinal Plant Preserve" (MPP) or a "Medicinal Biodiversity Preserve (MBP)," since the intention is to refer to the protected medicinal plant components of these preserves. Also, for simplicity, each of these preserves is referred to by the name of the province where the preserve is located.

ITM vehicles were used to drive from Vientiane to the preserves where fieldwork was conducted. Except for Bolikhamxay MBP, Dong Natat Protected Forest and Sekong MBP, fieldwork was conducted by three scientists: two were staff of the ITM (Sydara and either Xayvue or Souliya), and one was from UIC (Soejarto). The team usually stayed in a local guesthouse or hotel, which served as a home base for the duration of each trip.

5.4.2 PERMISSION TO ENTER A PRESERVE TO CARRY OUT ETHNOBOTANICAL INTERVIEWS AND COLLECT PLANTS

In every MBP or MPP location, permission to carry out field interviews in the preserve and collect voucher herbarium specimens was obtained from the FDD of the Provincial Health Department. If a preserve was under the administration of a District Office, such as Luang Prabang and Dong Natat preserves, additional permissions were obtained from these offices. In this process, these offices have also been generous in aiding with the recruitment of healers to be interviewed.

As a matter of protocol, the ITM-UIC researchers conducted the interviews, supported by staff of the provincial TMS, as well as locally recruited workers.

5.4.3 RECRUITMENT OF HEALERS TO BE INTERVIEWED AND HEALER COMPENSATION. FIGURE 5.8

Prior to every field trip, the Director of the ITM contacted the TMS staff of each province to explain the ITM's proposal to carry out a field survey in the medicinal plant preserve. TMS assistance was requested to help provide staffing support, connect with the Village Head in or near the preserve, and recruit a healer to be interviewed.

Upon arrival in the province, the ITM team went directly to the Provincial Health Department, then to TMS headquarters to discuss the proposal. With recommendation of the TMS Head, the team traveled to the village in separate vehicles. Once in the village, the healer was identified. Because a good relationship already existed between the healer to be recruited and the TMS Head and staff, the field team was able to visit the healer's home.

During each stop in a village, the TMS Head accompanied the research team to assist with explaining the purpose of the visit, obtaining permission for conducting the interview, and describing how the interview would be conducted. The entire chain of permissions and clearances, including the ITM, the Provincial Health Department, the FDD, the TMS, the Head of the village, down to the healer, attest to the legal procedures and cultural norms the researchers strictly adhered to during the entire informed consent process. Healers were monetarily compensated for their time and effort in collaborating with the interviewing team.

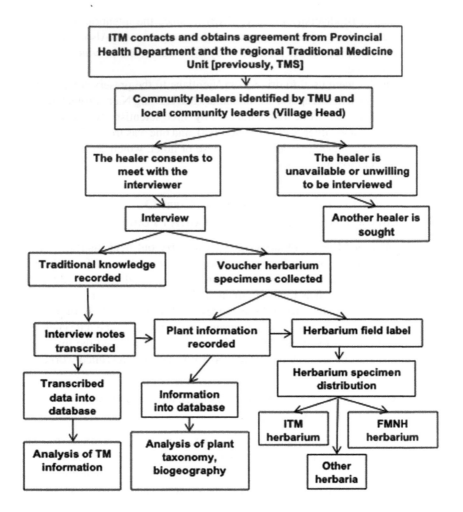

FIGURE 5.8 Workflow for recruiting and interviewing healers and documenting the ethnomedical data and the plants collected.

It should be noted that in this specific project, only male healers were interviewed. This was unintentional, but typical, as previous research projects in Laos have noted more male than female healers (Libman et al. 2009; Elkington et al. 2012, 2014; Soejarto et al. 2012).

The entire process of healer recruitment and data documentation is depicted in Figure 5.8.

5.4.4 INTERVIEWS. FIGURES 5.9 AND 5.10

The rapid ethnobotanical field interviews for this research were conducted under a research protocol approved by the UIC Institutional Review Board's approval

FIGURE 5.9 A walk-in-the-forest interview with a healer. A healer in the Oudomxay MBP pointed to a tree of important medicinal value (*Pellacalyx yunnanensis*; Soejarto et al. 15114) and explained the use of the tree for treating intestinal infection. Photo credit: D. Soejarto.

no. 2013-0763, and the Lao National Ethics Committee for Health Research No. 001 (2013–2014) and No. 087 (2019–2023). All interviews were conducted using the "walk-in-the-forest" method (Figures 5.9 and 5.10). The ITM research scientists conversed with the healer in the Lao language, while the UIC research scientist exchanged questions and answers with the healer in the English language, interpreted by the ITM colleagues. The researchers recorded the information on the plant and its medicinal use in their field notebooks.

5.4.5 Voucher Herbarium Specimen Documentation. Figure 5.11

Each plant pointed out by the healer was photographed and notes recorded in field notebooks about the habitat, date collected, geographic location (including GPS reading), local name, part of plant used, traditional medicinal use, preparation of

FIGURE 5.10 A healer in Xieng Khouang MBP explained the use of a plant (*Leea indica*; Soejarto et al. 14995; for the treatment of rheumatism) to the field team; the field researchers recorded the information in their notebooks. Photo credit: Villat.

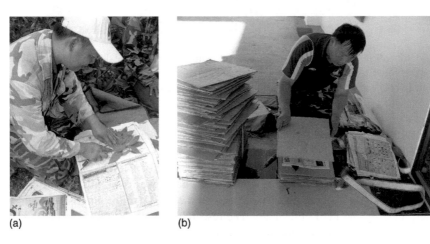

(a)   (b)

FIGURE 5.11 Taxonomic documentation. A field assistant presses a numbered intact twig bearing leaves and flowers/fruits between newspapers (a). On arrival at the fieldwork base station, the field assistant places the pressed plant specimens between corrugated cardboards into a stack (b), which will be tied and placed in a portable, electric-powered, specimen drying box during the night and throughout the fieldwork period. Photo credit: D. Soejarto.

the medicine and dosage. A set of three voucher herbarium specimens was collected to document each interview. These specimens serve as the primary record of the plant and healer interview and are imperative for accurate taxonomic identification. Each of these specimens consisted of a clipping of around 20 cm of a plant part with intact leaves and fruits and/or flowers. A tag with a collection number was attached to each specimen, the specimen was pressed flat into newspapers, and then it was placed between wooden presses (Figure 5.11a). Every attempt was made to find and collect fertile specimens bearing a flower and/or a fruit.

After returning to the field home base at the end of the day, each specimen was placed between corrugated cardboard between wooden presses and tied with a rope (Figure 5.11b). The presses were then placed inside a portable electric drier.

Upon arrival at the ITM herbarium, specimens that were not fully dry were separated to be dried in the sun or in the herbarium dryer at ITM until completely dry.

One set of fully dried herbarium specimens was shipped to the John G. Searle Herbarium of the Field Museum (Field Museum 2022) in Chicago, one set was retained at the ITM Herbarium, and one set is kept at ITM as a standby to be sent to a taxonomic specialist for assistance with the identification process or for confirmation of identity.

5.4.6 Taxonomic Determination. Figure 5.12

Taxonomic determination was initially performed at the ITM Herbarium (Figure 5.12) through consultation with the literature and by comparison with identified specimens at the herbarium. This process continued at the Field Museum Herbarium in Chicago.

At the ITM Herbarium, common names in the ITM's in-house compilations, together with those in published works (Inthakoun and Delang 2011; Vidal 1959), were consulted and found to be useful in eventually coming to the identification of a species. Floristic treatises of Asian plants were also consulted, especially the book *Cây cỏ Việt Nam* (Ho 2000). In Chicago, the Field Museum's Herbarium collections and the library resources were consulted. Searches were also performed on the online databases that house photographic images of identified specimens, including Tropicos® (Tropicos.org 2023), the National Herbarium of the Smithsonian Institution (Smithsonian Institution 2022), the Muséum national d'Histoire naturelle (2023), Plants of the World Online (Kew 2022b), and Global Plants at JSTOR (which houses photographic images of type specimens) (JSTOR® 2022). *A Checklist of the Vascular Plants of Lao PDR* (Newman et al. 2007) was consulted to confirm the previous occurrence of certain species in the country.

In the catalog of medicinal plant species presented below, the nomenclature of plant family names follows the APG system (Chase et al. 2016), while that of the species follows names adopted in the Plants of the World Online (Kew 2022b) and in the WFO Plant List (WFO 2022), formerly, The Plant List (Kew 2022a).

FIGURE 5.12   Research team in the process of taxonomic identification of plants collected. Initial taxonomic identification of the medicinal plant specimens collected was performed at the ITM Herbarium in Vientiane. Photo credit: K. Sydara.

5.5   RESULTS

5.5.1   TAXONOMIC DIVERSITY OF MEDICINAL PLANTS COLLECTED AS USED BY THE HEALERS

A total of 336 medicinal plant collections were made from the six medicinal plant preserves and one protected forest, comprising 279 taxa, of which 98.2% (274 taxa) comprise the angiosperms, 1.1% gymnosperms (three species of *Gnetum*), and 0.7% pteridophytes (a species of *Lygodium* and a species of *Selaginella*). Of the angiosperms, 251 species (91.6%) comprise the dicots and 23 species (8.4%) the monocots. Clearly, the angiospermous plants represent the major portion and the most diverse medicinal plants that the healers used.

5.5.2   PART OF THE PLANT USED

As stated above, the greater percentage (>98%) of plants used by the healers comprises the angiosperms. Consequently, the part of the plant used by healers in disease treatment refers to parts of a tree, a shrub, or an herb. In this regard, we can say that every part of the plant, but especially the leaf, the stem (stem bark or stem wood), and the root—or a combination of these—is used by the healer in disease treatment, as accounted in the species catalog below.

5.5.3   DIVERSITY OF THERAPEUTIC USES

A great diversity of therapeutic uses as communicated by the healers was recorded. Based on the order of frequency of medicinal use mentioned by the healers, as determined by the number of voucher herbarium specimens collected, the most common diseases treated are "pain" and afflictions related to pain (55 herbarium records), digestive system-related afflictions (45 herbarium records), female-related afflictions (37 herbarium records), rheumatism- and joint pain-related afflictions (36 herbarium records), fever and fever-related afflictions (34 herbarium records), and skin-related afflictions (25 herbarium records). The least mentioned body afflictions or diseases treated are metabolic diseases (4 herbarium records), eye (vision) afflictions (4 herbarium records), polio and paralysis afflictions (4 herbarium records), tooth-related afflictions (4 herbarium records), and treatment of thyroid gland (3 herbarium records) (Table 5.1). See also Chapter 6.

TABLE 5.1
Therapeutic Diversity of the Medicinal Plants Inventoried

Types of disease or afflictions treated	Number of herbarium specimens for which medicinal use is recorded
Pain, bodily pain, and other body weaknesses	55
Digestive system-related diseases	45
Women-related diseases	37
Rheumatism and joint pain	36
Fever and fever-related diseases	34
Skin infection and related diseases	25
Liver-related diseases	20
Kidney- and urination-related diseases	18
Dysentery and undefined infections	15
Lung afflictions and related diseases	13
Nervous system-related diseases (disorders)	13
Poisoning and poison antidote	12
Sexual intercourse-related diseases	12
Heart-related afflictions (diseases)	7
Metabolic diseases	4
Eye and vision afflictions (diseases)	4
Paralysis and polio	4
Tooth-related diseases	4
Thyroid gland-related diseases	3
Total	361 herbarium records (many collections have a record with more than one disease)

5.5.4 METHOD OF USE (PREPARATION, DOSAGE)

The most common mode of preparation, route of administration, and dosage of applications is "boil and drink" (boil the plant material and drink the decoction as needed). This method of use is comparable to that obtained by Delang (2007), who found that more than half of 57 medicinal plant materials purchased from a marketplace in Laos were used by boiling and drinking the decoction. It is also comparable to methods used by healers and communities of Bach Ma National Park, Vietnam, where it was recorded that boiling and drinking the decoction was the most common practice (Thiên and Ziegler 2001).

A sampling of method and route of administration is provided by the following (as noted for the cited taxon collection):

"Cut the root into small pieces, boil in water, and drink the decoction (ຝັກຊອຍຮາກເປັນບ່ຽງນ້ອຍໆ, ຕົ້ມໃນນ້ຳ ແລະ ດື່ມນ້ຳຕົ້ມ)" – Soejarto et al. 14925.
(*Polyalthia evecta* – Annonaceae)

"Crush the red fruits and rub it on the affected part of the skin (ຍ່ອງໝາກສຸກ ແລ້ວ ໃນທຸໃສ່ບ່ອນທີ່ເປັນພະຍາດ)" – Soejarto et al. 14964.
(*Lobelia nummularia* – Campanulaceae).

"Macerate the stem wood in water and use the liquid as a bath (ແຊ່ວ່າຕົ້ນທີ່ພັກ ຊອຍເປັນຕ່ອນນ້ອຍໆໃນນ້ຳ, ໃຊ້ນ້ຳຕົ້ມອາບ)" – Soejarto et al. 14985.
(*Elaeocarpus ovalis* – Elaeocarpaceae)

"The leaf or root is used as a medicine to treat goiters occurring mainly in women with thyroid gland disorders by chewing the leaves or root and swallowing the watery liquid (ໃຊ້ເປືອກບໍ່ຄໍໝປຽກກີນເລືອດ ໂດຍການຫຍ້ຳກີນນ້ຳຈາກເປືອກ ມັກເກີດກັບຜູ້ຍິງ)" – Soejarto et al. 14969.
(*Callicarpa macrophylla* – Lamiaceae)

"Cut the whole plant into small pieces, boil in water, and drink the decoction (ຝັກ ຊອຍ ວ່າຕົ້ນເປັນບ່ຽງນ້ອຍໆ, ຕົ້ມໃນນ້ຳ ແລະ ດື່ມນ້ຳຕົ້ມ)" – Soejarto et al. 15022.
(*Antidesma* cf. *hainanensis* – Phyllanthaceae)

"Take one handful of fresh leaves and pound to make a paste (ເອົາໃບສົດປະມານ ໜຶ່ງກຳມື ຕຳ ແລະ ເຮັດເປັນຍາທ້ອນປຸງກ), then apply on the anus areas for hemorrhoids (ເອົາຍາທ້ອນປຸງກປັ້ນແປະໃສ່ຮູຫະວານເພື່ອບໍ່ວິດສິດວງ), and for fever apply on the body (ສຳລັບບໍ່ໄຂ້ແມ່ນແປະໃສ່ຕົນໂຕ)" – Soejarto et al. 15007.
(*Oldenlandia microcephala* – Rubiaceae)

"Boil entire plant and drink the decoction (kidney stones, backaches) or use as a gargle with salt (tonsilitis) (ຕົ້ມພິດທັງໝົດໃນນ້ຳ ແລະ ດື່ມນ້ຳ) (ສຳລັບບໍ່ຫິ່ນ ໝາກໄຂ່ຫຼັງ, ເຈັບແອວ) ຫຼື ໃຊ້ກໍຄໍ ໂດຍບະສົມກັບເກືອລ້າງນ້ອຍ (ບໍ່ອີກ ເຊບວ່າຄໍ)" – Soejarto et al. 14961.
(*Rubia argyi* – Rubiaceae)

"Cut the root into small pieces and dry, then take a handful and boil in two liters of water (ຕັດຮາກເປັນຕ່ອນນ້ອຍໆ ແລະ ຕາກໃຫ້ແຫ້ງ); drink as needed in cases of kidney stones or as an antidote (ດື່ມບ່າຕົ້ມສ່າລັບບໍ່ໝື່ອໝາກຫໄຂ່ໜ້າ ແລະ ຖອນພິດ); boil the leaves and use as sauna for dizziness (ຕົ້ມໃບເພື່ອຮົມອາຍບ່າ ສ່າລັບບໍ່ອື່ນວຽນ)" – Soejarto et al. 15055.

(*Micromelum minutum* – Rutaceae)

"Boil one handful of dried small pieces of the liana/stem or leaves in one liter of water, drink the decoction (ຕົ້ມຕ່ອນນ້ອຍໆທິໝໄຂ້ກ ຽອຍມາຈາກເຄືອ (ລ່າຕົ້ນ) ຫຼື ໃບປະມານໜຶ່ງກ່າມີໃນນ້ຳຫຶ່ງລິດ ແລະ ດື່ມບ່າຕົ້ມ)" – Soejarto et al. 15067.

(*Pothos scandens* – Araceae)

"Boil a handful of small, dried pieces of the whole plant in one liter of water and drink the decoction (ຕົ້ມໝົດເຄືອທິໝໄຂ້ກ ຽອຍ ເປັນຕ່ອນນ້ອຍໆ ແລະ ຕາກແຫ້ງ ໃນນ້ຳຫຶ່ງລິດ ແລະ ດື່ມບ່າຕົ້ມ)" – Soejarto et al. 14916.

(*Lygodium flexuosum* – Lygodiaceae)

5.6 CATALOG OF INVENTORIED MEDICINAL PLANTS

A catalog of the medicinal plant species acquired through healers' interviews is presented below. For each species, the following data are provided in one paragraph, in the following order:

1. The Latin binomial
2. Common name in the Lao script and in the Roman script.
3. Location of the interview and collection of the voucher specimen.
4. Field characteristics of the plant, as recorded in the field notebook.
5. Medicinal use (and plant part used), as recorded in the field notebook.
6. Method of use (and dosage if given), as recorded in the field notebook.
7. Voucher herbarium specimen collection number.
8. Photographic image(s) for selected species.

As to the habitat of the plant, the reader is referred to the habitat description for each MBP/MPP described above under 5.3.

GROUP 1: ANGIOSPERMAE – DICOTYLEDONAE

Acanthaceae

Asystasia chelonoides Nees. Figure 5.13
ໝວດແມວປ່າ; ໝາກເພັດນ້ຳ – Nuat Meo Pa; Mak Pet Nam. Oudomxay MBP. An extensively spreading herb, flowers white to light pink, the lower lip with deep pink smudges adaxially. The entire plant is used to treat wounds and to stop bleeding (ໃຊ້ໝົດຕົ້ນເພື່ອບໍ່ບາດແຜ ແລະ ຫ້າມເລືອດ). Crush the whole plant and apply it on the injured skin (ຍ່ອງໝົດຕົ້ນ ແລະ ແປະໃສ່ບາດແຜເທິງຜິວໜ້າ). Soejarto et al. 15086.

(a)   (b)

FIGURE 5.13 A flowering herbaceous stem of *Asystasia chelonoides* in its natural habitat bearing open flowers borne on a terminal inflorescence (a) and a close-up view of the fully open zygomorphic flowers of *Asystasia chelonoides* (b). Photo credit: D. Soejarto.

ໜວດແມວປ່າ; ໝາກເພັດນ້ຳ – Nuat Meo Pa; Mak Pet Nam. Sekong MBP. Shrub, 1 m in height, stem with protuberances, leaves solitary, alternate, margins slightly crenate, flower funnel-shaped, white to lavender. The whole plant is employed as a medicine to treat infections and kidney stones. Boil a handful of dried pieces of the whole plant in one liter of water and drink the decoction (ໃຊ້ໝົດຕົ້ນປິ່ນປົວການຕິດຊື້ອ, ບໍ່ໝື້ອໝາກໄຂ່ຫຼັງ. ຕົ້ນພົດທີ່ຕັດເປັນ ຕ່ອນນ້ອຍໆ ແລະ ຕາກແຫ້ງແລ້ວ, ໃຊ້ປະມານ 1 ກຳມືໃນນ້ຳ 1 ລິດ ແລະ ດື່ມນ້ຳຕົ້ມ). Elkington et al. 304.

Barleria cf. *obtusa* Nees
ເຂົ້າວິບ - Khao Liib. Sekong MBP. Small herb, 20–30 cm in height, leaves with smooth margins, glabrous, flowers purple. The whole plant is used as a health tonic (ຮາກເປັນຍາບຳລຸງສຸຂະພາບ). Boil the whole plant in water, filter, then drink the decocted water (ຕົ້ນໃນນ້ຳ, ຕອງ, ແລ້ວດື່ມນ້ຳຕົ້ມ). Elkington et al. 269.

Barleria strigosa Willd.
ເຂົ້າວິບ – Khao Liib. Sekong MBP. Small herb, around 20–30 cm in height. Leaves glabrous, with smooth margins, flowers purple. The whole plant is used to treat body weakness (ໃຊ້ໝົດຕົ້ນເພື່ອບໍ່ອສຸຂະພາບອ່ອນເພຍ). Boil the plant with water, filter the liquid, drink the decoction (ຕົ້ນໃນນ້ຳ, ຕອງ, ແລ້ວດື່ມນ້ຳຕົ້ມ). Elkington et al. 261.

FIGURE 5.14   An inflorescence of *Justicia ventricosa*, standing erect, with some flowers opening, each subtended by a brown, veiny bract. The habitat in the background is water-logged terrains at the base of a hill. Photo credit: D. Soejarto.

Justicia ventricosa Wall. ex Hook.f. Figure 5.14.

ຮູຮາຂາວ – Hou Ha Khao. Luang Prabang MBP. Shrub on water-logged terrains at base of a slope, inflorescences upright, bracts olive-brown, flowers white with fine, purple striations along the inner surface of the labium. The root is used to prepare a medicine to treat internal rashes (ໃຊ້ຮາກເພື່ອປ້ອຕຸ່ມພາຍໃນ). Rub the roots with trickles of water on a rock and paste the slurry liquid on the affected part of the body (ຝົນຮາກໃສ່ຫົນຝົນ ແລ້ວເອົານ້ຳຊຸ້ງໆທິ່ໄດ້ແປະບ່ອນທີ່ມີອາການ). Soejarto et al. 15464.

Phlogacanthus curviflorus (Wall.) Nees. Figure 5.15

ຫອມຊ້າງ – Hom Xang. Oudomxay MBP. Large shrub, inflorescence erect, terminal, pink-purple when young. Fruit a capsule, dry, set in a racemose cluster.

(a)   (b)

FIGURE 5.15 A flowering twig of *Phlogacanthus curviflorus* (a) in its natural habitat and a close-up view of *Phlogacanthus curviflorus* flowers borne in a racemose inflorescence (b). Photo credit: D. Soejarto.

The root is used to make a medicine to treat malaria and other cases of fever (ໃຊ້ຮາກເພື່ອບໍ່ໄຂ້ມາລາເຣຍ ແລະ ກໍລະນີເປັນໄຂ້ອື່ນໆ). Macerate the root in water and drink the macerate (ແຊ່ຮາກ ແລ້ວດື່ມນ້ຳ). Soejarto et al. 15081.

Thunbergia grandiflora Roxb.
ເຄືອນ້ຳແນ່ – Kheua Nam Neh. Luang Prabang MBP. A vine, flowers light blue to pinkish blue. Very common plant. Stem to treat cases of poisoning (intoxication) (ເປັນພິດທີ່ພົບເຫັນທົ່ວໄປ. ໃຊ້ລ່າຕົ້ນເພື່ອຖອນພິດ). Boil the vine (stem) and drink the liquid as needed (ຕົ້ມເຄືອ (ລ່າຕົ້ນ) ແລະ ດື່ມນ້ຳຕົ້ມຕາມຕ້ອງການ). Soejarto et al. 15443.

Actinidiaceae
Saurauia napaulensis A. DC. Figure 5.16

ກົກໝາກເໝົ້າ; ສົ້ມເບົາ - Kok Mak Mao; Som Bao. Xieng Khouang MBP. Treelet 6 m tall, stem bark gray, roughish, lightly fissured, flower buds gray, open flowers pink, corolla lobes deep pink. The stem or the root is used to prepare a medicine to treat cases of nervous disorders (ໃຊ້ລ້າ ຫຼື ຮາກເພື່ອບໍ່ເສັ້ນປະສາດພິການ). Boil the root or stem and drink the liquid (ຕົ້ມລ່າຕົ້ນ ຫຼື ຮາກ ແລ້ວດື່ມນ້ຳ). Soejarto et al. 14988.

Adoxaceae
Viburnum sambucinum Reinw. ex Bl.

ກົກຄ້າມບຸ; ໄມ້ຄ່າຕາຍ - Ko Kam Pou; Mai Kam Tai. Xieng Khouang MBP. Treelet to 4 m tall, flower buds brownish green, profuse, on terminal, wide-spreading corymbose clusters. The root is used to promote pregnancy (ໃຊ້ຮາກເພື່ອຊ່ວຍໃຫ້ຖືພາ

(a)   (b)

FIGURE 5.16 A flowering branchlet (a) of a *Saurauia napaulensis* tree showing axillary paniculate inflorescence and a fully open flower (b) of *Saurauia napaulensis*. Photo credit: D. Soejarto.

- ມີລູກ). Boil a handful of small, dried pieces of the root in one liter of water and drink the decoction (ຕຶ້ມຮາກທີ່ຝັກ ຊອຍເປັນຕ່ອນນ້ອຍໆປະມານ 1 ກຳມື ໃນນ້ຳ 1 ລິດ ແລະ ດື່ມນ້ຳຕົ້ມ). Soejarto et al. 14967.

Anacardiaceae
Mangifera laurina Bl.

ມ່ວງໃຊ່ – Muang Khai. Luang Prabang MBP. Tree 20 m tall, dbh 30 cm, bark gray, finely fissured and cracked, flowers drying up dirty white, young fruits green, 1-seeded. Crushed fruitlets smell of mango. The stem bark is used to make a medicine to treat cases of swelling of body parts (edema) (ໃຊ້ເປືອກຕົ້ນເພື່ອພະ ລິດເປັນຍາບໍ່ອາການຕົ້ນໂຕໄຕ່ (ບວມ). The bark is mixed with white buffalo horn; macerate and use the liquid as a bath (ປະສົມເປືອກຕົ້ນໃສ່ກັບເຂົາຄວາຍ ດ່ອນ, ແຊ່ນ້ຳ ແລ້ວໃຊ້ນ້ຳອາບ). Soejarto et al. 15433.

Ancistrocladaceae
Ancistrocladus tectorius (Lour.) Merr.

ຫາງກວາງເຄືອ – Hang Kouang Kheua. Savannakhet MBP. A liana, long leaves, flowers densely clustered, perianth brownish red. The liana (stem) is used to treat fever and weakness (as a tonic) (ໃຊ້ເຄືອ (ລ່າຕົ້ນ) ບໍ່ໄຂ້, ບໍ່ສຸຂະພາບອ່ອນເພຍ (ເປັນຍາບຳລຸງກຳລັງ). Slice the liana into small pieces, then dry them. Take one handful of the dried pieces and put to a boil in about one liter of water for 10–15 minutes. Drink the cooled decoction as needed (ຝັກ ຊອຍເຄືອເປັນບ່ວງນ້ອຍໆ, ຕາກໃຫ້ແຫ້ງ. ໃຊ້ພິດແຫ້ງ ໜຶ່ງກຳມື ຕົ້ມໃນນ້ຳບະມານ 1 ລິດ ເປັນເວລາ 10–15 ນາທີ. ດື່ມນ້ຳຕົ້ມຕາມຕ້ອງການ). Sydara et al. 19035.

Annonaceae

Fissistigma cf. *acuminatissima* Merr.

ກຳວັງເສືອໂຄ່ງ – Kamlang Seua Kong. Sekong MBP. Liana, stem approx. 3 cm diam., leaves glabrous on top, underside tomentose, inflorescence umbellate, fruits globular, 2 cm across. The leaves and the stem are used to treat body weakness (as a tonic for strength) (ໃຊ້ໃບ, ລ່າຕົ້ນ ບໍ່ສຸຂະພາບອ່ອນເພຍ (ເຮັດໃຫ້ສຸຂະພາບແຂງແຮງ). Boil dried pieces of leaf or stem in one liter of water and drink the decoction (ຕົ້ມໃບ ແລະ ລ່າແຫ້ງຫນຶ່ງກຳມືໃນນ້ຳຫນຶ່ງລິດ ແລະ ດື່ມນ້ຳຕົ້ມ). Elkington et al. 297.

Goniothalamus cf. *gabriacianus* (Baill.) Ast.

ເຂົ້າຫຼາມດົງ – Khao Lam Dong. Savannakhet MBP. Small tree 2–4 m high, leaves alternate, flowers pale white, disposed along twigs. Stem wood, root to treat rheumatism and cases of gastrointestinal disease (ໃຊ້ແກ່ນລ່າ, ຮາກ ບໍ່ປະດົງຂໍ້ ແລະ ພະຍາດກ່ຽວກັບກະເພາະລ່າໃຊ້). Cut the stem wood or root into small pieces and dry them, then take a handful and boil in two liters of water. (ພັກ ຂອຍແກ່ນລ່າ ຫຼື ຮາກເປັນບ່ຽງນ້ອຍໆ, ຕາກໃຫ້ແຫ້ງ, ຕົ້ມປະມານຫນຶ່ງກຳມືໃນນ້ຳສອງລິດ). Drink the liquid as needed (ດື່ມນ້ຳຕົ້ມຕາມຕ້ອງການ). Sydara et al. 19026.

Melodorum cf. *vietnamense* Ban

ລ່າດວນ; ນ້ຳເຕົ້ານ້ອຍ – Lam Douan; Nam Tao Noy. Savannakhet MBP. Treelet with drooping, liana-like branches, flower buds light green. Only one seen. The root is used to prepare a medicine for back pain (ໃຊ້ຮາກເພື່ອບຸງແຕ່ງເປັນຍາ ບໍ່ເຈັບແອວ). Boil a handful of small pieces in a bowl of water and drink the decoction (ໃຊ້ພິດແຫ້ງທີ່ ພັກ ຂອຍເປັນບ່ຽງປະມານຫນຶ່ງກຳມື ຕົ້ມໃນນ້ຳຫນຶ່ງ ຖ້ວຍ ແລະ ດື່ມນ້ຳຕົ້ມ). Soejarto et al. 14904.

Monocarpia kalimantanensis P.J.A. Kessler. Figure 5.17

ຕົ້ນຊາເດນ Tohn Sa Den. Sekong MBP. Small tree, approx. 0.5 m high. Leaves ~30 cm long, 10 cm wide. Fruits orange and slightly tomentose, ca. 4 cm in diameter, set on a roundish umbel. The stem bark is used as a medicine to treat liver disease and cases of nervous disorders (ໃຊ້ເປືອກຮາກເພື່ອປິ່ນບໍ່ພະຍາດຕັບ ແລະ ເສັ້ນປະສາດພິການ). Boil the stem bark and drink the liquid (ຕົ້ມລ່າຕົ້ນ ແລະ ດື່ມນ້ຳ). Elkington et al. 274.

Monocarpia maingayi (Hook.f. & Thoms.) L.M. Turner

ກົກດ້ານຂວານ, ເຕັ້ມຕາກໍ່ຍູງ – Kok Dam Kwan; Teum Ta Koh Joung. Sekong MBP. Tree approximately 6 m high, trunk approximately 25 cm dbh. Leaves alternate. Flowers small, white, fruits white. Root to treat joint pain (ໃຊ້ຮາກບໍ່ເຈັບ ຂໍ້). Boil a handful of small, dried pieces of root in one liter of water and drink

FIGURE 5.17 A leafy branch of *Monocarpia kalimantanensis* (a) an intact fruit (b) and a cross section of the fruit (c) showing several seeds. Photo credit: B. Elkington.

the decoction (ຕົ້ມຮາກແຫ້ງທີ່ພັກ ຊອຍເປັນບ່ຽງນ້ອຍໆ ປະມານທັ້ງຍກ່ານມໍ ໃນນ້ຳ ທັ້ງ ລົດ ແລະ ດື່ມນ້ຳຕົ້ມ). Elkington et al. 303.

Monoon thorelii (Pierre) B. Xue & R.M.K. Saunders
ນ້ຳເຕົ້ານ້ອຍ; ຕາມກາຕຸງ – Nam Tao Noy; Tam Ka Toung. Sekong MBP. Small tree, 1.25 m in height, trunk diam. 1.5 cm, leaves glabrous, alternate, margin smooth, 30 cm long, 10 cm wide, fruits green turning white. Root to treat fever, rashes (ໃຊ້ຮາກບໍ່ໄຂ້, ບໍ່ໄຂ້ອອກຕຸ່ມ). Rub on stone into a powder, and mix into water, then give to patient to drink (ຝົນຮາກໃສ່ຫິນຝົນ, ປະສົມກັບນ້ຳ, ດື່ມນ້ຳທີ່ຝົນໄດ້). Elkington et al. 277.

Polyalthia evecta (Pierre) Finet & Gagn.
ກະດອຍ; ນ້ຳເຕົ້ານ້ອຍ – Kadoy; Nam Tao Noy. Bolikhamxay MPP. Shrub 1 m tall, fruits light green, in bunches of 4 or more. The stem or root is used to make a medicine to treat cases of rheumatism, joint-pain, and body weakness in a nursing

mother (prepared as a tonic) (ໃຊ້ລ່າຕັ້ນ ຫຼື ຮາກເພື່ອບົວດ້ງຂໍ້, ເຈັບຂໍ້ຕໍ່ ແລະຮ່າງ ກາຍບໍ່ແຂງແຮງໃນເວລາອອກລູກໃໝ່ (ເປັນຍາບໍາລຸງ). Cut the stem or root into pieces, take 30–40 g and boil in 2 liters of water and drink the liquid as needed (ຟັກ ຊອຍລ່າຕັ້ນເປັນຕ່ອນບ້ອຍໆ, ຕົ້ມ ແລະ ດື່ມນໍ້າຕາມຕ້ອງການ). Also take 20–30 g and place in alcohol; drink 2 times a day. The pieces may also be placed in alcohol in a container and drunk 2 times a day (ອາດຈະໃຊ້ດອງເຫຼົ້າ ແລະ ດື່ມ 2 ເທື່ອຕໍ່ມື້). Soejarto et al. 15006.

ນ້າເຕົ້ານ້ອຍ – Nam Tao Noy. Savannakhet MBP. Shrub 1 m tall, fruits green. Root to treat weakness in a nursing mother (as a tonic) (ໃຊ້ຮາກເປັນຍາບໍາລຸງສຸຂະພາບ (ເປັນຍາບໍາລຸງແມ່ກໍາ); also diarrhea (ປົວທ້ອງຖອກ). Cut the root into small pieces, boil in water, and drink the decoction (ຟັກຊອຍ ຮາກເປັນບ່ຽງນ້ອຍໆ, ຕົ້ມໃນນໍ້າ ແລະ ດື່ມນໍ້າຕົ້ມ). Soejarto et al. 14925.

ກົກນ້າເຕົ້ານ້ອຍ – Kok Nam Tao Noy. Savannakhet MBP. Small shrub about 1 m high, leaves alternate, fruits reddish brown. Root to treat rheumatism (ໃຊ້ຮາກເພື່ອປົວປະດົງຂໍ້). Cut the root into small pieces and dry, take a handful and boil in two liters of water, then drink the liquid as needed (ຟັກ ຊອຍຮາກ ເປັນບ່ຽງນ້ອຍໆ,ຕົ້ມບະມານເຫຍົ່ງກໍາມືໃນນໍ້າສອງລິດ,ດື່ມນໍ້າຕົ້ມຕາມຕ້ອງການ). Sydara et al. 19056.

Uvaria tonkinensis Finet & Gagn.
ກ້ອຍເຫັນ – Koy Hen. Bolikhamxay MPP. Liana, flowers pale greenish white, perianth of 6 tepals, leathery. Stem to treat cases of infections (external) (ໃຊ້ລ່າຕັ້ນບໍ່ການຕິດເຊື້ອ (ໃຊ້ພາຍນອກ). Take one handful and boil with an adequate amount of water (2 liters) and drink as needed (ໃຊ້ບະມານເຫຍົ່ງກໍາມື ຕົ້ມໃນນໍ້າບະມານສອງລິດ ແລະ ດື່ມຕາມຕ້ອງການ). Soejarto et al. 15037.

Apocynaceae
Alstonia scholaris (L.) R. Br.
ຕິນເປັດ – Tin Pet. Luang Prabang MBP. Treelet, latex white, sterile. Stem to treat liver disease (ໃຊ້ລ່າປົວພະຍາດຕັບ). Cut the stem into small pieces and dry, then take a handful and boil in two liters of water, cool the liquid and drink as needed (ຟັກ ຊອຍລ່າເປັນບ່ຽງນ້ອຍໆ ແລະ ຕາກໃຫ້ແຫ້ງ, ໃຊ້ບະມານເຫຍົ່ງກໍາມື ຕົ້ມໃນນໍ້າ ສອງລິດ, ປະໃຫ້ເຍັນ ແລະ ດື່ມຕາມຕ້ອງການ). Soejarto et al. 15462.

Ceropegia candelabrum L.
ເຄືອຂົນ – Kheua Khon. Luang Prabang MBP. Pubescent vine, latex white, scanty, plant sterile. Whole plant to treat paralysis (ໃຊ້ໝົດຕັ້ນບໍ່ວເປັ້ຍລ່ອຍ). Cut the whole plant into small pieces and dry, then take a handful and boil in two liters of water (ຟັກຊອຍ ພົດໝົດຕັ້ນ ແລະ ຕາກແດດໃຫ້ແຫ້ງ, ໃຊ້ພົດແຫ້ງບະມານເຫຍົ່ງກໍາມື ຕົ້ມໃນນໍ້າສອງລິດ). Drink as needed (ດື່ມຕາມຕ້ອງການ). Soejarto et al. 15431.

(a)

(b)

(c)

FIGURE 5.18 A flowering vine of *Cryptolepis dubia* (a) and (b) showing white latex exuding from the pricked stem and a close-up view of the flowers of *Cryptolepis dubia* (c). Photo credit: D. Soejarto.

Cryptolepis dubia (Burm.f.) M.R. Almeida. Figure 5.18

ເຄືອເອັນອ່ອນ – Kheua En Ohn. Xieng Khouang MBP. Vine, flowers greenish yellow to pale yellow, latex milky white, sticky. The vine (stem) is used to prepare a medicine to treat cases of nervous afflictions or disorders (ໃຊ້ເຄືອ (ລຳ) ເປັນຍາບົ່ວເສັ້ນບະສາດອັກເສບ ຫຼື ເຈັບເສັ້ນພິການ). Cut the dried stem into pieces, boil in water, and drink the liquid as needed (ຟັກ ຊອຍເຄືອແຫ້ງເປັນຕ່ອນນ້ອຍໆ, ຕົ້ມນ້ຳ ແລະ ດື່ມນ້ຳຕາມຕ້ອງການ). Soejarto et al. 14971.

Cynanchum corymbosum Wight

ເຄືອໝາກຍາງຂາວ; ໝາກຊິມຂົນ – Kheua Mak Yang Khao; Mak Siim Khon. Luang Prabang MBP. The vine is employed to treat kidney infection and TB cases (ໃຊ້ລຳບົ່ວໝາກໄຂ່ຫຼັງຕິດເຊື້ອ ແລະ ບົ່ວວັນນະໂລກ). Boil the vine and drink the liquid after cooling, as needed (ຕົ້ມເຄືອ ແລະ ດື່ມນ້ຳຕົ້ມທີ່ເຢັນແລ້ວ ຕາມຕ້ອງການ). Soejarto et al. 15447.

Hoya oblongacutifolia Costantin

ເຄືອປ້າງນ້ອຍ – Kheua Pang Noy. Savannakhet MBP. A small vine attached to a tree, flower buds pinkish. The whole plant is employed for treating liver disease; also cases of pain (ໃຊ້ໝົດຕົ້ນເພື່ອບົ່ວພະຍາດບອດ, ແລະ ບົ່ວການເຈັບບອດ). Cut the whole plant into small pieces and dry, then take a handful and boil in two liters of water (ຟັກ ຊອຍໝົດຕົ້ນເປັນບ່ຽງນ້ອຍໆ ແລະ ຕາກແດດໃຫ້ແຫ້ງ, ຕົ້ມບະມານໜຶ່ງກຳມືໃນນ້ຳ ສອງລິດ). Drink as needed (ດື່ມຕາມຕ້ອງການ). Sydara et al. 19039.

Hoya cf. *parasitica* (Wall. ex Hornem.) Wight

ດອກຕ້າງ – Dok Tang. Luang Prabang MBP. Epiphytic vine on a sloping tree trunk, flowers on a dense axillary umbel, brownish green. The leaf is used to prepare a medicine to treat wounds and cases of liver afflictions (disease); also used to treat spleen disease and jaundice (ໃຊ້ໃບເພື່ອບົ່ວບາດແຜ ແລະ ຕັບອັກເສບ, ແລະຍັງໃຊ້ບົ່ວພະຍາດມ້າມ ແລະ ຂີ້ໝາກເຫຼືອງຕົ້ມອິກ). Take 7 leaves and make a salad like papaya salad and eat it (ໃຊ້ 7 ໃບຕຳ ຄືຕຳໝາກທຸ່ງ ແລະ ກິນ). Soejarto et al. 15457.

Kibatalia macrophylla (Pierre ex Hua) Woodson

ມູກໃຫຍ່; ເໝືອດຂົນ – Mouk Nyai; Meuat Khon. Savannakhet MBP. Small tree 4–5 m high, leaves large, dark green, latex white, fruits in pairs, 10–20 cm long. Root or stem is used to treat dysentery (ໃຊ້ຮາກ ຫຼື ລຳຕົ້ນ ບົ່ວທ້ອງບິດ). Cut the root or stem into small pieces and dry, then take a handful and boil in two liters of water (ຟັກ ຊອຍຮາກ ຫຼື ລຳເປັນບ່ຽງນ້ອຍໆ, ຕົ້ມບະມານໜຶ່ງກຳມືໃນນ້ຳ ສອງລິດ). Drink the decoction as needed (ດື່ມຕາມຕ້ອງການ). Sydara et al. 19028.

Marsdenia sp.

ມ້າກະທືບໂຮງ, ຂົນຟືມ – Ma Ka Teub Hong; Khon Feum. Savannakhet MBP. Vine, fruits large, conical in shape, light green. Liana (stem) is used to treat rheumatism; also body weakness (as a tonic) (ໃຊ້ເຄືອ (ລ່າຕົ້ນ) ບໍ່ບະດົ່ງຂໍ້, ບໍ່ ສຸະພາບອ່ອນເພຍ). Boil a handful of small pieces of the stem in a bowl of water and drink the decoction (ຕົ້ມບະມານຫນຶ່ງກຳມືໃນບ້າ ຫນຶ່ງຖ້ວຍ ແລະ ດື່ມ ນ້ຳຕົ້ມ). Soejarto et al. 14950.

Marsdenia tonkinensis Costantin

ມູກເຄືອ – Mouk Kheua. Savannakhet MBP. A vine climbing on a tree, leaves opposite, fruits conical, green, in pairs with white latex. Vine (stem) to treat dysentery (ໃຊ້ເຄືອ (ລ່າຕົ້ນ) ບໍ່ທ້ອງບິດ). Cut the vine into small pieces and dry; take a handful and boil in water, then drink the cooled solution as needed (ຟັກ ຂອຍ ເຄືອ (ລ່າຕົ້ນ) ເປັນບ່ຽງນ້ອຍໆ ແລະ ຕາກໃຫ້ແຫ້ງ, ໃຊ້ບະມານຫນຶ່ງກຳມືໃນບ້າ, ດື່ມນ້ຳຕົ້ມ ທີ່ເຢັນແລ້ວຕາມຕ້ອງການ). Sydara et al. 19042.

Myriopteron extensum (Wight & Arn.) K. Schum.

ຫມາກຊິມ – Mak Siim. Luang Prabang MBP. Vine, fruits light green, follicles covered with prominent, deep, lengthwise ridges. Latex white. The root or the fruit is used as a medicine to treat persistent coughs and a condition of low feeling (ໃຊ້ຮາກ ຫຼື ຫມາກບໍ່ໄອຊ້າເຮື້ອ ແລະ ຄວາມຮູ້ສຶກຕ່ຳ). Chew the fresh fruit and swallow or boil and drink as a tonic (ຫຍ່າຫມາກສົດ ແລະ ກືນກົນບ້າ ຫຼື ຕົ້ມບ້າ ແລະ ດື່ມນ້ຳຕົ້ມ ເປັນຢາບ່າລຸງ). Soejarto et al. 15471.

Tabernaemontana bufalina Lour.

ຫມາກເປັດປ່າ – Mak Pet Pa. Bolikhamxay MPP. Treelet, latex profuse, milky white, sticky, fruit a follicetum with two out-curving follicles, their tips touching each other. The root is employed for postnatal care (as a tonic); also used to treat weakness in men (as a tonic) (ໃຊ້ຮາກເພື່ອບໍ່ສຸະພາບຫຼັງອອກລູກ ໃໝ່ (ເປັນຢາບ່າລຸງແມ່ກ່າ); ຍັງໃຊ້ເພື່ອບ່າລຸງສຸະພາບຂອງຜູ້ຊາຍບ່າອືກ). Boil one handful of dried small pieces of the root in one liter of water, drink the decoction (ຕົ້ມຮາກທີ່ພັກ ຂອຍ ແລະ ຕາກແຫ້ງບະມານຫນຶ່ງກຳມືໃນບ້າຫນຶ່ງລິດ, ດື່ມນ້ຳຕົ້ມ). Soejarto et al. 15045.

Tabernaemontana cf. bovina Lour.

ຫມາກເປັດປ່າ – Mak Pet Pa. Sekong MBP. Small tree, 1.5 m in height, leaves 8 cm in length, flowers yellow, bilateral, fruits bilateral. Root or stem to treat gonorrhea (ໃຊ້ຮາກ ຫຼື ລ່າຕົ້ນບໍ່ໂລກທນອງໃນ). Cut the root and stem, dry and then boil in water and drink the decoction (ຟັກ ຂອຍຮາກ ຫຼື ລ່າຕົ້ນ, ຕາກໃຫ້ແຫ້ງ ແລະ ຕົ້ມໃນບ້າ, ດື່ມນ້ຳຕົ້ມ). Elkington et al. 270.

Tabernaemontana peduncularis Wall.

ກົກມູກໃຫຍ່ – Kok Mouk Nyai. Bolikhamxay MPP. Treelet, latex profuse, milky white, sticky, flowers dull pale yellow to cream-colored. Stem wood to treat diarrhea (ໃຊ້ແກ່ນລ່າບໍ່ຍທ້ອງຖອກ). Boil one handful of dried small pieces of the stem wood in one liter of water, drink the decoction (ຕົ້ມພິດແຫ້ງທີ່ພັກ ຊອຍເປັນປ່ຽງນ້ອຍໆ ບະມານ ໜຶ່ງກ່າມຶໃນນ້ຳໜຶ່ງລິດ, ດື່ມນ້ຳຕົ້ມ). Soejarto et al. 15048.

Urceola rosea (Hook. & Arn.) D.J. Middl.

ມູກເຄືອ – Mouk Kheua. Savannakhet MBP. Woody vine, 15 m above ground, fruits a follicetum, of two long, cylindrical, brown follicles with white dots. The stem is used to make a medicine for the treatment of paralysis (ໃຊ້ຮາກເພື່ອບໍ່ພະຍາດເປັ້ຍລ່ອຍ). Boil the stem and drink the liquid (ຕົ້ມຮາກ ແລະ ດື່ມນ້ຳ). Soejarto et al. 14946.

Aquifoliaceae

Ilex umbellata (Wall.) Loes.

ແກ້ມອົ້ນ; ຕັບເຕົ່າ – Kehm Ohn; Tap Tao. Luang Prabang MBP. Treelet 8 m tall, dbh 15 cm, bark gray, slash light brown, turning olive-brownish with brick brown rings. Stem with intermittent rings of marking. Stem to treat asthma (ໃຊ້ລຳບໍ່ຫຶດ). Cut the stem into small pieces and dry, then take a handful and boil in two liters of water and drink the decoction after cooling (ພັກ ຊອຍລຳເປັນປ່ຽງນ້ອຍໆ ແລະ ຕາກແຫ້ງ, ໃຊ້ບະມານໜຶ່ງກ່າມິ ຕົ້ມໃນນ້ຳສອງລິດ ແລະ ດື່ມນ້ຳຕົ້ມທີ່ເຍັນແລ້ວ). Soejarto et al. 15456.

Araliaceae

Aralia finlaysoniana (Wall. ex G. Don) Seem.

ຕ້າງໜາມຄັນ; ຕ້າງຄອນໜາມ – Tang Nam Khanh; Tang Khone Nam. Oudomxay MBP. Ground level liana, leaves large, about 1 m long, bi- to tripinnate, densely and sharply spiny along stem, petiole and inflorescence, flowers greenish white, set in long-peduncled umbels, the umbels arranged into a loose, spreading pani-cles. Whole plant to treat polio (ໃຊ້ໝົດຕົ້ນບໍ່ເປ້ຍລ່ອຍ). Cut the whole plant into small pieces and dry, then take a handful and boil in about two liters of water; drink the liquid as needed (ພັກ ຊອຍ ໝົດຕົ້ນເປັນຕ່ອນນ້ອຍໆ ແລະ ຕາກແຫ້ງ, ຕົ້ມບະມານໜຶ່ງກ່າມິ ໃນນ້ຳບະມານສອງລິດ, ດື່ມນ້ຳຕົ້ມຕາມຕ້ອງການ). Soejarto et al. 15100.

Heteropanax fragrans (G. Don) Seem.

ອ້ອຍຊ້າງ – Oy Xang. Oudomxay MBP. Tree 8 m tall, stem slender, gray. Plant sterile. The stem bark and the leaf are used as a medicine for the treatment of backache, and cases of bodily swellings (edema) — ໃຊ້ເປືອກຕົ້ນ ແລະ ໃບບໍ່

ເຈັບຫຼັງ, ແລະກໍລະບິດຳປໂຕໄຕ່ (ບວມ). For backache, boil the stem bark and drink the liquid (ຕົ້ມເປືອກຕົ້ນ, ດິ່ມ). For bodily swellings, boil the leaves and take a sauna as needed (ຕົ້ມໃບ ແລະ ຮົມຕາມຕ້ອງການ). Soejarto et al. 15083.

Macropanax schmidii C.B. Shang

ເລັບມິນາງຕົ້ນ; ຕ້າງຕໍ່ຕົ້ນ – Lep Meu Nang Ton; Tang Toh Ton. Tree 8 m tall, leaves palmate, on a long petiole, flowers yellowish green, disposed in an umbel on a long peduncle, in turn, these umbels are held on a long, terminal raceme. Stem bark to treat rheumatism (ໃຊ້ເປືອກຕົ້ນບໍ່ປະດົງຂໍ້). Boil one handful of dried small pieces of the stem bark in one liter of water, drink the decoction (ຕົ້ມບະມານໜຶ່ງກຳມື ພິດທິ່ພັກຊອຍເບັ້ນຕອນບ້ອຍໆ ໃນນ້ຳໜຶ່ງລິດ, ດິ່ມນ້ຳຕົ້ນ). Soejarto et al. 15069.

Schefflera sp.

ຕ້າງໃຫຍ່ – Tang Nyai. Oudomxay MBP. Small tree with large, palmate leaves on a long petiole. Sterile. Stem, leaves are used to treat hepatitis and swollen liver (ໃຊ້ລຳຕົ້ນ, ໃບບໍ່ອພະຍາດຕັບອັກເສບ ແລະ ຕັບໄຕ່). Boil one handful of dried small pieces of the stem or leaves in one liter of water, then drink the decoction as needed (ຕົ້ມບະມານໜຶ່ງກຳມື ລຳ ຕົ້ນ ຫຼື ໃບທິ່ພັກຊອຍເບັ້ນຕອນບ້ອຍໆ ໃນນ້ຳໜຶ່ງລິດ, ດິ່ມນ້ຳຕົ້ນຕາມຕ້ອງການ). Soejarto et al. 15064.

Asteraceae

Acmella oleracea (L.) R.K. Jansen

ຜັກຄາດ – Pak Khat. Sekong MBP. Cultivated herb, 20 cm in height, leaves opposite, cordate-crenate, flowers small, yellow. Leaf to treat tooth disease (ໃຊ້ໃບບໍ່ເຈັບແຂ້ວ). Wash the fresh leaves, crush to make a paste and put the paste on the sore tooth (ລ້າງໃບດິບໃຫ້ສະອາດ, ຍ່ອງໆ ແລ້ວເອົາມາຕຳໃສ່ແຂ້ວທິ່ເຈັບ). Elkington et al. 288.

Ageratum houstonianum Mill.

ຫຍ້າດອກຂາວ – Nya Dok Khao. Sekong MBP. Herbaceous plant, leaves opposite, pilose, margins dentate, very common. The whole plant is used to treat hemorrhoids (ໃຊ້ໝົດຕົ້ນບໍ່ວິດສິດວງຫະວານ). Boil a handful of small, dried pieces of the whole plant in one liter of water and drink the decoction (ຕົ້ມບະມານໜຶ່ງກຳມື ສ່ວນທິ່ພັກຊອຍ ຈາກໝົດຕົ້ນ ເບັ້ນຕອນບ້ອຍໆ ໃນນ້ຳໜຶ່ງລິດ, ດິ່ມນ້ຳຕົ້ນ). Elkington et al. 313.

Cyanthillium cinereum (L.) H. Rob.

ຫຍ້າຫຼອດພິດ – Nya Toht Piit. Sekong MBP. Herbaceous plant, 0.5–1 m in height, leaves alternate, flowers purple, ca. 25 cm across. Very common. Whole plant is employed for treatment of fever (ໃຊ້ໝົດຕົ້ນບໍ່ໄຂ້). Boil a handful of small quantity of dried pieces of the whole plant in one liter of water and drink the decoction (ຕົ້ມປະມານຫນຶ່ງກໍາມື ສ່ວນທີ່ຟັກຂອຍ ຈາກຫມົດຕົ້ນ ເປັນຕອນ ນ້ອຍໆ ໃນນ້ຳທັ່ງ1ລິດ, ດື່ມນ້ຳຕົ້ມ). Elkington et al. 312.

Elephantopus mollis HBK.
ຄີໄຟນົກຄຸ້ມ – Khi Fai Nok Koum. Xieng Khouang MBP. Herb on an open patch of forest edge, flowerheads green, soon turning brown on drying, flowers white. Common. The root is used to prepare a medicine to treat persistent coughs; it is also used to treat liver diseases (ໃຊ້ຮາກເພື່ອບໍ່ໄອຢ່າເຮື້ອ ແລະ ພະຍາດຕັບ). In either case, boil the root and drink the liquid as needed (ຕົ້ມຮາກ ແລະ ດື່ມນ້ຳຕາມຕ້ອງການ). Soejarto et al. 14991.

Elephantopus scaber L.
ຄີໄຟນົກຄຸ້ມ – Khi Fai Nok Koum. Savannakhet MBP. Herb on forest floor, inflorescence terminal, flowerheads green, subtended by large green bracts. Common. The whole plant is to treat diabetes (ໃຊ້ຫມົດຕົ້ນບໍ່ເບົາຫວານ). Boil a handful of small, dried pieces of the whole plant in one liter of water and drink the decoction (ຕົ້ມປະມານຫນຶ່ງກໍາມືສ່ວນທີ່ຟັກ ຂອຍ ຈາກຫມົດຕົ້ນ ເປັນຕອນນ້ອຍໆ ໃນນ້ຳທັ່ງ1ລິດ, ດື່ມນ້ຳຕົ້ມ). Soejarto et al. 14914.

Balanophoraceae
Balanophora abbreviata Bl.
ປີດິນຂາວ; ຕ້າງດິນ – Pii Din Khao; Tang Diin. Oudomxay MBP. Parasitic chlorophyll-less male *Balanophora*, cones gray. Whole plant is used to prepare a tonic for body weakness (ໃຊ້ຫມົດຕົ້ນເປັນຢາບຳລຸງສຸະພາບອ່ອນເພຍ). Cut the whole plant into small pieces and dry, then take a handful and boil in two liters of water; drink the liquid as needed (ຟັກ ຂອຍ ຫມົດຕົ້ນເປັນຕ່ອນນ້ອຍໆ ແລະ ຕາກແຫ້ງ, ຕົ້ມປະມານຫນຶ່ງກໍາມື ໃນນ້ຳສອງລິດ, ດື່ມນ້ຳຕົ້ມຕາມຕ້ອງການ). Soejarto et al. 15132.

Campanulaceae
Lobelia nummularia Lam. Figure 5.19
ຫຍ້າເກັດຫອຍ – Nha Ket Hoi. Xieng Khouang MBP. Delicate ground-matted herb, flowers pink, fruits green, turning purplish red. Ground level in a secondary forest. The fruit is used to treat scabies and other diseases of the skin (ໃຊ້ຫມາກ ເພື່ອບໍ່ຂີ້ທິດ ແລະ ພະຍາດຜິວຫນັງອື່ນໆ). Crush the red fruits and rub it on the affected part of the skin (ຍ່ອງຫມາກສຸກ ແວ້ວ ໄປຖູໃສ່ບ່ອນທີ່ເປັນພະຍາດ). Soejarto et al. 14964.

(a)

(b)   (c)

FIGURE 5.19 A flowering-fruiting creeper of *Lobelia nummularia* (a) with close-up view of the flower of *Lobelia nummularia* (b) and fruits of *Lobelia nummularia* (c). Photo credit: D. Soejarto.

Cannabaceae

Trema orientalis (L.) Bl.

ປໍຫູ – Po Hou. Xieng Khouang MBP. Treelet 7 m tall, flowers in axillary glomerules, greenish white. Stem bark is used to treat mushroom poisoning (as an antidote) (ໃຊ້ເປືອກຕົ້ນເພື່ອແກ້ພິດເຫັດ (ແກ້ພິດເບື່ອ)). Boil the stem bark,

FIGURE 5.20   A flowering twig of *Capparis pubiflora*. Photo credit: K. Sydara.

cool the liquid and drink as needed (ຕົ້ມເບິ່ອກຕົ້ນ, ປະໃຫ້ເຍັນ ແລະ ດື່ມ). Soejarto et al. 14983.

Capparaceae
Capparis acutifolia Sw.

ຊາຍຊຸ້ເຄືອ – Sai Sou Kheua. Savannakhet MBP. Vine, fruits green. The liana (stem) is used to treat liver disease (ໃຊ້ເຄືອ (ລຳຕົ້ນ) ບໍ່ພະຍາດຕັບ); also fever (ແລະ ບໍ່ໄຂ້). Boil a handful of dried pieces of the liana/stem in one liter of water and drink the decoction (ຕົ້ມປະມານທັ້ງໆກຳມືເຄືອ (ລຳຕົ້ນ) ທີ່ຟັກຽອຍເປັນຕອນນ້ອຍໆ ໃນນ້ຳທັ້ງໆວິດ, ດື່ມນ້ຳຕົ້ມ). Soejarto et al. 14938; Sydara et al. 19021.

Capparis pubiflora DC. Figure 5.20

ຊາຍຊຸ້ຕົ້ນ – Sai Sou Ton. Savannakhet MBP. Small shrub on forest floor, flowers white, in extra-axillary clusters along stem. Stem, root to treat amygdalitis. Boil the stem or root and drink the liquid to treat amygdalitis (ຕົ້ມລຳຕົ້ນ ຫຼື ຮາກ ແລະ ດື່ມ ເພື່ອບໍ່ເຈັບອາມິດານ). Sydara et al. 19017.

Capparis cf. *tomentosa* Lam.

ຊາຍຊູ້ເຄືອ – Sai Sou Kheua. Sekong MBP. Small woody vine, around 3 m long, leaf margins smooth, upper side glabrous, fruits dark red, ellipsoid, approx. 3.5 cm long and 4 cm in diam. Root is used to treat hepatitis, sore throat (ໃຊ້ຮາກບົ່ວຕັບອັກເສບ, ເຈັບຄໍ). Cut the root into small pieces and then boil in water and drink (ຟັກ ຊອຍ ຮາກເປັນບ່ຽງນ້ອຍໆ ແລະ ຕົ້ມນ້ຳດື່ມ). Elkington et al. 262.

Capparis trinervia Hook.f. & Thoms.

ຊາຍຊູ້ເຄືອ – Sai Sou Kheua. Bolikhamxay MPP. Large liana, stem very spiny; a pendulous, globular-ellipsoid fruit is set on a long pedicel and peduncle; young branches brown-pubescent. Liana (stem) or stem bark to treat rheumatism (ໃຊ້ເຄືອ (ວ່າຕົ້ນ) ຫຼື ເປືອກວ່າ ບໍ່ປະດົງຂໍ້). Take one handful of liana or stem bark and boil with an adequate amount of water (about 2 liters) and drink the solution as needed (ຕົ້ມປະມານຫນຶ່ງກຳມືເຄືອ(ວ່າຕົ້ນ)ຫຼືເປືອກວ່າໃນນ້ຳປະມານສອງລິດ, ດື່ມນ້ຳຕົ້ມຕາມຕ້ອງການ). Soejarto et al. 15053.

Celastraceae

Celastrus paniculatus Willd.

ເຄືອຫມາກແຕກ – Kheua Mak Tek. Luang Prabang MBP. Large liana, stems coiled into ropes, lower part of stem has a diameter of 5 cm; bark fissured. The liana (stem) is employed to treat stomachache (ໃຊ້ເຄືອ ຫຼື ວ່າ ບໍ່ເຈັບກະເພາະ). Cut the stem into small pieces and dry, then take a handful and boil in two liters of water (ຟັກ ຊອຍ ຮາກເປັນບ່ຽງນ້ອຍໆ ແລະ ຕົ້ມປະມານຫນຶ່ງກຳມືໃນນ້ຳສອງລິດ). Drink the decoction as needed (ດື່ມນ້ຳຕົ້ມຕາມຕ້ອງການ). Soejarto et al. 15421.

Salacia cochinchinensis Lour.

ກົກຕາໄກ້ – Kok Ta Kai. Savannakhet MBP. Woody vine, flowers dull orange, small, fruits bright red. Only this one seen. Liana to treat body weakness (as a general tonic) — ໃຊ້ເຄືອບໍ່ສຸະພາບອ່ອນເພຍ (ເປັນຍາບຳລຸງ). Boil a handful of small pieces of liana in a bowl of water and drink the decoction (ຕົ້ມເຄືອທີ່ຟັກຊອຍເປັນຕອນນ້ອຍໆ ປະມານຫນຶ່ງກຳມືໃນນ້ຳຫນຶ່ງຖ້ວຍ ແລະ ດື່ມນ້ຳຕົ້ມ). Soejarto et al. 14909; 14948.

Salacia verrucosa Wight

ເຄືອຕາກວາງ; ສົ້ມຊື່ນໃຫຍ່ – Kheua Ta Kouang; Som Seun Nyai. Savannakhet MBP. Tree, flowers minute, profuse, light green, long-pedicelled, set in loose axillary globular clusters. Stem is used to treat flatulence (ໃຊ້ວ່າຕົ້ນບໍ່ທ້ອງຍ້ຽ ທ້ອງເບັ່ງ). Boil a handful of small pieces of stem in a bowl of water and drink the decoction (ຕົ້ມວ່າຕົ້ນທີ່ຟັກຊອຍເປັນຕອນນ້ອຍໆ ປະມານຫນຶ່ງກຳມືໃນນ້ຳຫນຶ່ງຖ້ວຍ ແລະ ດື່ມນ້ຳຕົ້ມ). Soejarto et al. 14943; Sydara 19027.

Chloranthaceae

Chloranthus spicatus (Thunb.) Makino

ຕົ້ນກຳລັງມ້າກຳ – Tohn Kam Lang Ma Kam. Xieng Khouang MBP. Shrub, inflorescence a spike, sometimes branched, green with white spots (anthers). Stem or root is used to treat body weakness (as a tonic) and cases of rheumatism — ໃຊ້ລຳຕົ້ນ ຫຼື ຮາກ ບໍ່ຢ່າງກາຍອ່ອນເພຍ (ເປັນຢາບຳລຸງ) ແລະ ໃຊ້ບໍ່ບະດົງຂໍ້). Cut the root or stem into pieces and boil it and drink the decoction (ຟັກ ຊອຍຮາກ ຫຼື ລຳຕົ້ນເປັນຕ່ອນນ້ອຍໆ ແລະ ຕົ້ມ, ດື່ມນ້ຳຕົ້ມ). Soejarto et al. 14997; 15004.

Clusiaceae

Garcinia xanthochymus Hook.f. ex T. Anderson

ກົກໝາກດະ; ສົ້ມໂມງ – Kok Mak Da; Som Mong. Oudomxay MBP. A large tree, 15–20 m tall, dbh 30 cm, bark blackish gray, smoothish, slash white, with dots of white latex. Fruits profuse, large, globular with scar of stigma off center, slice of fruits shows yellow, thick latex. The stem bark (the yellow latex) is used to treat cases of diarrhea (ໃຊ້ເປືອກຕົ້ນ (ຢາງສີເຫຼືອງ) ບໍ່ທ້ອງຖອກ). Boil the bark and drink the liquid as needed (ຕົ້ມເປືອກ ແລະ ດື່ມຕາມຕ້ອງການ). Soejarto et al. 15121.

Garcinia cf. *xanthochymus* Hook.f. ex T. Anderson

ຕົ້ນລິກຶກ; ສົ້ມບົງ – Ton Lii Kuek; Som Bohng. Bolikhamxay MPP. Treelet 4 m tall, latex yellow. Plant sterile. Stem to treat rheumatism and liver affliction (ໃຊ້ລຳຕົ້ນບໍ່ບະດົງຂໍ້ ແລະ ພະຍາດຕັບ). Take one handful and small pieces of stem, dry, and boil in an adequate amount of water (about 2 liters) and drink the decoction as needed (ຕົ້ມລຳຕົ້ນທີ່ຟັກຊອຍເປັນຕ່ອນນ້ອຍໆ ບະມານໜຶ່ງກຳມືໃນນ້ຳບະມານສອງລິດ ແລະ ດື່ມນ້ຳຕົ້ມຕາມຕ້ອງການ). Soejarto et al. 15042.

Garcinia sp.

ກົກໝາກແປມ – Kok Mak Pem. Bolikhamxay MPP. Small tree, fruits deep crimson-red, pulp white, sour-sweet. The root is used to treat boils and other abscesses (ໃຊ້ຮາກບໍ່ຕຸ່ມຝີ ແລະ ຝີຫນອງ). Cut the root into small pieces and dry and decoct one handful in a bowl of water; drink the decoction as needed (ຕົ້ມລຳຕົ້ນທີ່ຟັກຊອຍເປັນຕ່ອນນ້ອຍໆ ບະມານໜຶ່ງກຳມືໃນນ້ຳບະມານສອງລິດ ແລະ ດື່ມນ້ຳຕົ້ມຕາມຕ້ອງການ). Soejarto et al. 15038.

Combretaceae

Combretum deciduum Collett & Hemsl.

ເຄືອກະແດ້ງໃຫຍ່; ເຄືອວິ້ນປີ – Kheua Ka Deng Nyai; Kheua Liin Pii. Savannakhet MBP. A small shrub with opposite leaves, flowers white. The stem is used to treat dengue fever and hemorrhagic conditions (ໃຊ້ລຳເຄືອເພື່ອບໍ່ໄຂ້ເລືອດອອກ ແລະ

ໃບກ່ວະນິທີ່ເລືອດໄຫຼ). Boil the stem in water and drink the decoction (ຕົ້ມວ່າເຄືອ ແລະ ດື່ມນ້ຳຕົ້ມ). Sydara et al. 19006.

Connaraceae

Cnestis palala (Lour.) Merr.

ເຄືອຫອນໄກ່ແດງ; ຂາມເຄືອ – Kheua Hone Kai Deng; Kham Kheua. Luang Prabang MBP. Large spreading liana with shrub-like stem, flowers brown in small paniculate clusters, on burrs along stem and branches. The stem is used to treat syphilis by boiling cut pieces of stem and drinking the decoction. All parts of the plant together are used to prepare a medicine for women in cases when menstrual bleeding does not stop (ໃຊ້ໝົດຕົ້ນ ເພື່ອບໍ່ວຜູ້ຍິງມີປະຈຳເຄືອນ ລົງເລືອດບໍ່ຕຸດ). Cut the plant parts and mix the pieces with a species of *Smilax* (*S. glabra*) together with unpeeled sweet rice. Boil the mixture (ພັກ ຊອຍໝົດຕົ້ນ ປະສົມຍາທ່ໍໃສ່, ແລະ ຕົ້ມເຂົ້າເປືອກເຂົ້າໜຽວ, ຕົ້ມສ່ອນປະສົມທັງໝົດ). Drink the decoction (ດື່ມນ້ຳຕົ້ມ). Soejarto et al. 15430; 15460.

ເຄືອຫອນໄກ່ແດງ; ສະຄາມເຄືອ; ຂາມເຄືອ – Kheua Hone Kai Deng; Sa Kham Kheua; Kham Kheua. Savannakht MBP. Liana, flowers dull greenish white on terminal or axillary inflorescence, or ramiflorous, flowers profuse. Common. Liana (stem) is used to treat cases of gastritis and rheumatism (ໃຊ້ເຄືອ ຫຼື ວ່າ ບໍ່ກະເພາະອັກເສບ ແລະ ປະດົງຂໍ). Slice the liana into small pieces, dry them, take one handful and put to a boil in about one liter of water for 10–15 minutes (ຕົ້ມເຄືອທ່ີພັກຊອຍເປັນຕອນບໍ່ຢໆ ປະມານໜຶ່ງກຳມືໃນນ້ຳປະມານໜຶ່ງລິດ, ຕົ້ມ 10–15 ນາທີ). Drink the cooled decoction as needed (ດື່ມນ້ຳຕົ້ມທ່ີເຍັນແລ້ວ ຕາມຕ້ອງການ). Soejarto et al. 14911; Sydara et al. 19044.

Connarus cf. *bariensis* Pierre

ເຄືອຂົາປ້ອກ – Kheua Khao Peuak. Sekong MBP. Shrub, leaves opposite, glabrous, inflorescence a spike, flowers small, white, fruits green, tomentose. Liana to treat stomachaches (ໃຊ້ເຄືອບໍ່ວຈັບກະເພາະ). Boil the dried liana and drink the decoction (ຕົ້ມເຄືອແຫ້ງທ່ີພັກ ຊອຍເປັນຕ່ອນບໍ່ຢໆ ແລະ ດື່ມນ້ຳຕົ້ມ). Elkington et al. 283.

Connarus paniculatus Roxb. Figure 5.21

ເຄືອໝາວໍ; ເຄືອສອບແສບ – Khua Ma Vo; Kheua Sop Seb. Bolikhamxay MPP. Liana, flower buds greenish yellow with a pink top, fruit a follicetum of 2 follicles, orange-yellow to orange-red, each with a black glossy seed sticking out. Stem is used as a medicine for postpartum care in nursing mother and to treat rheumatism and joint pains (ໃຊ້ວ່າເຄືອບໍ່ວຜູ້ຍິງອອກລູກໃໝ່ ແລະ ບໍ່ປະດົງຂໍ ແລະ ເຈັບຂໍ). Cut the stem into pieces, dry and boil one handful in 2 liters of water (ພັກ ຊອຍວ່າເຄືອເປັນຕ່ອນບໍ່ຢໆ, ຕາກໃຫ້ແຫ້ງ ແລະ ຕົ້ງປະມານໜຶ່ງກຳມືໃນນ້ຳ 2

FIGURE 5.21   A flowering specimen of *Connarus paniculatus*. Photo credit: D. Soejarto.

ວິດ). Drink the decoction as needed (ດື່ມນ້ຳຕົ້ມຕາມຕ້ອງການ). Soejarto et al. 15043; 15052.

ເຄືອແສງຄຳ; ເຄືອຊອບແຊບ; ເຄືອຕັບໄກ່ – Kheua Sang Kham; Kheua Sop Seb; Kheua Tap Kai. Savannakhet MBP. Shrub with compound leaves, white flowers, red fruits. Stem to treat rheumatism (ໃຊ້ລຳບໍ່ວະດັ່ງຂໍ້). Cut the stem into small pieces, dry, then boil in water and drink the decoction (ພັກ ຊອຍລຳ ເຄືອເປັນຕ່ອນນ້ອຍໆ, ຕາກໃຫ້ແຫ້ງ ແລະ ຕົ້ມໃນນ້ຳ ແລະ ດື່ມນ້ຳຕົ້ມ). Sydara et al. 19011.

FIGURE 5.22   A fruiting twig of *Alangium kurzii* showing the underside view of the leaves. Photo credit: D. Soejarto.

Cornaceae

Alangium kurzii Craib. Figure 5.22

ກໍ່ຕູດ – Ko Tout. Xieng Khouang MBP. Treelet 3 m tall, stem with swollen bumps, flowers white with yellow conical-shaped anthers, fruits green. Common on this site. The stem is used to treat cases of nervous breakdown (ໃຊ້ລຳຕົ້ນປົວເສັ້ນປະສາດພິການ). Boil the cut-up stem and drink as needed (ຕົ້ມລຳຕົ້ນ ແລະ ດື່ມຕາມຕ້ອງການ). Soejarto et al. 14962.

Dilleniaceae

Dillenia turbinata Roxb.

ສ້ານໃຫຍ່ – San Nyai. Savannakhet MBP. Tree 10 m tall, fruits yellow, growing in the vicinity of visitor center building. The stem is used to treat cases of paralysis (ໃຊ້ລຳຕົ້ນປົວອ່ອຍ). Boil a handful of small pieces of stem in a bowl of water and drink the decoction (ຕົ້ມລຳຕົ້ນແຫ້ງທີ່ຫັ່ນ ຂອຍເປັນຕ່ອນນ້ອຍໆ ໃນນ້ຳໜຶ່ງຖ້ວຍ ແລະ ດື່ມນ້ຳຕົ້ມ). Soejarto et al. 14930.

Ebenaceae

Diospyros cf. *malabarica* (Desr.) Kostel.

ກົກຮ້າງຮ້ວນ – Kok Hang Hone. Savannakhet MBP. Tree 8 m tall, stem slender, fruits green, calyx cup blackish green. Stem is employed for treatment of skin diseases (itching, boils) and parasitic infection (ໃຊ້ວ່າຕັ້ນບໍ່ອພະຍາດຜິວໜັງ (ຄັນ, ຕຸ່ມຝີ) ແລະ ຕິດເຊື້ອກາຝາກ). Boil a handful of small pieces of stem in a bowl of water and drink the decoction (ຕົ້ມວ່າຕັ້ນແຫ້ງກຳມື ຊອຍເປັນຕ່ອນ ນ້ອຍໆບະນານໜຶ່ງກຳມື ໃນນ້ຳ ໜຶ່ງຖ້ວຍ ແລະ ດື່ມນ້ຳຕົ້ມ). Soejarto et al. 14944.

Elaeocarpaceae
Elaeocarpus ovalis Miq.
ກົກສົ້ມມຸ້ນ – Kok Som Moun. Xieng Khouang MBP. Treelet 8 m tall, trunk 20 cm dbh, bark hard, cracked, flower buds green, along racemose inflorescences arising from leafless branch. The stem wood is used to prepare a medicine for the treatment of jaundice (ໃຊ້ວ່າຕັ້ນເພື່ອບໍ່ຂີ້ໜາງເຫຼືອງ). Macerate the stem wood in water and use the liquid as a bath (ແຊ່ວ່າຕັ້ນທີ່ຜ່າກ ຊອຍເປັນຕ່ອນນ້ອຍໆໃນນ້ຳ, ໃຊ້ນ້ຳຕົ້ມອາບ). Soejarto et al. 14985.

Euphorbiaceae
Croton caudatus Geiseler
ເຄືອກະດອຍເຕົ່າ; ຫັດສະຄືນ – Kheua Ka Doy Tao; Had Sa Kheun. Xieng Khouang MBP. Liana, leaves cordate, fruits gray-brown, globular. The stem is used for detoxification in cases of poisoning (ໃຊ້ວ່າເຄືອເພື່ອຖອນພິດ). Cut the stem into pieces and boil in water, then drink the decoction as needed (ຜ່າ ຊອຍ ວ່າເຄືອເປັນຕ່ອນນ້ອຍໆ ແລະ ຕົ້ມໃນນ້ຳ, ດື່ມນ້ຳຕົ້ມຕາມຕ້ອງການ). Soejarto et al. 15090.

Croton cf. *dongnaiensis* Pierre ex Gagn.
ກົກເປົ້ານ້ອຍ; ເປົ້າທອງ – Kok Pao Noy; Pao Tong. Savannakhet MBP. Shrub, fruits brown. Root to treat flatulence (ໃຊ້ຮາກບໍ່ທ້ອງຢັ້ງ ທ້ອງເປັ່ງ). Cut the root into small pieces, boil in water and drink the decoction (ຜ່າ ຊອຍຮາກເປັນຕ່ອນ ນ້ອຍໆ ແລະ ຕົ້ມໃນນ້ຳ, ດື່ມນ້ຳຕົ້ມ). Soejarto et al. 14929.

Croton thorelii Gagn.
ກົກເປົ້ານ້ອຍ; ເປົ້າທອງ – Kok Pao Noy; Pao Tong. Savannakhet MBP. Small tree 1–2 m high, leaves alternate. The leaf is used to treat internal infections and to treat injuries from a fall (such as from climbing a tree) (ໃຊ້ໃບບໍ່ການຕິດເຊື້ອພາຍໃນ ແລະ ບໍ່ການບາດເຈັບຈາກການຕົກຈາກທີ່ສູງ (ເຊັ່ນຕົກກົກໄມ້). Burn the leaves and put the patient over the smoke (ຈຸດໃບ ແລະ ໃຫ້ຄົນເຈັບນອນຢູ່ທາງເທິງນັ້ນ ເພື່ອຮົມຄວັນ). Sydara et al. 19004.

Mallotus macrostachyus (Miq.) Muell.-Arg.
ກໍຕັ້ງໂກບ; ຕອງເທົ້າຄົນ – Ko Tang Kop; Tong Tao Khon. Xieng Khouang MBP. Treelet, flowers disposed along slender racemes, yellowish brown. Root to treat athus, muquet (ໃຊ້ຮາກບໍ່ເປັນເມັ້ງ ເປັນກາງ). Cut the root into small pieces, boil

(a) (b)

FIGURE 5.23 A fruiting liana of *Omphalea bracteata* (a) and an enlarged image of the fruit of *Omphalea bracteata* showing the intact fruit (b) and the cross section of the fruit, revealing 3 cut seeds. Photo credit: D. Soejarto.

in water and drink the decoction (ຝັກ ຂອຍຣາກເບັນຕ່ອນນ້ອຍໆ ແລະ ຕົ້ມ ໃນນ້າ, ຕື່ມນ້າຕົ້ມ). Soejarto et al. 14980.

Omphalea bracteata (Blanco) Merr. Figure 5.23

ຄິງໄກ່ – Khing Kai. Bolikhamxay MPP. Large climber, fruits globular, large, light green, glabrous, 3-seeded. The stem is used as a purgative in cases of constipation (ໃຊ້ລຳຕົ້ນເພື່ອເປັນຢາລະບາຍໃນກໍລະນີທ້ອງຜູກ). Cut the stem into pieces and boil, then drink the liquid (ຝັກ ຂອຍ ລຳຕົ້ນເປັນຕ່ອນນ້ອຍໆ, ຕົ້ມໃນນ້າ 1 ລິດ, ຕື່ມນ້າຕົ້ມ). Soejarto et al. 15031.

Suregada multiflora (A.Juss.) Baill.

ກົກຜ່າສາມ, ດູກໃສ – Kok Pha Sam; Douk Sai. Savannakhet MBP. Tree 6–8 m high, alternate leaves dark green, flowers greenish yellow. Root is used as a poison antidote and to treat gastric disorders (ໃຊ້ຮາກເພື່ອແກ້ອາການພິດເບື່ອ ແລະ ບໍ່ໂລກກະເພາະ). Cut the root into small pieces, boil in one liter of water, then drink the decoction (ຝັກ ຂອຍຮາກເບັນຕ່ອນນ້ອຍໆ, ຕົ້ມໃນນ້າ 1 ລິດ, ຕື່ມນ້າຕົ້ມ). Sydara et al. 19045.

ກົກປ່າສາມ; ດູກໃສ – Kok Pa Sam; Douk Sai. Luang Prabang MBP. Treelet
2.5 m tall, sterile. The root is used as a poison antidote (ໃຊ້ຮາກເພື່ອຖອນພິດ).
Cut the root into small pieces and dry, then take a handful and boil in two liters
of water (ຟັກ ຂອຍຮາກເປັນຕ່ອນນ້ອຍໆ ແລະ ຕາກແຫ້ງ, ຕົ້ມປະມານຫນຶ່ງກຳມື
ໃນນ້ຳສອງລິດ). Cool the decoction and drink as needed (ດື່ມນ້ຳຕົ້ມທີ່ເຢັນແລ້ວ
ຕາມຕ້ອງການ). Soejarto et al. 15426.

Trigonostemon cochinchinensis Gagn.
ກົກໝາກສ້ານນ້ຳຢອດແດງ – Kok Mak San Nam Yod Deng. Oudomxay MBP.
Small tree, branches liana-like, stem 8 cm in diameter, fruits profuse, green, turning black, disposed on a spreading, axillary, terminal panicles. Stem wood (no
bark) is used to make a blood tonic by boiling the wood and drinking the decoction (ໃຊ້ວ່າ (ບໍ່ມີເປືອກ) ປຸງແຕ່ງເປັນຢາບຳລຸງເລືອດ ໂດຍການຕົ້ມ, ດື່ມນ້ຳຕົ້ມ).
Soejarto et al. 15073.

Trigonostemon sp.
ຕິນສິ້ນບໍ່ຮີ – Tin Sin Boh Hii. Savannakhet MBP. Shrub, 2 m tall, fruits brown,
hairy. Root to treat body weakness (as a tonic; as a tincture) (ໃຊ້ຮາກບໍ່
ສຸຂະພາບທີ່ອ່ອນເພຍ (ເປັນຢາບຳລຸງ ໃນຮູບເຫຼົ້າຢາ). Cut the root into small
pieces, boil in water, and drink the decoction (ຟັກ ຂອຍຮາກເປັນຕ່ອນນ້ອຍໆ ແລະ
ຕາກແຫ້ງ, ຕົ້ມປະມານຫນຶ່ງກຳມືໃນນ້ຳ ແລະ ດື່ມນ້ຳຕົ້ມ). Soejarto et al. 14932.

Fabaceae-Caesalpinioideae
Afzelia xylocarpa (Kurz) Craib. Figure 5.24
ໄມ້ແຕ້ຄ່າ – Mai Tae Kha. Luang Prabang MBP. A large tree, ca. 25 m tall, dbh
50 cm, fruits large, dry, black. Young leaves light green. Fruit (crushed seed)
is used to treat diarrhea (ໃຊ້ໝາກ (ທຸບໃຫ້ແຕກ) ປິ່ນປົວທ້ອງຖອກ). Boil the
crushed seeds and drink the liquid (ຕົ້ມໝາກທີ່ທຸບແຕກການນັ້ນຕົ້ມໃນນ້ຳ, ດື່ມນ້ຳ
ຕົ້ມ). Soejarto et al. 15472.

Bauhinia acuminata L.
ກົກສ້ວງງ່າງ– Kok Siao Ngang. Luang Prabang MBP. Treelet 6 m tall, crown
spreading, fruits green, drying brown. Stem bark is used to prepare women's
drink after giving birth (as a tonic), by boiling and drinking the decoction (ໃຊ້
ເປືອກຕົ້ນປຸງແຕ່ງເປັນຢາບຳລຸງແມ່ກຳໂດຍການຕົ້ມ ແລະ ດື່ມນ້ຳຕົ້ມ). Soejarto
et al. 15436.

Bauhinia glauca (Benth.) Benth.
ບານເຄືອດອກຂາວ; ສ້ຽວເຄືອດອກຂາວ – Ban Kheua Dok Khao; Siao Kheua
Dok Khao. Oudomxay MBP. Liana to 7 m above ground, leaves gray-white
underneath, flowers white, anthers dark red, legumes green. The stem is used to

FIGURE 5.24   A fruiting specimen of *Afzelia xylocarpa*, showing opened legume and seeds. Photo credit: D. Soejarto.

treat leucorrhea (ໃຊ້ລຳເຄືອ ເພື່ອບໍ່ວລົງຂາວ). Pieces of stem are boiled in water and the decoction drunk (ຕົ້ມຕ່ອນວຳເຄືອໃນນ້ຳ ແລະ ດື່ມນ້ຳຕົ້ມ). Soejarto et al. 15068.

Bauhinia sp. 1
ສົ້ມສ້ຽວຕົ້ນ – Som Siao Ton. Luang Prabang MBP. Tree 15 m tall, dbh 15 cm, bark gray, finely fissured, slash pinkish brown, fruits green, drying brown. Stem bark is used for treatment of cases of fever and swelling of body parts (ໃຊ້ເປືອກຕົ້ນເພື່ອບໍ່ວໄຂ້ ແລະ ສ່ວນຕ່າງໆໃນຮ່າງກາຍໄໝ່ ບວມ). Boil the stem bark and drink the liquid (ຕົ້ມເປືອກຕົ້ນ ແລະ ດື່ມນ້ຳຕົ້ມ). Soejarto et al. 15422.

Bauhinia sp. 2
ສ້ຽວເຄືອ – Siao Kheua. Savannakhet MBP. Liana, fruits green with reddish purple tinge at one end. Common species throughout the preserve. Liana (stem) is used to treat rheumatism (ໃຊ້ເຄືອ (ວ່າ) ເພື່ອບໍ່ວປະດົງ). Boil a handful of small pieces of liana in a bowl of water and drink the decoction (ຕົ້ມວຳເຄືອທິ່ຊອຍເປັນຕ່ອນ ນ້ອຍໆປະມານທັ່ງໆກຳມືໃນນ້ຳທັ່ງໆຖ້ວຍ ແລະ ດື່ມນ້ຳຕົ້ມ). Soejarto et al. 14903.

Caesalpinia digyna Rottl.

ขนามกะจาย – Nam Ka Jai. Luang Prabang MBP. Huge liana with prominent and large prickles on the stem, spiny throughout; the large, raised prickles pyramidal, terminated by a very sharp spine; prickled stem to 20 cm in diameter at base. Stem or root to treat intestinal infection (ใຊ້ລ່າตົ້ນ ຫຼື ธาภ ບ່ອภาบตິດเຊื້ອຢູ່ລ່າใສ້). Cut the root or stem into small pieces and dry, then take a handful and boil in two liters of water and drink the solution as needed (ພ້ກ ຊອຍธาภ ຫຼື ລ່າตົ້ນ ເປັນต่อบบ້อຍໆ ແລະ ຕາກແຫ້ງ, ຕົ້ມປະມານໜຶ່ງກໍາມືใນບ້າສອງລິດ ແລະ ດື່ມບ້າຕົ້ນ ຕາມต้องภาบ). Soejarto et al. 15429.

เคือขนามกะจาย; ขนามกะจาย – Kheua Nam Ka Jai; Nam Ka Jai. Luang Prabang MBP. Liana, spiny throughout, fruits light green, brown when dry. Stem or root is used to treat syphilis, intestinal infection, and also cases of stomachache (ใຊ້ລ່າเคือ ຫຼື ธาภบ่อโລกຊິພິລິດส, ອັກເສບລ່າใສ້, เจับภะเພาะอาฑาบ). Boil the stem or root, cool and drink the decoction (ຕົ້ມລ່າเคือ ຫຼື ธาภ, ປະใຫ້ເย็บ ແລະ ດື່ມບ້າຕົ້ນ). Soejarto et al. 15449.

Caesalpinia pulcherrima (L.) Sw.

เคือขนามกะจาย – Kheua Nam Ka Jai. Savannakhet MBP. Small shrub, compound leaves, red and pink flowers, disposed along a dense inflorescence. The root is employed to treat syphilis; also, dysentery (ใຊ້ธาภบ่อล่อย ແລະ ท้องບິດ). Cut the root into small pieces and dry, then take a handful and boil in 2 liters of water; drink the liquid as needed (ພ້ກ ຊອຍธาภ ຫຼື ລ່าตົ้ນ ເປັນต่อบบ้อຍໆ ແລະ ຕາກແຫ້ງ, ຕົ้ມปะมาบໜຶ່ງກໍາມືใນບ້າສອງລิด ແລะ ດื่มบ้าตົ้ນ ตามต้อງภาบ). Sydara et al. 19007.

Dialium cochinchinense Pierre

ขมากเค้ง – Mak Kheng. Savannakhet MBP. A big tree 15–20 m high, fruits subglobular, black. Stem bark or root is used to treat infection and cases of pain (ใຊ້ເປືອກຕົ້ນ ຫຼື ธาภບ່ອภาบตິດเຊື້ອ ແລະ ภาบเจັບບอด). Cut stem bark or roots into small pieces, boil in 1 liter of water, drink the liquid (ພ້ກ ຊอย ເປืອກຕົ້ນ ຫຼື ธาภເປັນต่อบบ้อຍໆ, ຕົ້ມใນບ້າ 1 ลิด, ดื่มบ้าตົ້ນ). Sydara et al. 19043.

Peltophorum dasyrhachis (Miq.) Kurz

อะลาງ; ສະคาม – Ah Lang; Sa Kham. Savannakhet MBP. Large tree, to 25 m tall, fruits dry. Very common throughout the preserve. Stem bark to treat fever and swelling of body parts (ใຊ້ເປືອກຕົ້ນบ่อไຊ້ ແລະ ອາภาบบอมส่อบต่างๆຂອງຮ่างภาย). Boil a handful of small pieces of stem in a bowl of water and drink the decoction (ຕົ້ມລ່າຕົ້ນ ທີ່ພ້ກ ຊอยເປັນต่อบบ้อຍໆ ปะมาบໜຶ່ງກໍາມືใນບ້າໜຶ່ງຖ້ວย ແລະ ดื่มบ้าตົ້ນ). Soejarto et al. 14951.

Piliostigma malabaricum (Roxb.) Benth.

ສ້ຽວເຄືອ – Siao Kheua. Savannakhet MBP. Liana, fruits green with reddish purple tinge at one end. Common species throughout the preserve. Liana (stem) to treat rheumatism (ໃຊ້ເຄືອ ຫຼື ວ່າ ຕົ້ນບໍ່ປະດົງຂໍ້). Boil a handful of small pieces of the stem in a bowl of water and drink the decoction (ຕົ້ມວ່າຕົ້ນ ທີ່ພັກ ຂຸອຍ ເປັນ ຕ່ອນນ້ອຍໆ ບະມານຫນຶ່ງກຳ ງມໍ ໃນນ້ຳ ຫນຶ່ງຖ້ວຍ ແລະ ດຶ່ມນ້ຳຕົ້ມ). Soejarto et al. 14903.

ສົ້ມສ້ຽວຕົ້ນ – Som Siao Ton. Luang Prabang MBP. Tree 15 m tall, dbh 15 cm, bark gray, finely fissured, slash pinkish brown, fruits green, drying brown. The stem is used to treat cases of fever, swelling of body parts, and cases of rheumatism (ໃຊ້ວ່າຕົ້ນບໍ່ໄຂ້, ຮ່າງກາຍໄຄ່ບອມ, ເຈັບປະດົງຂໍ້). Boil the stem and drink the liquid (ຕົ້ມນ້ຳ ແລະ ດຶ່ມ). Soejarto et al. 15422.

Senna sophera (L.) Roxb.

ຊ້າຄາມ – Sa Kham. Luang Prabang MBP. Tree 8 m tall, dbh 15 cm, bark gray, lenticellate, fruits dry, grayish brown. Stem or root is used for the treatment of rheumatism (ໃຊ້ວ່າຕົ້ນ ຫຼື ຮາກບໍ່ປະດົງຂໍ້). Boil the stem or root and drink the liquid (ຕົ້ມວ່າຕົ້ນ ຫຼື ຮາກ ແລະ ດຶ່ມນ້ຳຕົ້ມ). Soejarto et al. 15448.

Senna timorensis (DC.) H.S. Irwin & Barneby

ຂີ້ຫຼຶກປ່າ – Khi Lek Pa. Luang Prabang MBP. Treelet 3 m tall, low-branched, grows to a large tree; sterile. Mix with leaves of other plants (*Croton* and *Eclipta*) and the macerated liquid of rice, then apply on the paralyzed part; also used to treat cases of skin infections (ເປັນຫນຶ່ງໃນສ່ວນປະກອບຂອງຢາເປົ້ຍລ່ອຍ: ປະສົມກັບໃບບົ້າ, ຫຍ້າຫອມແກ່ວ ແລະ ນ້ຳເຂົ້າມອກ ແລ້ວໂປະໃສ່ບ່ອນເປັນລ່ອຍ). Soejarto et al. 15418.

Fabaceae-Mimosoideae

Albizia lucidior (Steud.) I.C. Nielson ex H. Hara

ກົກສະແຄ – Kok Sa Kheh. Luang Prabang MBP. Tree 8 m tall, stem brown, bark finely fissured and granular, leaves bipinnate, fruits light brown, profuse, flat. The stem bark is used to prepare a medicine for liver disease (hepatitis) (ໃຊ້ເປືອກຕົ້ນບໍ່ພະຍາດຕັບ (ຕັບອັກເສບ). One of the ingredients for liver disease and jaundice (ເປັນຫນຶ່ງໃນສ່ວນປະກອບຂອງຢາບໍ່ຕັບ ແລະ ຂີ້ ເໝາະ(ຫຼຶອງ)). Boil the stem bark and drink the liquid after cooling, as needed (ຕົ້ມເປືອກຕົ້ນ, ປະໃຫ້ເຢັນ, ດຶ່ມນ້ຳຕົ້ມຕາມຕ້ອງການ). Soejarto et al. 15445.

Fabaceae-Papilionoideae

Campylotropis pinetorum (Kurz) Schindl.

ກາຕາປິກ; ກາສາມປິກ; ຫົ້ວແຮະເຮືອ – Ka Ta Piik; Ka Sam Piik; Tua Heh Heua. Luang Prabang MBP. Large herb, flowers white, purplish pink streaks on the inner surface of the banner, wings purplish pink, keels light yellowish green. Root is used as a medicine for diarrhea (ໃຊ້ຮາກບ່ວທ້ອງຖອກ). Boil the root and drink the cooled liquid (ຕົ້ມຮາກ, ດື່ມນ້ຳຕົ້ມທີ່ເຢັນ). Soejarto et al. 15444.

Dalbergia cf. rimosa Roxb.

ເຄືອປະດົງເຄືອໄກ່; ດູ່ເຄືອ – Kheua Pa Dong Deua Kai; Dou Kheua. Bolikhamxay MBP. A liana; fruits dry, winged. The leaf or the stem is used as a medicine for rheumatism and to treat cases of bone injuries (fractured bones) (ໃຊ້ໃບ ຫຼື ລ່າເຄືອບໍ່ປະດົງຂໍ້ ແລະ ບໍ່ກະດູກຫັກ). For rheumatism, cut the stem into small pieces, then dry. Decoct 1 handful by boiling with 2 liters of water. Drink as needed (ຟັກ ຊອຍລ່າເຄືອເປັນຕ່ອນນ້ອຍໆ, ຕາກໃຫ້ແຫ້ງ. ຫຼຸງຈາກນັ້ນຈິ່ງໃຊ້ຢາບະມານ 1 ກຳມື ຕົ້ມໃນນ້ຳ 2 ລິດ, ດື່ມຕາມຕ້ອງການ). Take 2–3 fresh leaves, burn into ashes, and apply the ashes on the bone injuries (ໃຊ້ໃບສົດ 2–3 ໃບ, ທຸບ ແລະ ເຜົາໄຟໃຫ້ເປັນຂີ້ເຖົ່າ, ເອົາຂີ້ເຖົ່າໃສ່ບ່ອນທີ່ກະດູກຫັກ). Soejarto et al. 15051.

Dalbergia volubilis Roxb.

ເຄືອພອດຟາວນ້ຳ – Kheua Fot Fao Nam. Luang Prabang MBP. Large liana, flowers white with bluish-pink tinge set on terminal panicles. The leaf is used as an eye medicine in cases of blurred vision or other afflictions (ໃຊ້ໃບບໍ່ວຕາ ໃນ ກໍລະນີທີ່ຕາມົວ ຫຼື ຕິດເຊື້ອພະຍາດອື່ນໆ). Crush the leaves with water, then filter and use the clean liquid to wash the blurred eyes (ຍ່ອງໃບໃນນ້ຳສະອາດ, ຕອງ ແລະ ໃຊ້ນ້ຳຕອງທີ່ໃສ ລ້າງຕາທີ່ມົວ). Soejarto et al. 15440.

Derris scandens (Roxb.) Benth.

ອ້ອຍສາມສວນເຄືອ – Oy Sam Suan Kheua. Xieng Khouang MBP. Liana, flowers greenish white, profuse, in spreading, loose terminal paniculate clusters. The stem is used to prepare a tonic for debility in nursing mothers by boiling and drinking the solution (ໃຊ້ລ່າເຄືອເປັນຢາບຳລຸງແມ່ກ່າ ໂດຍການຕົ້ມ ແລະ ດື່ມນ້ຳຕົ້ມ). Soejarto et al. 14977.

ອ້ອຍສາມສວນເຄືອ – Oy Sam Suan Kheua. Luang Prabang MBP. Liana, stems twining, forming ropes, plant sterile. Masticated stem produces a sweetish sensation. Liana (stem) is used to treat constant coughing (ໃຊ້ເຄືອ (ລ່າ) ບໍ່ໄອຊ່າເຮື້ອ). Cut the liana (stem) into small pieces and dry, take a handful and boil in two liters of water, then cool and drink the decoction as needed. (ຟັກ ຊອຍເຄືອ (ລ່າ) ເປັນຕ່ອນນ້ອຍໆ ແລະ ຕາກໃຫ້ແຫ້ງ, ເອົາຫນຶ່ງກຳມືຕົ້ມໃນນ້ຳສອງລິດ, ດື່ມນ້ຳຕົ້ມ ທີ່ເຢັນແລ້ວຕາມຕ້ອງການ) Soejarto et al. 15432.

Fordia cauliflora Hemsl.

ປະດົງເຫຼືອງ – Pa Dong Leuang. Luang Prabang MBP. Treelet 8 m tall, stem crooked, fruits brownish green, densely and softly pubescent. The stem bark

(a)   (b)

FIGURE 5.25 A fruiting branchlet of *Millettia pachyloba* bearing several brown fruits at the tip of the branchlet (a) and the fruit of *Millettia pachyloba* splitting open, revealing 3 dark brown seeds (b). Photo credit: D. Soejarto.

is used to treat rheumatism by boiling it and drinking the solution as needed (ໃຊ້ເປືອກວ່າບໍ່ປະດັງຂໍ້ ໂດຍການຕົ້ມ ແລະ ດື່ມນ້ຳຕົ້ມຕາມຕ້ອງການ). Soejarto et al. 15454.

Millettia pachyloba Drake. Figure 5.25

ເຄືອຫໍາງົວ – Kheua Ham Ngua. Bolikhamxay MPP. A large liana, fruits covered with brown, velvety pubescence. The stem (without the bark) or root is used to prepare drink (by boiling) for body weakness during postpartum (puerperium) and as an antidote for food allergic reactions (ໃຊ້ວ່າເຄືອ (ບໍ່ມີເປືອກ) ຕົ້ມດື່ມ ເພື່ອບໍາລຸງແມ່ກໍາ ແລະ ໃຊ້ເພື່ອຖອນພິດ ໃນກໍລະນີ ແພ້ອາຫານ). Soejarto et al. 15029.

ເຄືອຫໍາງົວ – Kheua Ham Ngua. Oudomxay MBP. Liana, leaves pinnate, leaflets 5–9, legume brown, short-hairy, seeds deep brown, shiny. Vine (stem without bark) is used to treat cases of intoxication from allergic foods during postpartum (puerperium) — (ໃຊ້ເຄືອ (ບໍ່ມີເປືອກ) ເພື່ອຖອນພິດ ທີ່ເກີດຈາກການແພ້ອາຫານ ໃນເວລາຢູ່ໄຟ (ຜິດກໍາ). Cut the vine (stem; without the bark) into small pieces and dry, then take a handful and boil in two liters of water (ຫັ່ນ ຢອຍ ເຄືອ (ທີ່ ບໍ່ມີເປືອກ) ເປັນຕ່ອນນ້ອຍໆ, ຕາກແຫ້ງ, ຕົ້ມ ປະມານຫນຶ່ງກໍາມືໃນນ້ຳສອງລິດ). Drink the liquid as needed (ດື່ມນ້ຳຕົ້ມຕາມຕ້ອງການ). Soejarto et al. 15097.

ເຄືອຂີ້ຫ່າຄວາຍ – Kheua Khi Ha Khuay. Bolikhamxay MPP. Liana, inflorescences and branches brown-pubescent, fruits covered with brown velvety pubescence. Root to treat body weakness (as a tonic) (ໃຊ້ຮາກບໍ່ອສຸຂະພາບອ່ອນເພຍ (ເປັນຍາບໍາລຸງກໍາລັງ). Cut the root into small pieces and dry, then take a handful and boil in two liters of water (ຫັ່ນ ຢອຍຮາກເປັນຕ່ອນນ້ອຍໆ, ຕາກແຫ້ງ, ຕົ້ມ ປະມານຫນຶ່ງກໍາມືໃນນ້ຳສອງລິດ). Drink the liquid as needed (ດື່ມນ້ຳຕົ້ມຕາມ ຕ້ອງການ). Soejarto et al.15009.

Ormosia cambodiana Gagn.

ກົກຂີ້ໝູ – Kok Khi Mou. Luang Prabang MBP. Tree 10 m tall, dbh 15 cm, bark gray, flowers pinkish white, banner and keel pinkish, young fruits purplish brown. The stem is used to prepare a medicine to treat cases of indigestion and constipation (ໃຊ້ລຳຕົ້ນບໍ່ອາຫານບໍ່ຍ່ອຍ ແລະ ທ້ອງຜູກ). Cut the stem into small pieces, then boil in water and drink the solution (ຟັກ ຊອຍລຳຕົ້ນເປັນຕ່ອນນ້ອຍໆ, ຈາກນັ້ນຈິ່ງຕົ້ມ ແລະ ດື່ມນ້ຳຕົ້ມ). DDS et al. 15458.

Pterocarpus macrocarpus Kurz

ໄມ້ດູ່ເລືອດ – Mai Dou Leuat. Luang Prabang MBP. Tree 15 m tall, dbh 20 cm, bark gray, fissured and peeling off, fruits dry. Very common. The stem bark is used to treat human and animal wounds and other diseases of the skin (ໃຊ້ເປືອກຕົ້ນບໍ່ບາດແຜໃນຄົນ ແລະ ສັດ ແລະ ພະຍາດຜິວໜັງອື່ນໆ). Boil the bark with 9 parts of water until a slurry is formed. Use the slurry to rub on the wounds (for person) or to wash the mouth and the foot of the animal (ຕົ້ມເປືອກກັບນ້ຳ 9 ສ່ວນ ຈົນກວ່າຈະໄດ້ນ້ຳຂຸ້ນໆ. ໃຊ້ນ້ຳຂຸ້ນງນັ້ນທາບາດແຜໃນຄົນ ຫຼື ໃຊ້ລ້າງປາກ ແລະ ຕີນສັດ). Soejarto et al. 15434.

Pueraria alopecuroides Craib. Figure 5.26

ຫາງເສືອເຄືອ – Hang Seua Kheua. Luang Prabang MBP. Large, extensive scrambling vine, bearing prominent, showy upright inflorescences of 50 cm long—a

(a)   (b)

FIGURE 5.26 A scrambling vine of *Pueraria alopecuroides* showing several upright inflorescences (a) and a close-up view of several open flowers of *Pueraria alopecuroides* exhibiting the deep violet keels (b). Photo credit: D. Soejarto.

dense, cylindrical, softly brown-haired raceme; flowers white, keel bluish pink. The root is used to treat internal rashes. Boil and drink the cooled decoction (ໃຊ້ຮາກເພື່ອປິ່ນບົວຕຸ່ມພາຍໃນ. ຕົ້ມ ແລະ ດື່ມ ນ້ຳຕົ້ມທີ່ເຢັນແລ້ວ). Soejarto et al. 15453.

Tadehagi triquetrum (L.) H. Ohashi
ຫຍ້າຍອດແດງ; ຫຍ້າຫນອນຫນ່າຍ – Nya Nyot Deng; Nya Nohn Nyai. Xieng Khouang MBP. Herb on open patch of forest edge, ground-level, young leaves copper-brown to brownish red. The whole plant is used to make a medicine to treat kidney stones and liver disease by boiling and drinking the decoction (ໃຊ້ຫມົດຕົ້ນບໍ່ຫນື່ອຫມາກໃສ່ຫຼັງ ແລະ ພະຍາດຕັບ ໂດຍການຕົ້ມ ແລະ ດື່ມນ້ຳຕົ້ມ). Soejarto et al. 14992.

ຫຍ້າຫນອນຫນ່າຍ – Nya Nohn Nyai. Sekong MBP. Herbaceous plant, 20 cm high, flowers purple. Whole plant to treat liver disease (ໃຊ້ຫມົດຕົ້ນບໍ່ພະຍາດຕັບ). Boil a handful of small pieces in a bowl of water and drink the decoction (ຕົ້ມ ບະມານຫນຶ່ງກຳມືໃນນ້ຳຫນຶ່ງຖ້ວຍ ແລະ ດື່ມນ້ຳຕົ້ມ). Elkington et al. 293.

Gesneriaceae
Rhynchotechum ellipticum (Wall. ex D. Dietr.) A. DC.
ກົກເຫມົ້ານ້ຳ – Kok Mao Nam. Xieng Khouang MBP. Small shrub along stream bank, flowers brownish white, pubescence brown. The root is used to treat nervousness by boiling it and drinking the decoction as needed (ໃຊ້ຮາກບໍ່ໂລກປະສາດ ໂດຍການຕົ້ມ ແລະ ດື່ມນ້ຳຕົ້ມຕາມຕ້ອງການ). Soejarto et al. 14998.

Hernandiaceae
Illigera celebica Miq. Figure 5.27
ເຄືອຟອດຟາວຂຽວ – Kheua Fot Fao Khiao. Xieng Khouang MBP. Vine, leaves glabrous, flowers pinkish red, set on a 40 cm, terminal panicles. The stem and the leaf are used to treat scabies and other skin diseases (ໃຊ້ລຳເຄືອ ແລະ ໃບບໍ່ຂີ້ຫິດ ແລະ ພະຍາດຜິວຫນັງອື່ນໆ). Crush the leaves and stem in water and take a bath with the filtered solution (ຍ່ອງໃບ ແລະ ເຄືອໃບນ້ຳ ແລະ ໃຊ້ນ້ຳຕອງອາບ). Soejarto et al. 15091.

Hydrangeaceae
Dichroa febrifuga Lour.
ກົກບົ້ງເລັນ – Kok Bong Len. Xieng Khouang MBP. Shrub 2 m tall, flower buds light blue to bluish white, open flowers deeper blue. The root is used to treat kidney stone disease and cases of jaundice (ໃຊ້ຮາກບໍ່ຫນື່ອຫມາກໃສ່ຫຼັງ ແລະ ກໍລະນີ ເປັນຂີ້ຫມາກເຫຼືອງ). Boil the root and drink the cooled liquid (ຕົ້ມຮາກ ແລະ ດື່ມ ນ້ຳຕົ້ມທີ່ເຢັນແລ້ວ). Soejarto et al. 14972.

(a)   (b)

FIGURE 5.27   A flowering vine of *Illigera celebica* (a) showing several open flowers and a close-up view of a flower of *Illigera celebica* revealing prominent incurving stamens (b). Photo credit: D. Soejarto.

Icacinaceae
Gonocaryum lobbianum (Miers) Kurz

ຮົ່ມແຊງເມືອງ – Home Seng Meuang. Bolikhamxay MPP. Treelet, fruits pale greenish, ellipsoid to ellipsoid-ovoid, with perceptible longitudinal angles/ ridges. The stem wood is used as a medicine for post-natal care (as a tonic) (ໃຊ້ລຳຕົ້ນເປັນຍາບຳລຸງແມ່ກຳ). Cut the stem wood into small pieces and dry, then decoct 1 handful by boiling with 2 liters of water (ຟັກ ຢອຍ ລຳຕົ້ນເປັນຕ່ອນນ້ອຍໆ ແລະ ຕາກໃຫ້ແຫ້ງ, ຈາກນັ້ນຈິ່ງຕົ້ມບະມານ 1 ກຳມືໃນນ້ຳ 2 ລິດ). Drink the decoction as needed (ດື່ມຕາມຕ້ອງການ). Soejarto et al. 15020.

Lamiaceae
Callicarpa formosana Rolfe

ຟ້ານົກ – Fa Nok. Sekong MBP. Small tree, 2.5 m in height, trunk 10 cm dbh, leaves opposite, pubescent, slightly crenate, inflorescence a panicle, fruits emerging from leaf nodes, very small, round, ranging from pink to white. Stem to treat stomachache, GI ulcers (ໃຊ້ລຳຕົ້ນປົວເຈັບກະເພາະ). Boil a handful of small, dried pieces of stem in 1 liter of water and drink the decoction (ຕົ້ມລຳຕົ້ນທີ່ຟັກ ຢອຍ ເປັນຕ່ອນນ້ອຍໆໃນນ້ຳໜຶ່ງລິດ ແລະ ດື່ມນ້ຳຕົ້ມ). Elkington et al. 306.

Callicarpa macrophylla Vahl

ກົກເປືອກຜ້າ; ຜ້າຮ້າຍ – Kok Peuak Pa; Pa Hai. Xieng Khouang MBP. Tree, 8 m tall, leaves dark green above, grayish green below, flowers small, pink, profuse, set in corymbose clusters. The leaf or root is used as a medicine to treat

(a) (b)

FIGURE 5.28 A flowering shrub of *Callicarpa rubella* in its natural habitat (a) with a close-up view of the inflorescence, showing numerous open flowers (b). Photo credit: D. Soejarto.

goiters occurring mainly in women with thyroid gland disorders (e.g., Basedow disease) by chewing the leaves or root and swallowing the watery liquid (ໃຊ້ເບືອກບໍ່ຄໍ່ໜນຽງກ້ຳມເລືອດ (ບາຊີໂດ) ໂດຍການຫຍ້ຳກ້ຳບ້າຈາກເບືອກ (ມັກເກີດກັບຜູ້ຍິງ). Soejarto et al. 14969.

Callicarpa rubella Lindl. Figure 5.28
ກົກໝາກປັດ – Kok Mak Pat. Xieng Khouang MBP. Treelet 3 m tall, flowers deep purplish pink. All parts of the plant are harvested to prepare a medicine to treat cases of nerve disorders, including backache (ໃຊ້ໝົດຕົ້ນບໍ່ໂລກປະສາດ, ລວມທັງການເຈັບແອວ). It is also claimed to be used to treat syphilis, muscle spasms, and back pain (ມີການກ່າວເຖິງການນຳໃຊ້ບໍ່ຊິຟິລິດ, ຊື້ນບັ້ນ ແລະ ເຈັບຫຼັງ). Boil whole plant and drink the decoction after cooling (ຕົ້ມໝົດໝົດຕົ້ນ ແລະ ດື່ມນ້ຳຕົ້ມທີ່ເຢັນແລ້ວ). Soejarto et al. 14986.

Clerodendrum chinense (Osb.) Mabberley
ພວງພີ່ຂົນຂາວ; ຫິງຂົນ – Puang Pii Khon Khao; Ting Khon. Xieng Khouang MBP. Shrub, flowers pale pink to white, stamens exserted. Common in open and semi-open areas. The leaf or the root is used to treat headache and gonorrhea by boiling and drinking the tea (ໃຊ້ໃບ ຫຼື ຮາກບໍ່ເຈັບຫົວ ແລະ ໂລກຫນອງໃນ ໂດຍການຕົ້ມ ແລະ ດື່ມນ້ຳຕົ້ມ). Soejarto et al. 14955.

Clerodendrum godefroyi Kuntze

ພ້າຄີຕົ້ນ – Pang Khi Tohn. Sekong MBP. Shrub, 2 m tall, stem 3 cm dbh, leaves opposite, 8–10 cm long, 3–5 cm wide, flowers white with 4 petals. The root is used to treat baldness, rashes (ໃຊ້ຮາກເພື່ອບົ່ງຫົວລ້ານ, ບໍ່ໄຂ້ອອກຕຸ່ມ). Boil a handful of small pieces in a bowl of water and drink the decoction (ຕົ້ມຮາກທິ່ຫັກ ຂອຍ ເປັນຕ່ອນນ້ອຍໆໃນນ້າທັ່ງໆຖ້ວຍ ແລະ ດື່ມນ້າຕົ້ມ). Elkington et al. 290.

Clerodendrum japonicum (Thunb.) Sweet

ພວງພີຂົນ – Puang Pii Khon. Shrub to 3 m tall, fruits green, glossy, encircled by dirty green to white persistent sepals. The root is used to prepare a medicine to treat gonorrhea (ໃຊ້ຮາກບໍ່ໂລກໜອງໃນ). Boil the root and drink the decoction (ຕົ້ມຮາກທີ່ຫັກ ຂອຍເປັນຕ່ອນນ້ອຍໆ ແລະ ດື່ມນ້າຕົ້ມ). Soejarto et al. 15468.

Clerodendrum schmidtii C.B. Clarke. Figure 5.29

ພ້າຄີດົງ; ພວງພີຂາວ – Pang Khi Dong; Puang Pii Khao. Savannakhet MBP. Shrub, fruits green, calyx red, plant hairy. The root is used to treat body weakness (as a tonic) (ໃຊ້ຮາກບໍ່ສຸຂະພາບອ່ອນເພຍ (ເປັນຢາບຳລຸງ). Macerate the root in alcohol and drink it 20 ml each time, morning and evening before meals. (ແຊ່ຮາກໃນເຫຼົ້າ (ເຫຼົ້າບອງຢາ), ດື່ມ ເຊົ້າແລງເທື່ອລະ 20 ມິລິວິດກ່ອນອາຫານ). Soejarto et al. 14940.

Congea tomentosa Roxb. Figure 5.30

ເຄືອງວງຊຸ່ມ – Kheua Ngouang Soum. Luang Prabang MBP. Liana to 5 m above ground, flowers profuse, white, in a group of 3, surrounded by 3 bracts, bracts light pinkish above, gray green below. The root or the stem is used to prepare a medicine to treat excessive menstrual bleeding (ໃຊ້ຮາກ ຫຼື ລຳເຄືອ ປຸງແຕ່ງເປັນ ຢາບໍ່ວ່າງເລືອດຫຼາຍ (ບະຈຳເດືອນມາຫຼາຍຜິດປົກກະຕິ)). Boil and drink (ຕົ້ມ ແລະ ດື່ມ). Soejarto et al. 15451.

ເຄືອກະແດ້ງ; ເຄືອແມງກະເບື້ອ – Kheua Ka Deng; Kheua Meng Ka Beau. Savannakhet MBP. Liana, inflorescences whitish, profusely flowered, showy from a distance. Common species throughout the preserve, especially on the canopy area. Root to treat asthma (ໃຊ້ຮາກບໍ່ຫືດ). Boil the root and drink the liquid as needed (ຕົ້ມຮາກ ແລະ ດື່ມນ້າຕົ້ມຕາມທີ່ຕ້ອງການ). Soejarto et al. 14902.

ເຄືອງວງຊຸ່ມ – Kheua Ngouang Soum. Sekong MBP. Vine/liana, stem 3 cm dbh, leaves tomentose, margins smooth, flowers white, glaucous, set on paniculate inflorescence. Root, stem to treat dengue fever (ໃຊ້ຮາກ, ລຳຕົ້ນເພື່ອບໍ່ໄຂ້ເລືອດອອກ). Boil a handful of small pieces in a bowl of water and drink the decoction (ຕົ້ມຮາກ ຫຼື ລຳຕົ້ນ ທີ່ຫັກ ຂອຍ ເປັນຕ່ອນນ້ອຍໆໃນນ້າທັ່ງໆຖ້ວຍ ແລະ ດື່ມນ້າຕົ້ມ). Elkington et al. 291.

FIGURE 5.29 A fruiting shrub of *Clerodendrum schmidtii* in its natural habitat, showing green fruits, each subtended by dark red persistent calyx. Photo credit: D. Soejarto.

Premna latifolia Roxb.

ເຄືອໂມກ – Kheua Mok. Xieng Khouang MBP. Climber-like shrub, fruits green on a corymbose cluster. The stem is used to treat asthma (ໃຊ້ລຳຕົ້ນບໍ່ອຫຶດ). Cut the stem into small pieces, boil in water, and drink the decoction (ຫັກ ຊອຍ ລຳຕົ້ນເປັນຕ່ອນນ້ອຍໆ, ຕົ້ມໃນນ້ຳຫຶ່ງໆຖ້ວຍ ແລະ ດື່ມນ້ຳຕົ້ມ). Soejarto et al. 15001.

Rotheca serrata (L.) Steane & Mabberley

ກົກຕັບເຕົ່າ – Kok Tab Tao. Oudomxay MBP. Shrub, flowers white with dark pink bracts. Stem to treat hepatitis (liver disease) (ໃຊ້ລຳຕົ້ນບໍ່ຕັບອັກເສບ (ພະຍາດຕັບ)). Cut the stem into small pieces and dry, then take a handful and boil in two liters of water; drink the decoction as needed (ຫັກ ຊອຍລຳຕົ້ນໃຫ້ ເປັນຕ່ອນນ້ອຍໆ, ຕົ້ມບະມານຫຶ່ງໆກຳມືໃນນ້ຳສອງລິດ ແລະ ດື່ມນ້ຳຕົ້ມຕາມຕ້ອງການ). Soejarto et al. 15135.

(a)   (b)

FIGURE 5.30 A flowering vine of *Congea tomentosa* (a) showing numerous terminal bracts (in threes), each bearing flowers inside and a close-up view of the flowers of *Congea tomentosa* (b) each revealing prominent filament-like stamens arising above the zygomorphic corolla. Photo credit: D. Soejarto.

ຕາຮາເຣ; ຕັບເຕົ່າ – Ta Har; Tap Tao. Sekong MBP. Herb 30 cm high, stem squared, leaves opposite, 40 cm long and 15 cm wide, flowers purple, fruits small, green. The root is used to treat liver disease (ໃຊ້ລາກັບບໍ່ອພະຍາດຕັບ). Boil a handful of small pieces in a bowl of water and drink the decoction (ຕົ້ມລາກັບທີ່ພັກ ຊອຍເປັນຕ່ອນນ້ອຍໆ ປະມານໜຶ່ງກໍາມືໃນນ້າໜຶ່ງຖ້ວຍ ແລະ ດື່ມນ້າຕົ້ມ). Elkington et al. 289.

Sphenodesme griffithiana Wight
ເຄືອຫວານເຍັນ – Kheua Van Yen. Savannakhet MBP. Liana, leaves opposite, flowers green in clusters along twigs. Liana (stem) to treat malaria (ໃຊ້ເຄືອ (ລາຕົ້ນ) ບໍ່ໄຂ້ມາລາເຣຍ). Cut stem into small pieces, boil a handful in plenty of water, and drink the cooled decoction as needed (ພັກ ຊອຍ ເປັນຕ່ອນນ້ອຍໆ, ຕົ້ມປະມານໜຶ່ງກໍາມືໃນນ້າ ແລະດື່ມນ້າຕົ້ມທີ່ເຢັນແລ້ວຕາມຕ້ອງການ). Sydara et al. 19034.

Vitex cf. *pierrei* Craib
ສະຄ່າງ; ສະລ່າງຄ່າງ – Sa Khang; Sa Lang Khang. Savannakhet MBP. A medium tree 5–8 m high, trunk with white bark, flowers profuse, pale green, in terminal paniculate clusters. Stem to treat liver disease, jaundice (ໃຊ້ລາກັບບໍ່ອພະຍາດຕັບ, ຂີ້ໝາກເຫຼືອງ). Cut the stem into small pieces, dry, then boil in water and drink the decoction (ພັກ ຊອຍ ລາກັບເປັນຕ່ອນນ້ອຍໆ, ຕົ້ມປະມານໜຶ່ງກໍາມືໃນນ້າ ແລະດື່ມນ້າຕົ້ມ). Sydara et al. 19036.

Vitex quinata (Lour.) F.N. Williams

ກົກຂີ້ເຫງັນ – Kok Khi Ngen. Xieng Khouang MBP. Tree 6 m tall, trunk dbh 45 cm, bark gray, irregularly fissured and cracked, slash pale brown, flower buds gray-green, open flowers with pink labellum and yellow lateral petals. The stem is used to make a medicine for jaundice by boiling in water and drinking the decoction (ໃຊ້ລຳຕົ້ນບໍ່ອຂີ້ໝາກເຫຼືອງ ໂດຍການຕົ້ມ ແລະ ດື່ມນ້ຳຕົ້ມ). Soejarto et al. 14989.

Vitex tripinnata (Lour.) Merr.

ແຄພູ; ກົກຂີ້ເຫງັນ – Keh Pou; Kok Khi Ngen. Oudomxay MBP. Tree, low- and many-branched, bark gray, fruits dull green to greenish brown. The stem is used to treat women's irregular menstruation (profuse bleeding) (ໃຊ້ລຳຕົ້ນ ບໍ່ ວິງເລືອດ ໝິດບໍ່ກະຕິ). Boil 1 handful of dried small pieces of the stem in 1 liter of water, drink the decoction (ຕົ້ມລຳຕົ້ນທີ່ແຫ້ກ ຊອຍເປັນບ່ຽງນ້ອຍໆ ແລະ ຕາກແຫ້ງແລ້ວ ປະມານໜຶ່ງກຳມືໃນນ້ຳໜຶ່ງລິດ,ດື່ມນ້ຳຕົ້ມ). Soejarto et al. 15063.

Lauraceae

Actinodaphne rehderiana (C.K. Allen) Kosterm. Figure 5.31

ກໍເມືອກ – Ko Meuak. Xieng Khouang MBP. Tree 10 m tall, leaves large, upper surface dark green, lower surface gray-green, fruits profuse, green, turning deep

FIGURE 5.31 A fruiting branchlet of *Actinodaphne rehderiana* bearing many fruits; red when mature. Photo credit: D. Soejarto.

red, glossy. The stem is used to prepare medicine to treat liver disease by boiling in water and drinking the decoction (ໃຊ້ລຳຕົ້ນປົວພະຍາດຕັບ ໂດຍການຕົ້ມ ແລະ ດື່ມນ້ຳຕົ້ມ). Soejarto et al. 14968.

Litsea cambodiana Lecomte

ກົກໝີ່ຫອມ – Kok Mii Hom. Savannakhet MBP. Tree up to 10–15 m high, flowers in loose terminal panicles. Stem, root to treat liver disease (ໃຊ້ລຳຕົ້ນ, ຮາກປົວພະຍາດຕັບ). Cut the stem or root into small pieces, dry, then boil in water and drink the decoction (ຟັກ ຊອຍ ລຳຕົ້ນ ຫຼື ຮາກ ເປັນຕ່ອນນ້ອຍໆ, ຕາກແຫ້ງ, ຕົ້ມໃນນ້ຳ ແລະ ດື່ມນ້ຳຕົ້ມ). Sydara et al. 19020.

Litsea cubeba (Lour.) Pers.

ກໍໝາກຊາງ; ກົກສີໄຄຕົ້ນ – Ko Mak Sang; Sii Khai Ton. Xieng Khouang MBP. Treelet 4 m tall, fruits dark green, glossy with white dots. The stem or the root is used to make a medicine to treat smallpox (ໃຊ້ລຳຕົ້ນ ຫຼື ຮາກປົວໄຂ້ທໍລະພິດ). Boil the root or stem and drink the decoction. Soejarto et al. 14982.

Lecythidaceae

Barringtonia racemosa (L.) Spreng.

ກົກນົມຍານ – Kok Nom Nyan. Savannakhet MBP. Medium-sized tree 4–5 m high, leaves alternate, inflorescences racemose, long, pendulous, flower buds greenish white with pink streaks. Bark, root is used to treat postnatal weakness (as a tonic) (ໃຊ້ເປືອກ, ຮາກ ປົວແມ່ຢູ່ກຳ (ອາການອ່ອນເພຍຫຼັງຈາກອອກລູກ)). Cut the bark or root into small pieces and dry, then take a handful and boil in two liters of water (ຟັກ ຊອຍ ເປືອກ ຫຼື ຮາກ ເປັນຕ່ອນນ້ອຍໆ, ຕາກແຫ້ງ, ຕົ້ມໜຶ່ງກຳມື ໃນນ້ຳສອງລິດ). Drink the liquid as needed (ດື່ມນ້ຳຕົ້ມຕາມຕ້ອງການ). Sydara et al. 19031.

Linderniaceae

Torenia asiatica L.

ກົງສະເດັນ – Kong Sa Den. Xieng Khouang MBP. Delicate herb on forest floor, flowers bluish pink to violet. Stem, root for prenatal weakness (ໃຊ້ເປືອກ, ຮາກ ປົວແມ່ຢູ່ກຳທີ່ອ່ອນເພຍຫຼັງຈາກອອກລູກ). Cut the stem or root into small pieces and dry, then take a handful and boil in two liters of water; drink the liquid as needed (ຟັກ ຊອຍ ເປືອກ ຫຼື ຮາກ ເປັນຕ່ອນນ້ອຍໆ, ຕາກແຫ້ງ, ຕົ້ມໜຶ່ງກຳມື ໃນນ້ຳສອງລິດ, ດື່ມນ້ຳຕົ້ມຕາມຕ້ອງການ). Soejarto et al. 15003.

Loganiaceae

Strychnos nux-blanda A.W. Hill. Figure 5.32

ໝາກຕູມກາຂາວ – Mak Toum Ka Khao. Luang Prabang MBP. Tree 15 m tall, fruits globular, greenish yellow, many-seeded. The root is one of the ingredients

FIGURE 5.32 A leafy branchlet of *Strychnos nux-blanda* bearing a large, globular fruit. Photo credit: D. Soejarto.

used to make a medicine to treat body weakness (as a tonic) and cases of backache by boiling in water and drinking the cooled decoction (ใຊ້ຮາກເບັນໝົ້ງຂອງ ສ່ວນບະກອບຍາບົວເຈັບແອວ ໂດຍການຕົ້ມ ແລະ ດື່ມນ້ຳຕົ້ມທີ່ເຍັນ). Soejarto et al. 15419.

Loranthaceae

Macrosolen cochinchinensis (Lour.) van Tieghem

ພາກຕົ້ອ – Pak Tiao. Bolikhamxay MPP. Large, spreading parasitic plant on a *Cratoxylum* tree, fruits small, globular, greenish yellow. Whole plant to treat rheumatism (ใຊ້ໝົດຕົ້ນບໍ່ປະດົງຂໍ້). Cut the whole plant into small pieces, boil in water, and drink the decoction (ພັກ ຊອຍ ໝົດຕົ້ນ ເບັນຕ່ອນນ້ອຍໆ, ຕາກແຫ້ງ, ຕົ້ມໃນນ້ຳ, ດື່ມນ້ຳຕົ້ມ). Soejarto et al. 15050.

ພາກໝາກໝຸ້ນ – Pak Mak Moun. Savannakhet MBP. A parasite on tree, flower buds pale green, on bunches along stem. Whole plant for rheumatism (ใຊ້ໝົດຕົ້ນບໍ່ປະດົງຂໍ້). Cut the whole plant into small pieces and dry, then take a handful and boil in two liters of water. Drink the decoction as needed (ຜັກ ຊອຍ ໝົດຕົ້ນ ເປັນຕ່ອນນ້ອຍໆ, ຕາກແຫ້ງ, ຕົ້ມໃນນ້ຳສອງລິດ, ດື່ມນ້ຳຕົ້ມຕາມຕ້ອງການ). Sydara et al. 19038.

Malpighiaceae
Hiptage elliptica Pierre. Figure 5.33

ຄຳພະມ້າຜູ້ – Kham Pa Ma Pou. Luang Prabang MBP. Large liana, flowers white with the lip yellow on the inner surface, set on a spreading terminal inflorescence. Rare occurrence. The stem is an ingredient of a medicine to treat paralysis and also to treat heart disease (ໃຊ້ລຳເປັນສ່ວນປະກອບຂອງຢາບົ່ວເປ້ຍລ່ອຍ ແລະ ໂລກ ຫົວໃຈ). Cut the stem into pieces, then boil in water and the liquid is taken by mouth (ຜັກ ຊອຍລຳເປັນຕ່ອນນ້ອຍໆ ແລະ ຕົ້ມ, ດື່ມນ້ຳຕົ້ມ). Soejarto et al. 15420.

Malvaceae
Grewia tomentosa Roxb. ex DC.

ໝາກຄອມ – Mak Khom. Savannakhet MBP. Tree 15 m tall, fruits green, turning black. Only this one seen. Root, stem to treat weakness in nursing mothers (as a tonic) and in cases of gastritis (ໃຊ້ຮາກ, ລຳຕົ້ນ ບົວແມ່ລູກອ່ອນທີ່ມີອາການອ່ອນເພຍ ແລະ ບົວເຈັບກະເພາະ). Boil a handful of small pieces of root or stem in a bowl of water and drink the decoction (ຕົ້ມ ປະມານໜຶ່ງກຳມື ຮາກ ຫຼື ລຳຕົ້ນ ທີ່ ຜັກ ຊອຍເປັນຕ່ອນນ້ອຍໆ ໃນນ້ຳໜຶ່ງຖ້ວຍ ແລະ ດື່ມນ້ຳຕົ້ມ). Soejarto et al. 14908.

Helicteres viscida Bl.

ປໍຂີ້ຕຸ່ນ – Poh Khi Toun. Bolikhamxay MBP. Large shrub, 3 m all, frits dry, with dense, gray and wooly pubescence. The root is used for the treatment of gonorrhea. Women also use it to treat leucorrhea (ໃຊ້ຮາກບົວໂລກหนองใน. ພວກຜູ້ຍິງຍັງໃຊ້ເພື່ອບົວລົງຂາວຕົ້ມອິກ). Cut the root into small pieces and dry (ຜັກ ຊອຍ ຮາກເປັນຕ່ອນນ້ອຍໆ). Decoct 1 handful by boiling in 2 liters

(a)   (b)

FIGURE 5.33 A leafy branchlet of *Hiptage elliptica* (a) and a close-up view of the flower of *Hiptage elliptica* (b). Photo credit: D. Soejarto.

of water (ຕົ້ມປະມານ 1 ກຳມືໃນນ້ຳ 2 ລິດ) and drink the decoction as needed (ດື່ມຕາມຕ້ອງການ). Soejarto et al. 15021.

ບໍ່ຂີ້ກະເທີ່; ບໍ່ຂີ້ຕຸ່ນ – Poh Khi Keh Teu; Poh Khi Toun. Savannakhet MBP. Shrub 1–2 m high, fruits covered with fluffy hairs. The root to treat gonorrhea (ໃຊ້ຮາກປົວໂລກໜອງໃນ). Cut the root into small pieces and dry, then take a handful and boil in two liters of water and drink the solution as needed (ຟັກ ຊອຍ ຮາກ ເປັນຕ່ອນນ້ອຍໆ, ຕາກແຫ້ງ, ຕົ້ມປະມານໜຶ່ງກຳມືໃນນ້ຳສອງລິດ, ດື່ມນ້ຳຕົ້ມຕາມຕ້ອງການ). Sydara et al. 19055.

ບໍ່ຂີ້ຕຸ່ນ – Poh Khi Toun. Sekong MBP. Shrub, 1.5–2 m in height, the trunk dbh 1.5 cm, leaves cordate-crenate, flowers white, arising from leaf nodes. Root to treat gonorrhea (ໃຊ້ຮາກປົວໂລກໜອງໃນ). Cut the root into small pieces and dry, decoct one handful by boiling in 2 liters of water and drink the decoction as needed (ຟັກ ຊອຍ ຮາກ ເປັນຕ່ອນນ້ອຍໆ, ຕາກແຫ້ງ, ຕົ້ມປະມານໜຶ່ງກຳມື ໃນນ້ຳສອງລິດ, ດື່ມນ້ຳຕົ້ມຕາມຕ້ອງການ). Elkington et al. 281.

Microcos paniculata L.

ກົກໝາກຄອມ – Kok Mak Khom. Oudomxay MBP. Small tree 6 m tall, stem bark gray, fruits green, profuse, on terminal panicles. The root is used as a remedy for headache due to excessive drinking (alcoholic drink) (ໃຊ້ຮາກປົວເຈັບຫົວ ທີ່ເກີດຈາກການດື່ມເຫຼົ້າຫຼາຍ). Boil the root and drink the cooled decoction (ຕົ້ມ ຮາກ ແລະ ດື່ມນ້ຳຕົ້ມທີ່ເຢັນ). Soejarto et al. 15084.

Pterospermum argenteum Tardieu. Figure 5.34

ກົກແສງ; ຫຳອາວນ້ອຍ – Kok Seng; Ham Ao Noy. Bolikhamxay MBP. Treelet, fruits brown, drying splitting open into a star-like configuration, seeds winged. The root is a medicine for rheumatism and joint pain (ໃຊ້ຮາກເພື່ອປົວປະດົງຂໍ້ ແລະ ເຈັບຂໍ້). Cut the root into small pieces and dry (ຟັກ ຊອຍ ຮາກເປັນຕ່ອນ

FIGURE 5.34 A leafy twig of *Pterospermum argenteum* (a) and a splitting open fruit of *Pterospermum argenteum* revealing dark brown, winged seeds (b). Photo credit: D. Soejarto.

ນ້ອຍໆ ແລະ ຕາກໃຫ້ແຫ້ງ). Mix with Meuat Nyai (*Aporosa ficifolia*) and Meuat Noy (*Aporosa tetrapleura*) (ປະສົມກັບຕົ້ນເໝືອດໃຫຍ່ ແລະ ເໝືອດນ້ອຍ). Decoct 1 handful of the mixture by boiling in 2 liters of water (ຕົ້ມປະມານ 1 ກຳ ມີໃບນ້ຳ 2 ລິດ). Drink as needed (ດື່ມຕາມຕ້ອງການ). Soejarto et al. 15017.

Pterospermum lanceifolium Roxb.
ກົກແຮງ – Kok Seng. Savannakhet MBP. Tree 9 m tall, fruits brown, ellipsoid, pendulous. Only this tree seen. Stem or root are used to treat certain conditions of rigid nerves and rigid tendon (ໃຊ້ລຳຕົ້ນ ຫຼື ຮາກບໍ່ເອັ້ນຮານ ແລະ ເອັ້ນແຮງ). Boil the stem and root and drink the decoction (ຕົ້ມລຳ ຫຼື ຮາກ ແລະ ດື່ມນ້ຳຕົ້ມ). Soejarto et al. 14907.

Pterospermum semisagittatum Buch.-Ham. ex Roxb.
ຫຳອາວ – Ham Ao. Luang Prabang MBP. Tree 15 m tall, leaves dark olive green above, light brownish green below, fruits brown, splitting open lengthwise. Stem is used for the treatment of rheumatism (ໃຊ້ລຳຕົ້ນບໍ່ປະດົງຂໍ້). Cut the stem into small pieces, dry, then boil in water and drink the decoction (ຟັກ ຊອຍລຳຕົ້ນເປັນຕ່ອນນ້ອຍໆ, ຕົ້ມ ແລະ ດື່ມນ້ຳຕົ້ມ). Soejarto et al. 15473.

Sterculia coccinea Roxb.
ໝາກລິ້ນວາງ – Mak Lin Wang. Sekong MBP. Smal tree, 2 m tall, leaves simple, alternate, glabrous, with arcuate veins, flowers pentamerous, pink. Stem to treat kidney stone (ໃຊ້ລຳຕົ້ນບໍ່ເໜື້ອໝາກໄຂ່ຫຼັງ). Boil a handful of small, dried pieces of stem in one liter of water and drink the decoction (ຕົ້ມປະມານໜຶ່ງກຳມີ ລຳຕົ້ນ ທີ່ ຟັກ ຊອຍເປັນຕ່ອນນ້ອຍໆ ໃນນ້ຳໜຶ່ງລິດ ແລະ ດື່ມນ້ຳຕົ້ມ). Elkington et al. 307.

Sterculia cf. *hyposticta* Miq.
ລິ້ນວາງ; ປໍລິ້ນ – Lin Wang; Po Liin. Bolikhamxay MPP. Treelet, fruit deep velvety-red, a capsule that splits open to reveal five carpels, each with 1–3 subglobular glossy seeds. Stem wood to treat lung infection and fluid retention (edema) (ໃຊ້ແກ່ນລຳບໍ່ພະຍາດປອດ. ບໍ່ນ້ຳຄັ່ງ (ບວມ)). Boil 1 handful of dried small pieces of the stem wood in 1 liter of water, drink the decoction (ຕົ້ມປະມານໜຶ່ງກຳມີ ແກ່ນລຳຕົ້ນ ທີ່ ຟັກ ຊອຍເປັນຕ່ອນນ້ອຍໆ ໃນ ນ້ຳໜຶ່ງລິດ ແລະ ດື່ມນ້ຳຕົ້ມ). Soejarto et al. 15049.

Melastomataceae
Melastoma imbricatum Wall. ex Triana
ເບັນອ້າ – Ben Aa. Xieng Khouang MBP. Shrub, flowers pink, fruits green, turning brownish red with prominent covering trichomes. Common in open to semi-open areas. Root for treatment of stomachaches (ໃຊ້ຮາກປິ່ນບໍ່ເຈັບກະເພາະ).

(a)   (b)

FIGURE 5.35   An upright shrub of *Oxyspora curtisii* (a) in its natural habitat and a downward view of the stem of *Oxyspora curtisii* (b) showing a spreading terminal inflorescence bearing numerous flowers. Photo credit: D. Soejarto.

Cut the root into small pieces, boil in water, and drink the decoction (ຟັກ ຊອຍ ຮາກເປັນຕ່ອນນ້ອຍໆ, ຕົ້ມໃນນ້ຳຫຶ່ງລິດ ແລະ ດື່ມນ້ຳຕົ້ມ). Soejarto et al. 14957.

Oxyspora curtisii King. Figure 5.35

ຂະໂກາະຄານ – Ko Kheh Khan. Oudomxay MBP. Shrub; leaves ovate, petiole winged, blade-like; inflorescences spreading, paniculate, young fruits light green. Stem or root is used to treat cases of food allergy reactions during puerperium (ໃຊ້ລຳຕົ້ນ ຫຼື ຮາກ ເພື່ອທອນພິດທີ່ເກີດຈາກການກິນຂອງຜິດໃນເວລາຢູ່ກຳ). Boil the stem or the root and drink the decoction (ຕົ້ມລຳຕົ້ນ ຫຼື ຮາກ, ດື່ມນ້ຳຕົ້ມ). Soejarto et al. 15119.

Pseudodissochaeta septentrionalis (W.W. Sm.) M.P. Nayar

ສ້ອຍເງິນລຽງ; ໃບສົ້ມກ່ອງ; ສົ້ມປ່ອງ – Soy Ngeun Lieng; Bai Som Kong; Som Pong. Xieng Khouang MBP. Shrub, flower buds pinkish. The root is used to treat rheumatism (ໃຊ້ຮາກບໍ່ປະດົງຂໍ້). Boil the root and drink the decoction (ຕົ້ມຮາກ ແລະ ດື່ມນ້ຳຕົ້ມ). Soejarto et al. 14966.

Meliaceae

Aglaia sp.

ກ້ອງຕາເສືອ – Kong Ta Seua. Oudomxay MBP. Treelet, leaves large, pinnate, about 40 cm long, fruits large, green, subglobular, on a pendulous raceme. Stem to treat joint pain (ໃຊ້ລຳຕົ້ນບໍ່ເຈັບຂໍ້). Boil one handful of dried small pieces of the stem in 1 liter of water and drink the decoction (ຕົ້ມປະມານຫຶ່ງກຳມື ແກ່ນ ລຳຕົ້ນ ທີ່ ຟັກ ຊອຍເປັນຕ່ອນນ້ອຍໆ ໃນນ້ຳຫຶ່ງລິດ ແລະ ດື່ມນ້ຳຕົ້ມ). Soejarto et al. 15065.

Cipadessa baccifera (Roth) Miq.

ເພັ້ຍຟານ; ໝາກແຜນດິນ – Pia Fan; Mak Fen Diin. Oudomxay MBP. Tree 12 m tall, dbh 30 cm, bark gray, fissured, slash pale brown to white, fruits green, globular. Stem or root is used to treat cases of kidney infection by boiling them and drinking the decoction (ໃຊ້ລຳຕົ້ນ ຫຼື ຮາກ ປົວໝາກໄຂ່ຫຼັງຕິດເຊື້ອ ໂດຍການຕົ້ມ ແລະ ດື່ມນໍ້າຕົ້ມ). Soejarto et al. 15099, 15104.

ເພັຍຟານ – Pia Fan. Luang Prabang MBP. Large shrub 3 m tall, fruits greenish, turning pink, then red, profuse, on a long-pedunculate paniculate cluster. Stem is used to treat fever and nervous disorders (ໃຊ້ລຳຕົ້ນປົວໄຂ້ ແລະ ພະຍາດເສັ້ນປະສາດ). Cut the stem into small pieces and dry, then take a handful and boil in two liters of water, drink the liquid as needed (ຟັກຊອຍຮາກເປັນຕ່ອນນ້ອຍໆ, ຕາກແດດ, ຕົ້ມປະມານໜຶ່ງກຳມື ໃນນໍ້າສອງລິດ ແລະ ດື່ມນໍ້າຕົ້ມຕາມຕ້ອງການ). Soejarto et al. 15459.

Melia azedarach L. Figure 5.36

ກົກຮ່ຽນ, ກະເດົາຊ້າງ – Kok Hien, Ka Dao Xang. Luang Prabang MBP. Tree 15 m tall, dbh 30 cm, bark gray roughly fissured, flowers white, stamen column bluish purple to dark purple. Leaf and stem bark are used to treat skin diseases (ໃຊ້ໃບ ແລະ ເປືອກຕົ້ນປົວພະຍາດຜິວໜັງ); fruit is used to treat stomach worms (ໃຊ້ໝາກຂ້າແມ່ທ້ອງ). In either case, boil the plant part and drink the decoction (ທັງສອງກໍລະນີ, ຕົ້ມສ່ວນຕ່າງໆນັ້ນ ແລະ ດື່ມນໍ້າຕົ້ມ). Soejarto et al. 15438.

ກະເດົາຊ້າງ; ຫ້ຽນ – Ka Dao Xang; Hien. Oudomxay MBP. Tree 12 m tall, dbh 30 cm, bark gray, fissured, slash pale brown to white, fruits green, globular, edge of forest. Fruit to treat stomach worms; stem bark to treat skin disease (ໃຊ້ໝາກຂ້າແມ່ທ້ອງ; ເປືອກຕົ້ນໃຊ້ປົວພະຍາດຜິງໜັງ). Boil the fruits and drink the decoction for worms (ຕົ້ມໝາກ ແລະ ດື່ມນໍ້າຕົ້ມເພື່ອຂ້າແມ່ທ້ອງ); boil stem bark and use the decoction to wash the diseased skin (ຕົ້ມເປືອກຕົ້ນເພື່ອໃຊ້ລ້າງປົວພະຍາດຜິວໜັງ). Soejarto et al. 15060.

(a) (b)

FIGURE 5.36 A flowering-fruiting branch (a) of a *Melia azedarach* tree viewed from below and a close-up view of the flower of *Melia azedarach* (b) exhibiting radiating white petals and a ring of stamens (pink-yellow). Photo credit: D. Soejarto.

Sandoricum koetjape (Burm.f.) Merr.

ໝາກຕ້ອງ – Mak Tong. Savannakhet MBP. A big tree 15–20 m high, fruits sub-globular, green. Stem bark is used to prepare a medicine to treat dysentery and diarrhea (ໃຊ້ເປືອກຕົ້ນບໍ່ທ້ອງບິດ, ທ້ອງຖອກ). Boil the stem bark and drink the decoction (ຕົ້ມເປືອກຕົ້ນ ແລະ ດື່ມນ້ຳຕົ້ມ). Sydara et al. 19040.

Menispermaceae

Pericampylus glaucus (Lam.) Merr. Figure 5.37

ເຄືອສະໜາດເປາະ – Kheua Sa Nat Pao. Xieng Khouang MBP. Vine, flowers min-ute, pale green, fruits green. Upper surface of leaves dark green, lower surface grayish green. The whole vine is used as a medicine for stomach worms (ໃຊ້ໝົດ ເຄືອຂ້າແມ່ທ້ອງ). Boil the crushed vine and drink the decoction (ຍ່ອງເຄືອ ແລະ ຕົ້ມ, ດື່ມນ້ຳຕົ້ມ). Soejarto et al. 14987.

Moraceae

Ficus auriculata Lour.

ເດື່ອວ່າ – Deua Va. Luang Prabang MBP. A low-branched tree from the base; main branches are trees of 15–30 cm dbh, bark smoothish, gray, slash light brown, exuding beads of white latex; fruits globular, greenish brown, in bunches attached to the tree trunk and branches, mature fruits maroon red. The fruit is used to

FIGURE 5.37 A fruiting vine of *Pericampylus glaucus* bearing small, profuse fruits. Photo credit: D. Soejarto.

prepare a tonic (by boiling the fruit) for postpartum weakness and nutrient defi-ciencies (ໃຊ້ໝາກເປັນຢາບຳລຸງກຳລັງ (ໂດຍການຕົ້ມໝາກ) ສຳລັບແມ່ລູກອ່ອນ ທີ່ມີອາການອ່ອນເພຍ ແລະ ຂາດທາດບຳລຸງບາງຊະນິດ). The scar left by fallen fruits is scraped and macerated in water and given to women to stimulate milk flow (ແຊ່ປູດຂອງຂົ້ວໝາກທີ່ລົ່ນອອກ ໃນນ້ຳ ແລະ ໃຫ້ແມ່ລູກອ່ອນດື່ມນ້ຳ ເພື່ອໃຫ້ມີນ້ຳນົມ). Soejarto et al. 15450.

Ficus cf. *gaspariana* Miq.
ເດື່ອນ້ອຍ – Deua Noy. Bolikhamxay MPP. Treelet, sap white, fruits green to yellow. Stem, root to treat kidney infection and TB cases (ໃຊ້ລຳຕົ້ນ, ຮາກ ບົວພະຍາດໝາກໄຂ່ຫຼັງ). Cut the stem or root into small pieces, boil in water, and drink the decoction (ຟັກ ຈອຍ ລຳຕົ້ນເປັນປ່ຽງນ້ອຍໆ, ຕົ້ມ ໃນນ້ຳສອງລິດ ແລະ ດື່ມນ້ຳຕົ້ມ). Soejarto et al. 15028.

Ficus hispida L.f. Figure 5.38
ເດື່ອປ່ອງ – Deua Pong. Luang Prabang MBP. Tree 10 m tall, 25 cm diameter at base, bark gray-white, smooth, slash pinkish brown, exuding white latex; fruits green, on pendulous raceme or narrow paniculate clusters along stem and branches. The stem wood is used as a medicine for skin rashes due to fever (ໃຊ້ລຳຕົ້ນບົວໄຂ້ອອກໝາກ, ອອກຕຸ່ມ). Crush and macerate the stem wood in water with one earthworm and drink the slurry (ແຊ່ລຳຕົ້ນ ແລະ ຂີ້ກະເດືອນ 1 ໂຕໃນນ້ຳ, ດື່ມນ້ຳແຊ່ນັ້ນ). Soejarto et al. 15452.

Ficus sarmentosa Buch.-Ham. ex J.E. Smith
ໄຮເຄືອ – Hai Kheua. Oudomxay MBP. Liana, profuse stems and leaves, leaves gray underneath, latex milky white, fruits dull pale green, beset with brown lenticels. The

(a)   (b)

FIGURE 5.38 Profuse fruits hanging on lateral inflorescences of a *Ficus hispida* tree (a) and a twig showing two large leaves of the plant (b). Photo credit: D. Soejarto.

stem wood is used to prepare a tonic for postpartum weakness and in cases of stomachache (ໃຊ້ລຳຕົ້ນເປັນຍາບຳລຸງແມ່ກຳທີ່ອ່ອນເພຍ ແລະ ໃຊ້ບໍ່ເຈັບກະເພາະ). Cut the stem into small pieces and boil and drink the decoction (ຟັກ ຊອຍ ລຳຕົ້ນເປັນຕ່ອນນ້ອຍ, ຕົ້ມ ແລະ ຕິ່ມນ້ຳຕົ້ມ). Soejarto et al. 15116.

Ficus simplicissima Lour.
ເດື່ອດິນ; ຫ່າຮອກ; ເຂົ້າກວາງອ່ອນ — Deua Diin; Ham Hok; Khao Kuang Ohn. Oudomxay MBP. Small tree, fruits green, turning red, plant hairy, latex profuse, milky. Root to treat body weakness (as a tonic) (ໃຊ້ຮາກບໍ່ອສຸຂະພາບອ່ອນເພຍ (ເປັນຍາບຳລຸງກຳລັງ)). Cut the root into small pieces and dry, then take a handful and boil in 2 liters of water; drink the decoction as needed (ຟັກ ຊອຍ ຮາກເປັນປູ່ປຽງນ້ອຍໆ ແລະ ຕາກແດດ, ຕົ້ມບະມານໜຶ່ງກຳມື ໃນນ້ຳສອງລິດ ແລະ ຕິ່ມນ້ຳຕົ້ມຕາມຕ້ອງການ). Soejarto et al. 15106.

ເດື່ອດິນ – Deua Diin. Sekong MBP. Small tree 1 m tall, leaves alternate, variably shaped, fruits arising from leaf nodes. Entire plant covered with small hairs. Root to treat men's weakness (ໃຊ້ຮາກບິ່ນບໍ່ອສຸຂະພາບຂອງທ່ານຊາຍທີ່ ອ່ອນເພຍ (ຍາບຳລຸງສຸຂະພາບທ່ານຊາຍ)). Cut the root into small pieces, boil in water, and drink the decoction as needed (ຟັກ ຊອຍ ຮາກເປັນປ່ງນ້ອຍໆ ແລະ ຕາກແດດ, ຕົ້ມໃນນ້ຳ ແລະ ຕິ່ມນ້ຳຕົ້ມຕາມຕ້ອງການ). Elkington et al. 294.

Ficus subincisa J.E. Sm. Figure 5.39
ກົກກະດອຍ – Kok Ka Doy. Xieng Khouang MBP. Shrub 2 m tall, latex white, fruits green with white dots, upper leaf surface with fine white dots, sap whitish. The stem and the root are used to prepare a tonic for postpartum weakness by boiling them and drinking the decoction (ໃຊ້ລຳຕົ້ນ ແລະ ຮາກເປັນຍາບຳລຸງແມ່ກຳທີ່ອ່ອນເພຍ ໂດຍການຕົ້ມ ແລະ ຕິ່ມນ້ຳຕົ້ມ). Soejarto et al. 14970.

Ficus triloba Buch.-Ham. ex Wall.
ປໍຫູຊ້າງ, ກົກນົມງົວ – Po Hou Xang; Kok Nom Ngua. Oudomxay MBP. Treelet, stem slender, leaves large, rounded cordate, plant covered by rusty brown hairs. Stem wood (no bark) is boiled in water and the decoction taken by mouth by women as a tonic for postpartum care and when milk flow is scanty or absent (ໃຊ້ລຳຕົ້ນ (ບໍ່ເອົາເປືອກ) ຕົ້ມໃຫ້ແມ່ລູກອ່ອນຕິ່ມເພື່ອເປັນຍາບຳລຸງ ໃນກໍລະນີທີ່ບໍ່ມີນ້ຳນົມ ຫຼື ນ້ຳນົມມານ້ອຍ). It is also used for the treatment of liver disease and jaundice (ຍັງໃຊ້ເພື່ອບໍ່ອພະຍາດຕັບ ແລະ ຂີ້ໝາກເຫຼືອງບ່າອິກ). Soejarto et al. 15089.

Maclura tricuspidata Carr.
ກົກໜາມຂີ້ແຮດ – Kok Nam Khi Het. Savannakhet MBP. Tree 3–4 m high, flower buds in clusters along stem. Stem to treat rheumatism (ໃຊ້ລຳຕົ້ນບໍ່ອປະດົງຂໍ້). Cut the stem into small pieces, dry, then boil in water and drink the decoction (ຟັກ ຊອຍ ລຳຕົ້ນເປັນປ່ງນ້ອຍໆ ແລະ ຕາກແດດ, ຕົ້ມໃນນ້ຳ ແລະ ຕິ່ມນ້ຳຕົ້ມ). Sydara et al. 19018.

FIGURE 5.39 Fruiting branchlets of *Ficus subincisa* bearing dotted mature (red) and immature fruits. Photo credit: D. Soejarto.

Myristicaceae

Knema globularia (Lamk.) Warb. Figure 5.40

ກະທ່ອມເລືອດ; ເລືອດມ້າ; ຮາວເລືອດນ້ອຍ – Ka Tohm Leuat; Leuat Ma; Hao Leuat Noy. Savannakhet MBP. Treelet, 10 m tall, stem slender, flower buds brown, open flowers reddish brown. The stem bark is used to prepare a medicine to treat cases of white leucorrhea (ໃຊ້ເປືອກຕົ້ນບໍ່ວົງຂາວ). Boil the smashed stem bark and drink the decoction (ຕົ້ມເປືອກຕົ້ນ ແລະ ດື່ມນ້ຳຕົ້ມ). Soejarto et al. 14945.

Knema pierrei Warb. Figure 5.41

ກົກຮາວເລືອດ – Kok Hao Leuat. Oudomxay MBP. Tree 8–9 m tall, stem 10 cm diameter, bark gray, finely cracked, slash dark red, sap dark red, leaves large,

(a)   (b)

FIGURE 5.40   A leafy branch of a *Knema globularia* tree (a) and an enlarged view of the flower of a *Knema globularia* tree (b). Photo credit: D. Soejarto.

(a)   (b)

FIGURE 5.41   A leafy branch arising from the tree of *Knema pierrei* (a) and a leafy twig of *Knema pierrei* with sectioned fruits showing white seeds (b). Photo credit: D. Soejarto.

obovate to elliptic-obovate, gray green below, fruits orange, tubercled. The stem bark is used as a medicine to treat dysentery (ໃຊ້ເປືອກຕົ້ນບໍ່ທ້ອງບິດ). Cut the stem bark into pieces and boil and drink the decoction (ພັກ ຊອຍ ເປືອກຕົ້ນ ແລະ ຕົ້ມນ້ຳ, ດື່ມນ້ຳຕົ້ມ). Soejarto et al. 15111.

Myrtaceae
Syzygium cf. *chloranthum* (Duthie) Merr. & L.M. Perry
ຫວ້ານ້ອຍ – Wa Noy. Savannakhet MBP. Tree 20–25 m tall, dbh 30 cm, bark brown, smooth, slash light brown, flower buds light green, open flowers with prominent, numerous white stamens. Stem bark to treat diarrhea, dysentery (ໃຊ້ເປືອກຕົ້ນບໍ່ທ້ອງຖອກ, ທ້ອງບິດ). Cut the stem bark into small pieces, boil in water, and drink the decoction (ພັກຊອຍເປືອກຕົ້ນເປັນບ່ຽງນ້ອຍໆ ແລະ ຕາກແດດ, ຕົ້ມໃນນ້ຳ ແລະ ດື່ມນ້ຳຕົ້ມ). Soejarto et al. 14941.

Ochnaceae

Gomphia serrata (Gaertn.) Kanis

ກົກລິ້ນກວາງ – Kok Liin Kuang. Savannakhet MBP. Shrub 2 m tall, flowers light yellow on elongated clusters, fruitlets glossy green. Common species throughout the preserve. The root is used to make a tonic for body weakness (ໃຊ້ຮາກເປັນ ຍາບຳລຸງສຸຂະພາບທີ່ອ່ອນເພຍ). Cut the root into small pieces, bring to a boil in water, and give the decoction to drink. (ຫັ້ກ ຊອຍ ຮາກເປັນຕ່ອນນ້ອຍໆ, ຕົ້ມນ້ຳ ແລະ ດື່ມນ້ຳຕົ້ມ). Soejarto et al. 14905.

Oleaceae

Jasminum cf. *subtriplinerve* Bl.

ໄສ້ໄກ່ນ້ອຍ – Sai Kai Noy. Bolikhamxay MBP. Vine, persistent sepals light green, fruits black to blue-black, glossy. Leaves to treat toothache (ໃຊ້ໃບບົ່ນບ໋ອເຈັບແຂ້ວ). Wash fresh leaves, crush into a paste, and put on ached tooth (ລ້າງໃບໃຫ້ສະອາດ, ຍ່ອງໃຫ້ຂຸ້ນໆ ແລ້ວແປະໃສ່ແຂ້ວທີ່ເຈັບ). Soejarto et al. 15041.

Ligustrum robustum (Roxb.) Bl. Figure 5.42

ກະດູກຂຽດ – Ka Douk Khiet. Oudomxay MBP. Treelet, fruits green, turning black, opaque, peduncle and pedicel hairy. The root is used to prepare a medicine

FIGURE 5.42   A fruiting branch of *Ligustrum robustum* bearing mature (black and black-brown in color) and immature fruits (green in color). Photo credit: D. Soejarto.

for back pain and other body pains (ໃຊ້ຣາກບົວເຈັບຫຼັງ ແລະ ເຈັບຕົນໂຕ). Boil the root and drink the decoction (ຕົ້ມຣາກ ແລະ ດື່ມນ້ຳຕົ້ມ). Soejarto et al. 15107.

Ligustrum sinense Lour.

ກໍຫັດ – Ko Hat. Xieng Khouang MBP. Treelet 3 m tall, flowers white, young fruits green. Not common. The root is used to prepare women's postpartum tonic and also to stimulate lactation (as a galactagogue) (ໃຊ້ຣາກບັ່ນຢາບຳລຸງແມ່ກຳ ແລະ ກະຕຸ້ນໃຫ້ມີນ້ຳນົມ (ຊ່ວຍໃຫ້ນ້ຳນົມມາຫຼາຍ). Boil the root and drink the decoction (ຕົ້ມຣາກ ແລະ ດື່ມນ້ຳຕົ້ມ). Soejarto et al. 14959.

Pandaceae

Microdesmis caseariifolia Pl. ex Hook.

ຫຸ່ນໄຮ່; ອ້ອມຕໍ – Houn Hai; Om Toh. Bolikhamxay MPP. Treelet, fruits globular-ellipsoid. Root, leaf to treat heat inside the body; also, for skin rashes, skin inflammation (ໃຊ້ຣາກ, ໃບ ບໍ່ອາການຮ້ອນພາຍໃນຮ່າງກາຍ; ບໍ່ເບັ່ນຕຸ່ມຕາມຜິວໜັງ, ຜິວໜັງອັກເສບ). In all cases, cut the root or leaves into small pieces, boil in water, and drink the decoction (ໃບທຸກກະນິ ແມ່ນພັ້ກ ຊອຍ ຣາກ ຫຼື ໃບເບັ່ນບ່ຽງນ້ອຍໆ, ຕົ້ມໃບນ້ຳ ແລະ ດື່ມນ້ຳຕົ້ມ. Soejarto et al. 15016.

Peraceae

Chaetocarpus castanocarpus (Roxb.) Thw. Figure 5.43

ກົກບົກຄາຍ – Kok Bok Kai. Savannakhet MBP. Tree 15 m tall, fruits light green covered with soft bristles. Common species throughout the preserve. Branch and twigs are used to prepare medicine to treat persistent coughs and cough-related disease (ໃຊ້ກິ່ງ ແລະ ງ່າບໍ່ໄອຊ້ຳເຊື້ອ ແລະ ພະຍາດທີ່ກ່ຽວພັນກັບໄອ). Boil and drink the decoction (ຕົ້ມນ້ຳ ແລະ ດື່ມນ້ຳຕົ້ມ). Soejarto et al. 14901.

(a)   (b)

FIGURE 5.43 A branchlet of *Chaetocarpus castanocarpus* bearing minute flowers and fruits (a) and a twig of *Chaetocarpus castanocarpus* bearing mature fruits covered by dense spiny prickles (b). Photo credit: D. Soejarto.

Phyllanthaceae

Antidesma comptum Tul.

ໝາກເໝົ້າ – Mak Mao. Sekong MBP. Small tree, ca. 1.5 m tall, leaves alternate, glabrous, margins entire, inflorescence a spike bearing purple fruits. The root is used to treat fever accompanied by hot body (ໃຊ້ຮາກບໍ່ໄຂ້ ທີ່ມີອາການຮ້ອນໃນ). Boil a handful of small pieces in a bowl of water and drink the decoction (ຕົ້ມຮາກທີ່ຟັກ ຂອຍ ເປັນຕ່ອນນ້ອຍໆ ປະມານທົ່ງກຳມືໃນນ້ຳທົ່ງຖ້ອຍ ແລະ ດື່ມນ້ຳຕົ້ມ). Elkington et al. 292.

Antidesma cf. *hainanense* Merr.

ກົກເໝົ້າສ້ອຍ – Kok Mao Soy. Bolikhamxay MPP. Large shrub, inflorescence a slender, dense, terminal raceme. Whole plant to treat cases of higher fever (ໃຊ້ໝົດຕົ້ນບໍ່ໄຂ້ສູງແຮງ). Cut the whole plant into small pieces, boil in water, and drink the decoction (ຟັກ ຂອຍ ວ່າຕົ້ນເປັນບ່ຽງນ້ອຍໆ, ຕົ້ມໃນນ້ຳ ແລະ ດື່ມນ້ຳຕົ້ມ). Soejarto et al. 15022.

Antidesma japonicum Sieb. & Zucc.

ກົກເໝົ້າເຕ້ຍ – Kok Mao Tia. Savannakhet MBP. Shrub, fruits green, turning to black, on racemose, axillary infructescence. Common. The stem is used to make a remedy to treat cases of fever by boiling and drinking the decoction (ໃຊ້ລ່າຕົ້ນບໍ່ໄຂ້ ໂດຍການຕົ້ມ ແລະ ດື່ມນ້ຳຕົ້ມ). Soejarto et al. 14913.

Antidesma tonkinensis Gagn.

ໝາກເໝົ້າຕາດອາຍ – Mak Mao Ta Koy. Sekong MBP. Liana, ca. 3 m long, stem dbh 2 cm, leaves alternate, glabrous, inflorescence a spike, fruits small, white turning to red. Stem to treat fever, kidney infection (ໃຊ້ລ່າຕົ້ນບໍ່ໄຂ້, ບໍ່ພະຍາດໝາກໄຂ່ຫຼັງ). Boil a handful of small, dried pieces of stem in 1 liter of water and drink the decoction (ຕົ້ມລ່າຕົ້ນທີ່ຟັກ ຂອຍ ເປັນຕ່ອນນ້ອຍໆ ແລະ ຕາກແຫ້ງປະມານທົ່ງກຳມືໃນນ້ຳທົ່ງລິດ ແລະ ດື່ມນ້ຳຕົ້ມ). Elkington et al. 295.

Aporosa ficifolia Baill. Figure 5.44

ກົກເໝືອດໃຫຍ່ – Kok Meuat Nyai. Bolikhamxay MPP. Pubescent treelet 4 m tall, fruits green to gray-brown, globular, split open to reveal a yellow to red, glossy, and translucent seed. The stem (sometimes mixed with other plants) is used to prepare a medicine to treat rheumatism and joint pain (ໃຊ້ລ່າ (ບາງຄັ້ງ ກໍປະສົມໃສ່ກັບຕົ້ນເໝືອດນ້ອຍ ແລະ ຕົ້ງຄອນແຄນ) ເປັນຍາບໍ່ປະດົງຂໍ້ ແລະ ເຈັບຂໍ້ຕໍ່). Take 1 handful and boil with an adequate amount of water (2 liters) and drink as needed (ໃຊ້ປະມານ 1 ກຳມື ຕົ້ມໃນນ້ຳພໍປະມານ (2 ລິດ) ແລະ ດື່ມນ້ຳ ຕົ້ມຕາມຕ້ອງການ). Soejarto et al. 15015.

FIGURE 5.44   A terminal branch of *Aporosa ficifolia* shrub in its natural habitat, bearing mature fruits. Photo credit: D. Soejarto.

Baccaurea ramiflora Lour. Figure 5.45

ກົກໝາກໄພແດງ – Kok Mak Fai Deng. Xieng Khouang MBP. Many-branched tree 10 m tall, trunk diameter 10–20 cm dbh, bark light gray, slash pale yellowish white, fruits blackish purple, profuse, in pendulous racemes, along branches and trunk. The stem bark or the root is used as a remedy to treat inflammation and edema and body weakness (ໃຊ້ເປືອກຕົ້ນ ຫຼື ຮາກ ບໍ່ອັກເສບ ແລະ ຕົນໂຕໄຂ່ ບວມ ແລະ ຮ່າງກາຍອ່ອນເພຍ). Prepare a tonic solution and drink as needed (ປຸງແຕ່ງເປັນຍາບາລຸງ ແລະ ດື່ມນ້ຳຕາມຕ້ອງການ). Soejarto et al. 14953.

ກົກໝາກໄພ – Kok Mak Fai. Savannakhet MBP. A medium tree 5–6 m high with yellow, globular fruits. Root, stem to treat liver disease (ໃຊ້ຮາກ, ລຳຕົ້ນບໍ່ ພະຍາດຕັບ). Cut the root or stem into small pieces and dry, then take a handful and boil in 2 liters of water (ຟັກ ຊອຍ ລຳຕົ້ນເປັນຕ່ອນຫ້ອຍໆ ແລະ ຕາກແຫ້ງ, ເກັບບະມານໜຶ່ງກຳໃນນ້ຳໃນນ້ຳສອງລິດ). Drink the decoction as needed (ດື່ມນ້ຳຕົ້ມ ຕາມຕ້ອງການ). Sydara et al. 19025.

FIGURE 5.45 *Baccaurea ramiflora* tree showing branches loaded with purple-brown fruits, set on dense, hanging racemes. Photo credit: D. Soejarto.

Breynia fruticosa (L.) M.A.

ກົກໝາກຕິ່ງທອງ; ຕັ່ງຕິບ – Kok Mak Ting Tong; Tang Tip. Xieng Khouang MBP. Large shrub, flower buds green, open yellow. Root, stem to treat nerve afflictions, gonorrhea (ໃຊ້ຮາກ, ລຳຕົ້ນບໍ່ພະຍາດເສັ້ນປະສາດ, ໂກກໜອງໃນ). Cut the root or stem into small pieces, boil in water, and drink the decoction (ໝັກ ຊອຍ ຮາກ, ລຳຕົ້ນເປັນຕ່ອນນ້ອຍໆ, ຕົ້ມໃນນ້ຳ ແລະ ດື່ມນ້ຳຕົ້ມ). Soejarto et al. 14973.

Breynia grandiflora Beille

ຫວ້ານຂີ້ໄກ່, ຊ້າແປກຫວານ – Van Khi Kai; Xa Pek Van. Bolikhamxay MPP. Treelet, flowers small, on axillary clusters, young fruits dull green, pedicellate. Stem to treat heart disease, cough, asthma (ໃຊ້ລຳຕົ້ນເພື່ອບໍ່ພະຍາດຫົວໃຈ, ໄອ, ຫືດ). Boil one handful of dried small pieces of the stem in 1 liter of water and drink the decoction (ຕົ້ມ ລຳຕົ້ນທີ່ພັກ ຊອຍ ແລະ ຕາກແຫ້ງປະມານໜຶ່ງກຳມື ໃນນ້ຳໜຶ່ງລິດ, ດື່ມນ້ຳຕົ້ມ). Soejarto et al. 15047.

Cleistanthus oblongifolius (Roxb.) M.A. Figure 5.46

ກົກຂະຫນ່ອງໄກ່ – Kok Kha Nong Kai. Bolikhamxay MPP. Liana-like shrub with a slender stem, fruits green, turning black, profuse. Very common. The stem is

FIGURE 5.46   A downward view of a branchlet of *Cleistanthus oblongifolius* shrub showing green, immature fruits set on axillary clusters. Photo credit: D. Soejarto.

used to make a medicine to treat heart disease (ໃຊ້ລ່າຕົ້ນບໍ່ອພະຍາດຫົວໃຈ). Boil and drink the decoction (ຕົ້ມບ້າ ແລະ ດື່ມນ້ຳຕົ້ມ). Soejarto et al. 15010; 15036.

Phyllanthus eriocarpus (Champ. Ex Benth.) Müll-Arg. Figure 5.47
ກໍ່ຫືບຫັດ, ກົກຂີ້ມົດ – Ko Hueb Hat; Kok Khi Mot. Xieng Khouang MBP. Treelet 6 m tall, flowers pale yellowish, fruits pinkish red. The stem is used to prepare a medicine to treat uterine infection and cases of gonorrhea (ໃຊ້ລ່າຕົ້ນ ບໍ່ມົດລູກອັກເສບ ແລະ ໂລກໜອງໃນ). In either case, boil and drink the decoction (ຕົ້ມບ້າ ແລະ ດື່ມນ້ຳຕົ້ມ). Soejarto et al. 14999.

Phyllanthus urinaria L.
ຫຍ້າໝາກໃຕ້ໃບ, ຫຍ້າໄຂ່ຫ້ຼງ – Nya Mak Tai Bai; Nya Kai Lang. Sekong MBP. Cultivated herb, 20 cm in height, leaves alternate, fruits growing on the underside from leaf nodes, small, round. Whole plant to treat rashes or cancer in the mouth; also, for liver disease (ໃຊ້ໝົດຕົ້ນບໍ່ອຕຸ່ມ ຫຼື ມະເຮັງປາກ; ແລະພະຍາດຕັບ). In all cases boil the whole plant and drink the decoction (ໃບທຸກກຳລະນີ ແມ່ນຕົ້ມໝົດຕົ້ນ ແລະ ດື່ມນ້ຳຕົ້ມ). Elkington et al. 287.

FIGURE 5.47   A lower side view of the stem of *Phyllanthus eriocarpus* shrub showing an axillary cluster of pink-colored fruits and small young fruits and flowers, also in axillary clusters. Photo credit: D. Soejarto.

Piperaceae

Piper laetispicum C. DC.

ເຄືອຜັກອີ່ເລິດ; ອີ່ເລິດດົງ; ຜິກໄທປ່າ – Kheua Pak II Leut; II Leut Dong; Pik Tai Pa. Oudomxay MBP. Epiphytic climber, attaching to a large tree-trunk, fruits green, set in a helical infructescence. Whole plant to treat cases of gonorrhea (ໃຊ້ໝົດຕົ້ນບໍ່ໂລກໜອງໄນ). Cut the whole plant into small pieces and dry, then take a handful and boil in 2 liters of water (ຟັກ ຊອຍ ໝົດຕົ້ນເປັນຕ່ອນນ້ອຍໆ ແລະ ຕາກແຫ້ງ, ຕົ້ມປະມານໜຶ່ງກຳມືໃນນ້ຳສອງລິດ). Drink the decoction as needed (ດື່ມນ້ຳຕົ້ມຕາມຕ້ອງການ). Soejarto et al. 15096.

ໝາກຂີ້ໄກ່ປ່າ – Mak Khi Kai Pa. Sekong MBP. A vine. Root to treat urinary tract infection, kidney stones (ໃຊ້ຮາກເພື່ອປົວທໍ່ປັດສະອະຕິດເຊື້ອ, ໜ້ວຍໝາກໄຂ່ຫຼັງ). In both cases, boil the dried root and drink the decoction (ໃນທັງສອງກໍລະນີ, ຕົ້ມຮາກແຫ້ງ ແລະ ດື່ມນ້ຳຕົ້ມ). Elkington et al. 279.

Polygalaceae

Xanthophyllum bibracteatum Gagn. Figure 5.48

ກົກໝາກຂົມ – Kok Mak Khom. Oudomxay MBP. Treelet 4 m tall, fruits green, 1-seeded. The stem is used to make a medicine to treat body swellings and other cases of

(a)   (b)

FIGURE 5.48   A fruiting twig of a *Xanthophyllum bibracteatum* treelet (a) and a close-up view of a mature fruit of *Xanthophyllum bibracteatum* attached to a terminal inflorescence (b). Photo credit: D. Soejarto.

edema by boiling and drinking the cooled decoction (ໃຊ້ລຳຕົ້ນບໍ່ຕົ້ນໂຕໄຕໄບໝ ແລະ ອາການບອມອື່ນໆ ໂດຍການຕົ້ມນ້ຳ ແລະ ດື່ມນ້ຳຕົ້ມທີ່ເຍັນແລ້ວ). Soejarto et al.15112; 15123.

Primulaceae

Ardisia annamensis Pit.

ຕິນຈຳນ້ອຍ – Tin Cham Noy. Bolikhamxay MPP. Tree with liana-like branches, fruits dull green. The root is used to prepare a medicine to treat cases of fluid retention (dropsy) and cases of edema (ໃຊ້ຮາກບໍ່ອອາການສະສົມນ້ຳຢ່າງຜິດບົກກະຕິ ແລະ ກໍລະນີໄຕ່ບອມ). Cut the root into small pieces and dry, then decoct one handful by boiling in a bowl of water (about 2 liters) and drink as needed (ຟັກ ຊອຍ ຮາກໃຫ້ເປັນຕ່ອນນ້ອຍໆ ແລະ ຕາກໃຫ້ແຫ້ງ, ຫຼັງຈາກນັ້ນ ຕົ້ມບະຍານ 1 ກໍາມືໃນນ້ຳ (ບະຍານ 2 ລິດ) ແລະ ດື່ມຕາມຕ້ອງການ). Soejarto et al. 15025.

Ardisia crenata Sims

ຕາກວາງໃຫຍ່; ຕິນຈຳກ້ຽງ – Ta Kuang Nyai; Tin Cham Kieng. Oudomxay MBP. Shrub, fruits light green, edge of leaf blade glandulous. The root is used to treat cases of body weakness (as a tonic) (ໃຊ້ຮາກບໍ່ສຸຂະພາບອ່ອນເພຍ (ເປັນຢາບຳລຸງກຳລັງ)). Cut the root into small pieces and put to dry, then take a handful and boil in 2 liters of water; drink the filtered liquid as needed (ຟັກ ຊອຍ ຮາກ, ລຳຕົ້ນເປັນຕ່ອນນ້ອຍໆ ແລະ ຕາກໃຫ້ແຫ້ງ, ຕົ້ມບະຍານນັ້ນໆກຳມືໃນນ້ຳສອງລິດ ແລະ ດື່ມນ້ຳຕົ້ມທີ່ຕອງແລ້ວ). Soejarto et al. 15129.

Ardisia florida Gagn. Figure 5.49

ຕິນຈຳກ້ຽງໃບຍາວ – Tin Cham Kieng Bai Niao. Bolikhamxay MPP. Treelet, fruits pink to dull gray-green to pinkish purple. The root is used to make a medicine for kidney stones and cases of urine retention (ໃຊ້ຮາກບໍ່ຫນ້ວໝາກໄຕ່ຫຼັງ ແລະ ກໍລະນີຖ່າຍເບົາບໍ່ສະດອກ (ຂັບປັດສະອະ)). Cut the root into small pieces and dry

(a)   (b)

FIGURE 5.49 Profuse small pink fruits of an *Ardisia florida* shrub set on a narrow terminal panicle (a) and a leafy terminal part of a branch of *Ardisia florida* shrub showing the underside of the leaves (b). Photo credit: D. Soejarto.

and decoct one handful in a bowl of water; drink the decoction as needed (ພັກ ຊອຍ ຮາກເປັນຕ່ອນນ້ອຍໆ ແລະ ຕາກໃຫ້ແຫ້ງ, ຕົ້ມບະມານ 1 ກຳມືໃນນ້ຳໜຶ່ງຖ້ວຍ, ດື່ມນ້ຳຕົ້ມຕາມຕ້ອງການ). Soejarto et al. 15039.

Ardisia hanceana Mez

ກົກຕາກວາງ; ຕິນຈຳກ້ຽງ – Kok Ta Kuang; Tin Cham Kieng. Oudomxay MBP. Small tree with liana-like, pendulous branches, fruits green. The stem is used as an antidote to a poisoning caused by eating some food allergens/allergenic foods during puerperium (ໃຊ້ລຳຕົ້ນບໍ່ດພິດເບື່ອ ທີ່ເກີດການແພ້ອາຫານ ໃນເວລາຍູ່ກຳ (ແມ່ຍີງຜິດກຳ)). Boil the stem and drink the solution as needed (ຕົ້ມຮາກໃນນ້ຳ, ດື່ມນ້ຳຕົ້ມຕາມຕ້ອງການ). Soejarto et al. 15079.

Ardisia helferiana Kurz. Figure 5.50

ຕິນຈຳຂົນ – Tin Cham Khon. Bolikhamxay MPP. Treelet, plant hairy, fruits green, glossy, flower buds dull greenish white, set on a long pedicel, disposed on a long, spreading, peduncled umbel. Not common. The root is used to treat kidney stones and urine retention, and cases of high fever (ໃຊ້ຮາກບໍ່ຫນື້ວຫມາກໄຂ່ຫຼັງ ແລະ ຂັບບັດສະວະ ແລະ ບໍ່ໄຂ້ສູງ). In either case, boil pieces of root and drink the decoction (ຕົ້ມຮາກໃນນ້ຳ ແລະ ດື່ມນ້ຳຕົ້ມ). Soejarto et al. 15054.

FIGURE 5.50 The terminal end of a branch of *Ardisia helferiana* treelet, showing hairy, perpendicularly set axillary racemes, each bearing small young fruits and unopened flowers. Photo credit: D. Soejarto.

ຕິນຈຳຂົນ – Tin Cham Khon. Savannakhet MBP. Shrub, hairy, flower buds dull greenish white. Root to treat cases of high fever (as an antipyretic) (ໃຊ້ຮາກບໍ່ໄຂ້ຮ້ອນ). Cut the root into small pieces, boil in water, and drink the decoction (ຝັກ ຊອຍ ຮາກເປັນຕ່ອນນ້ອຍໆ, ຕົ້ມ ໃນນ້ຳສອງລິດ ແລະ ດື່ມນ້ຳຕົ້ມ). Soejarto et al. 14928.

ຕິນຈຳຂົນ – Tin Cham Khon. Sekong MBP. Herb, 50 cm in height, entire plant covered with small hairs, leaves alternate, margins slightly crenate, the fruits red, glabrous. The root to treat fever (ໃຊ້ຮາກບໍ່ໄຂ້). Boil the root in water and then drink the decoction as needed (ຕົ້ມຮາກໃນນ້ຳ ແລະ ດື່ມນ້ຳຕົ້ມຕາມຕ້ອງການ). Elkington et al. 268.

Ardisia humilis Vahl. Figure 5.51

ຕິນຈຳກ້ຽງ – Tin Cham Kieng. Bolikhamxay MPP. Treelet 4 m tall, flowers pink with green pedicel, disposed on axillary and terminal umbellate clusters, fruits green, plant glabrous. The root is used to make a medicine to treat cases of urine retention (as a diuretic) and kidney stones (ໃຊ້ຮາກບໍ່ໃນກໍລະນີທີ່ຖ່າຍເບົາບໍ່ ສະດວກ (ເປັນຍາຂັບປັດສະວະ) ແລະ ບໍ່ຫນ້ວອຫນາກໄຂ່ຫຼັງ). Boil pieces of root and drink the decoction (ຕົ້ມຮາກໃນນ້ຳ ແລະ ດື່ມນ້ຳຕົ້ມ). Soejarto et al. 15034.

FIGURE 5.51　The terminal end of a branch of *Ardisia humilis* treelet bearing pink flower buds set on umbellate, terminal clusters. Photo credit: D. Soejarto.

Ardisia polysticta Miq.

ກົກຕາກວາງ; ຕິນຈ່າກ້ຽງ – Kok Ta Kuang; Tin Cham Kieng. Xieng Khouang MBP. Shrub, flowers pink, fruits deep red, glossy. Root to treat rheumatism (ໃຊ້ ຮາກບໍ່ປະດົງຂໍ້). Cut the root into small pieces, boil in water and drink the decoction (ຜັກ ຊອຍ ຮາກເປັນຕ່ອນນ້ອຍໆ, ຕົ້ມ ໃນນ້ຳ ແລະ ດື່ມນ້ຳຕົ້ມ). Soejarto et al. 14993.

Ardisia villosa Roxb. Figure 5.52

ຕາຟານ; ຕິນຈ່າຂົນ – Ta Fan; Tin Cham Khon. Oudomxay MBP. Erect shrub, 0.3–0.5 cm tall, fruits green, plant hairy. The root or the stem is used to make a remedy to treat back pain (unspecified) (ໃຊ້ຮາກ ຫຼື ລຳຕົ້ນບໍ່ອອາການເຈັບບວດ (ບໍ່ໄດ້ ບອກວ່າເຈັບບ່ອນໃດ)). Boil the root or stem and drink the decoction as needed (ຕົ້ມຮາກ ຫຼື ລຳຕົ້ນໃນນ້ຳ ແລະ ດື່ມນ້ຳຕົ້ມຕາມຕ້ອງການ). Soejarto et al. 15108.

Embelia ribes Burm.f.

ເຄືອສົ້ມຂີ້ມ້ອນ – Kheua Som Khi Mohn. Xieng Khoang MBP. Liana, flowers greenish white, profuse, set on a spreading, loose terminal paniculate clusters. The root is used to prepare a medicine to treat flatulence, stomachache, and diarrhea (ໃຊ້ຮາກບໍ່ອຶ່ງທ້ອງ, ເຈັບກະເພາະອາຫານ ແລະ ທ້ອງຖອກ). Cut the root

(a)   (b)

FIGURE 5.52 The terminal end of the stem of *Ardisia villosa* shrub bearing green fruits set on axillary racemes (a) and a close-up view of the cluster of hairy fruits of *Ardisia villosa* (b). Photo credit: D. Soejarto.

into small pieces; boil and drink the decoction (ຜັກ ຂອຍ ຮາກບົດຕ່ອນນ້ອຍໆ, ຕົ້ມໃບນ້ຳ ແລະ ດົ່ມນ້ຳຕົ້ມ). Soejarto et al. 14976.

Maesa japonica (Thunb.) Moritz. ex Zoll.
ກົກຄັບນ້ອຍ – Kok Khab Noy. Oudomxay MBP. Shrub, 2 m tall, fruits brownish, profuse, in axillary, racemous clusters. Root, aerial parts to treat fever (ໃຊ້ຮາກ, ພາກສ່ວນເຫິ່ອດິນ ບໍ່ໄຂ້). Cut the root or aerial parts into small pieces and dry, then take a handful and boil in two liters of water (ຜັກ ຂອຍ ຮາກບົດຕ່ອນນ້ອຍໆ ແລະ ຕາກແຫ້ງ, ຕົ້ມປະມານໜຶ່ງກຳມື ໃນນ້ຳສອງລິດ). Drink the solution as needed (ດົ່ມນ້ຳຕົ້ມຕາມຕ້ອງການ). Soejarto et al. 15101.

Maesa perlaria (Lour.) Merr.
ກົກຄັບ - Kok Khab Xieng Khouang MBP. Shrub, fruits profuse, dull green. Common. The leaf or the root is used to prepare a tonic given to women suffering a weakness of the body and to treat cases of gonorrhea (ໃຊ້ໃບ ຫຼື ຮາກ ເປັນ ຢາບຳລຸງກຳລັງສຳລັບຜູ້ຍິງທີ່ອ່ອນເພຍ ແລະ ບໍ່ໂລກຫນອງໃນ). Boil the leaf or the root separately or together, and drink the liquid as needed (ຕົ້ມໃບ ຫຼື ຮາກ ຫຼື ປະສົມກັນ ແລະ ດົ່ມນ້ຳຕົ້ມຕາມຕ້ອງການ). Soejarto et al. 14958.

Rhamnaceae
Gouania leptostachya DC.
ເຄືອຟອດຟາວ – Kheua Fot Fao. Sekong MBP. A liana, leaves alternate, cordate, tomentose, fruits small, green, oblong in outline, 3-winged, ca. 1 cm in length

and set along slender racemes. Whole plant used as a remedy for pain and child's rashes (ໃຊ້ໝົດຕົ້ນບໍ່ອາການເຈັດບອດ ແລະ ບໍ່ໄຂ້ອອກຕຸ່ມໃນເດັກ). Boil the whole plant and use the cold decoction to wash the infected site (ຕົ້ມໝົດຕົ້ນ ແລະ ໃຊ້ນ້ຳຕົ້ມທີ່ເຢັນແລ້ວລ້າງບ່ອນທີ່ມີອາການ). Elkington et al. 284.

Ventilago denticulata Willd.

ເຄືອຂ້າແກາບ – Kheua Khao Kep. Luang Prabang MBP. Liana, stem green, with prominent multi-angled ridges lengthwise. Sterile. The liana (stem) is used to prepare medicine for diarrhea (ໃຊ້ເຄືອ (ວ່າ) ບໍ່ທ້ອງຖອກ). Cut the liana into small pieces and dry, then take a handful and boil in 2 liters of water (ຟັກ ຂວບ ເຄືອ (ວ່າຕົ້ນ) ເປັນຕ່ອນນ້ອຍໆ ແລະ ຕາກແຫ້ງ, ຕົ້ມບະມານໜຶ່ງກຳມືໃນນ້ຳສອງລິດ). Drink the decoction as needed (ດື່ມນ້ຳຕົ້ມຕາມຕ້ອງການ). Soejarto et al. 15461.

Rhizophoraceae

Carallia brachiata (Lour.) Merr.

ເຄືອຕິດຕໍ່ – Kheua Tiit Toh. Savannakhet MBP. Tree 15–20 m tall, flowers greenish white, profuse, in axillary clusters. Not common. Stem (stem bark and stem wood) is used to prepare a medicine to treat jaundice, rheumatism, and nervousness (ໃຊ້ລ່າເຄືອ (ທ້າງເປືອກ ແລະ ວ່າ) ບໍ່ຂີ້ໝາກເຫຼືອງ, ປະດົງຂໍ້ ແລະ ເສັ້ນ ປະສາດ). In all cases, cut the stem into pieces, and boil and drink the decoction (ຟັກ ຂວບ ລ່າເປັນຕ່ອນນ້ອຍໆ, ຕົ້ມໃນນ້ຳ, ດື່ມນ້ຳຕົ້ມ). Soejarto et al. 14937.

Pellacalyx yunnanensis Hu. Figure 5.53

ກົກຕັ້ງເສົາ – Kok Tang Sao. Oudomxay MPP. Tree 15–20 m tall, dbh 25 cm, bark gray-brown, coarsely fissured, slash light brown, wood white, young fruit solitary on the leaf axil, stamens epipetalous, stipules needle-shaped; prop roots present.

(a) (b)

FIGURE 5.53 The terminal end of a fruiting branch of *Pellacalyx yunnanensis* tree showing fruits set on an axillary position (a) and a close-up view of the long-pedicelled fruits of *Pellacalyx yunnanensis*, hanging from an axillary position on the twig (b). Photo credit: D. Soejarto.

A handful of small pieces of the stem (stem bark and stem wood) is decocted and the decoction given to drink as treatment of stomachache and intestinal infection (ໃຊ້ລຳຕົ້ນ (ລຳ ແລະ ເປືອກ) ທີ່ພັກ ຂຍຍແລ້ວບະມານ ທັ້ງງກຳມິ ຕົ້ມໃບນ້ຳ, ດື່ມ ນ້ຳຕົ້ມເພື່ອບໍ່ເຈັບກະເພາະ ແລະ ລຳໄສ້ຕິດເຊື້ອ). Soejarto et al. 15114.

Rosaceae
Rubus alceifolius Poir. Figure 5.54
ເຄືອມະຣູ້ຝົນ; ເຄືອກຸ່ມ – Kheua Meh Hou Fon; Keuah Toum. Xieng Khouang MBP. Straggling shrub, fruits red. Common. The root is used to prepare a medicine to treat rashes inside the throat (ໃຊ້ຮາກບໍ່ວິ້ນເປັນຕຸ່ມໃນລຳຄໍ). Macerate the root in water, take a gargle, and drink the macerate (ແຊ່ຮາກໃນນ້ຳ, ດື່ມນ້ຳ ແຊ່). Soejarto et al. 14960.

Rubus pluribrateatus L.T. Lu & Boufford. Figure 5.55a and b
ເຄືອມະຣູ້ໃຫຍ່ – Kheua Meh Hou Nyai. Xieng Khouang MBP. Straggling shrub, flower buds enclosed by a large pinnatifid bract, brownish green in color, the plant spiny. Not common. The root is used to make a medicine to treat rashes inside the throat (ໃຊ້ຮາກບໍ່ວິ້ນເປັນຕຸ່ມໃນລຳຄໍ). Macerate the root in water and drink the macerate (ແຊ່ຮາກໃນນ້ຳ, ດື່ມນ້ຳແຊ່). Soejarto et al. 14963.

FIGURE 5.54 The terminal end of a fruiting branch of *Rubus alceifolius* shrub showing red fruits set on terminal and axillary clusters. Photo credit: D. Soejarto.

(a)   (b)

FIGURE 5.55 The terminal end of a flowering branch of *Rubus pluribracteatus* shrub showing flower buds and open flowers set on a hanging, racemose cluster (a) and an enlarged portion of the cluster showing pinnatifid bracts subtending the flowers (b). Photo credit: D. Soejarto.

Rubiaceae

Chassalia curviflora (Wall.) Thw. Figure 5.56

ກົກກ້ານກ່າ; ປ້ອງຂຽວ – Kok Kan Kam; Pong Kiao. Oudomxay MBP. Shrub 2 m tall, fruits black, glossy, pedicel and peduncle pink. The root or the stem (no leaf) are used to prepare a medicine to treat cases of bleeding cough (ໃຊ້ຮາກ ຫຼື ລຳຕົ້ນ (ບໍ່ເອົາໃບ) ບໍ່ອາການໄອອອກເລືອດ). Boil the root or stem, and drink the liquid as needed (ຕົ້ມຮາກ ຫຼື ລຳຕົ້ນ, ດື່ມນ້ຳຕົ້ມ). Soejarto et al. 15070; 15133.

ປ້ອງຂຽວ; ຂີ້ໝິ້ນປູນ – Pong Kiao; Khi Min Poun. Bolikhamxay MPP. Shrub 2–3 m tall, flowers with white corolla tube and lobes, with a light-yellow ring on the throat inside. The whole plant is used as a remedy to treat postnatal weakness (as a tonic) (ໃຊ້ໝົດຕົ້ນບໍ່ສຸຂະພາບອ່ອນເພຍ (ເປັນຍາບຳລຸງກຳ ລັງ)). Cut the whole plant into small pieces, boil in water, and drink the decoction (ຟັກ ຊອຍ ໝົດຕົ້ນເປັນຕ່ອນນ້ອຍໆ, ຕົ້ມໃນນ້ຳ ແລະ ດື່ມນ້ຳຕົ້ມ). Soejarto et al. 15032.

Coffea arabica L.

ກາເຟປ່າ – Café Pa. Sekong MBP. Cultivated tree, ca. 10 m high, trunk ca. 30 cm dbh, bark gray with white spots, plant glabrous, fruits green. Stem is used to prepare a medicine to treat leprosy (ໃຊ້ລຳຕົ້ນບໍ່ພະຍາດຂີ້ທູດ). Boil a handful of small, dried pieces of stem in 1 liter of water and drink the decoction (ຕົ້ມລຳຕົ້ນ ທີ່ ຟັກ ຊອຍເປັນຕ່ອນນ້ອຍໆປະມານໜຶ່ງກຳມື ໃນນ້ຳໜຶ່ງລິດ ແລະ ດື່ມນ້ຳຕົ້ມ). Elkington et al. 302.

FIGURE 5.56   A leafy twig of *Chassalia curviflora* shrub, bearing a corymbose cluster of black shiny fruits set on pink pedicels. Photo credit: D. Soejarto.

Diplospora dubia (Lindl.) Masam.
ກົກແສນໄຊ; ຕັງເບິ່ນ້ອຍ – Kok Sen Xai; Tang Beu Noy. Oudomxay MBP. Treelet, fruits green, turning reddish red, then blackish red, glossy, set in small, extra-axillary cymes, stipule acicular-subulate. The stem or the leaf is used to make a medicine to treat cases of rheumatism (stem) or stomachache (leaf) by boiling and drinking the decoction (ໃຊ້ລຳຕົ້ນ ຫຼື ໃບບໍ່ອບະດົງຂັ້ (ລຳຕົ້ນ) ຫຼື ເຈັບ ກະເພາະ (ໃບ) ໂດຍການຕົ້ມ ແລະ ດື່ມນ້ຳຕົ້ມ). Soejarto et al. 15109.

Diplospora viridiflora DC.
ກົກແສນໄຊ – Kok Sen Xai. Oudomxay MBP. Shrub, fruits red. Root to treat rheumatism, stomachache (ໃຊ້ຮາກບໍ່ອບະດົງຂັ້, ເຈັບກະເພາະ). Cut the root into small pieces and dry, then take a handful and boil in 2 liters of water (ຫັ້ກ ຂອຍ ຮາກເປັນຕ່ອນບ້ອຍໆ, ຕົ້ມບະມານຫັ້ງໆກຳມົ ໃບນ້ຳສອງລິດ). Drink the liquid as needed (ດື່ມນ້ຳຕົ້ມຕາມຕ້ອງການ). Soejarto et al. 15130.

Gaertnera cf. *vaginans* (DC.) Merr.
ເຂັມແດງປ່າ – Khem Deng Pa. Savannakhet MBP. Small shrub, fruits red, 2-lobed, somewhat compressed laterally. Forest floor. Root to treat body weakness (as a general tonic) (ໃຊ້ຮາກບໍ່ສຸຂະພາບອ່ອນເພຍ (ເປັນຍາບຳລຸງກຳລັງທົ່ວໄປ)).

Cut the root into small pieces, boil in water, and drink the decoction (ຟັກ ຂອຍຮາກເປັນຕ່ອນນ້ອຍໆ, ຕົ້ມໃນນ້ຳ ແລະ ດື່ມນ້ຳຕົ້ມ). Soejarto et al. 14917.

Gardenia coronaria Buch-Ham.
ໄຂ່ເນົ່າ, ກົກມອກ – Khai Nao; Kok Mok. Luang Prabang MBP. Tree 15 m tall, dbh 25 cm, bark gray, smoothish, finely cracked, forming rings encircling the stem, slash light brown, fruits green, many-seeded. The stem bark is used to make a medicine for diabetes and to treat wounds (to accelerate the skin formation) (ໃຊ້ ເປືອກຕົ້ນບໍ່ອບາດແຜ (ເວົ່າໃຫ້ຫນັງບົ່ງໄວ)). Macerate small pieces of stem bark and place the macerate on the affected part (ແຊ່ເປືອກຕົ້ນທີ່ຕັດເປັນຕ່ອນນ້ອຍໆ ໃນນ້ຳ ແລະ ເອົານ້ຳແຊ່ນັ້ນໃສ່ບ່ອນທີ່ເປັນບາດແຜ). Soejarto et al. 15455.

Gardenia stenophylla Merr. Figure 5.57
ກະດອມ – Ka Dom. Bolikhamxay MPP. Treelet, fruits ellipsoid to ellipsoid-ovoid, green, with longitudinal striations. The root is used to make a medicine to treat muscle inflammation and muscle pain; it is also used to treat bad smell from the mouth (ໃຊ້ຮາກບໍ່ກ້າມຊີ້ນອັກເສບ ແລະ ເຈັບປວດກ້າມຊີ້ນ; ຍັງໃຊ້ສ່າ ລັບບໍ່ກິ່ນບາກປໍ່ດີຕົມອິກ). In either case, a handful of the dried small pieces of

FIGURE 5.57 The terminal end of a fruiting branch of *Gardenia stenophylla* treelet showing a fruit borne on a terminal cluster. Photo credit: D. Soejarto.

the root are boiled in a bowl of water and the decoction given to drink as needed (ໃນທຸກກໍລະນີ, ໃຊ້ຮາກທີ່ພັກ ເປັນຕ່ອນນ້ອຍໆ ນັ້ນຕົ້ມໃນນ້ຳ ໜຶ່ງຖ້ວຍ, ດື່ມນ້ຳ ຕົ້ມຕາມຕ້ອງການ). Soejarto et al. 15018.

Geophila repens (L.) I.M. Johnst.
ເຈັດນ້ອຍ, ຕົດໝາຕົ້ນ – Jet noy; Tod Ma Ton. Sekong MBP. Herbaceous ground cover. Approximately 2 cm in height. Leaves cordate. Fruits red. The whole plant is used to make a medicine to treat uterine infections by boiling and drinking the cooled liquid (ໃຊ້ໝົດຕົ້ນບໍ່ມົດລູກອັກເສບ ໂດຍການຕົ້ມນ້ຳ ແລະ ດື່ມນ້ຳ ຕົ້ມທີ່ເຢັນແຄ້ວ). Elkington et al. 273.

Hedyotis sp. 1.
ກົກມຸ້ງກະຕ່າຍ – Kok Moung Ka Tai. Savannakhet MBP. Herb 20–30 cm, leaves simple, opposite. Root to treat stomachache (ໃຊ້ຮາກບໍ່ເຈັບແຄ້ວ). Cut the root into small pieces and dry, then take a handful and boil in 2 liters of water and drink the solution as needed (ຟັກ ຊອຍຮາກເປັນຕ່ອນນ້ອຍໆ ແລະ ຕາກແຫ້ງ, ຕົ້ມບະມານໜຶ່ງກຳມື ໃນນ້ຳສອງລິດ ແລະ ດື່ມນ້ຳຕົ້ມຕາມຕ້ອງການ). Sydara et al. 19048.

Hedyotis sp. 2.
ຫຍ້າມຸ້ງກະຕ່າຍ – Nya Moung Ka Tai. Savannakhet MBP. Prostrate herb, fruits green. The whole plant is used to prepare a medicine to treat cases of gastritis, yellow urine, and yellow eyes (ໃຊ້ໝົດຕົ້ນບໍ່ກະເພາະອັກເສບ, ບັດສະອະຫິ ເຫຼືອງ ແລະ ຕາເຫຼືອງ). Boil the plant material and drink the liquid (ຕົ້ມພິດໝົດຕົ້ນ ແລະ ດື່ມນ້ຳຕົ້ມ). Soejarto et al. 14927.

Hedyotis sp. 3.
ເຄືອຫຍ້າມຸ້ງກະຕ່າຍ – Keuah Nya Moung Ka Tai; Nya Si Liem. Oudomxay MBP. Herbaceous plant forming an extensive ground cover, flowers pinkish white. Whole plant to treat tinea versicolor (skin disease) — ໃຊ້ໝົດຕົ້ນບໍ່ຂີ້ກ້ຽນ (ພະຍາດຜິວໜັງ). Cut the whole plant into small pieces and dry, then take a handful and boil in two liters of water; drink the liquid as needed (ຟັກ ຊອຍໝົດຕົ້ນເປັນຕ່ອນນ້ອຍໆ ແລະ ຕາກແຫ້ງ, ຕົ້ມບະມານໜຶ່ງກຳມື ໃນນ້ຳສອງລິດ ແລະ ດື່ມນ້ຳຕົ້ມຕາມຕ້ອງການ). Soejarto et al. 15087.

Ixora chinensis Lam.
ເຂັມຂາວໃຫຍ່ – Khem Khao Nyai. Luang Prabang MBP. Treelet 4 m tall, flowers pinkish white, set on dense, rounded terminal panicles. The stem or root is used to prepare a remedy to treat nervous breakdown, also cases of diarrhea (ໃຊ້ລຳຕົ້ນ ຫຼື ຮາກບໍ່ທ້ອງຖອກ). In either case, boil small pieces of the stem and drink the cooled liquid as needed (ຕົ້ມລຳຕົ້ນທີ່ພັກ ເປັນຕ່ອນນ້ອຍໆ ແລະ ດື່ມນ້ຳຕົ້ມທີ່ ເຢັນແຄ້ວຕາມຕ້ອງການ). Soejarto et al. 15423.

Ixora delpyana Pierre ex Pit.

ກົກຮາກດຽວ – Kok Hak Diao. Oudomxay MBP. Large shrub, with terminal, spreading inflorescences, fruits green. The root is used to make a medicine to treat cases of fever (ໃຊ້ຮາກບົ່ວໄຂ້). Boil dried pieces of the root and drink the decoction (ຕົ້ມຮາກແຫ້ງທີ່ຕັດເປັນຕ່ອນນ້ອຍໆ ຕົ້ມ ແລະ ດື່ມນ້ຳຕົ້ມ). Soejarto et al. 15080.

Lasianthus attenuatus Jack

ຄັນແຮ້ວນົກຂໍ່ແມ່ – Khan Heo Nok Kho Meh. Oudomxay MBP. Shrub, plant hairy, fruits blue. The root is used to prepare a medicine for dysentery (ໃຊ້ຮາກ ບົ່ວທ້ອງບິດ). Boil small pieces of root in water and drink the liquid as needed (ຕົ້ມຮາກທີ່ຕັດເປັນຕ່ອນນ້ອຍໆ, ດື່ມຕາງນ້ຳ). Soejarto et al. 15098; 15127.

Lasianthus cf. *caeruleus* Pit.

ຄັນແຫ້ວນົກຂໍ່ – Khan Heo Nok Kho. Sekong MBP. Shrub 1 m high, flowers growing from leaf nodes, pentamerous, white, pubescent, fruits blue. Root, stem to treat diabetes and many other afflictions (ໃຊ້ຮາກ, ລຳຕົ້ນ ບົ່ວເບົາຫວານ ແລະ ຫຼາຍພະຍາດ). Boil a handful of small dried pieces of root or stem in 1 liter of water and drink the decoction (ຕົ້ມຮາກ ຫຼື ລຳຕົ້ນທີ່ຜັ້ກ ຈອຍ ເປັນຕ່ອນນ້ອຍໆ ປະມານຫນຶ່ງກຳມື ໃນນ້ຳຫນຶ່ງລິດ ແລະ ດື່ມນ້ຳຕົ້ມ). Elkington et al. 301.

Lasianthus chinensis (Champ. ex Benth.) Benth.

ຄັນແຮ້ວນົກຂໍ່; ໂຄຍງົວຕົ້ນ – Khan Heo Nok Kho; Khoy Ngou Ton. Bolikhamxay MPP. Treelet, flower buds greenish white, open flowers white. The stem or the root is used to make a medicine for cases of uterine weakness (ໃຊ້ລຳຕົ້ນ ຫຼື ຮາກບົ່ວ ມົດລູກບໍ່ແຂງແຮງ). Boil the root or stem and drink the cooled liquid as needed (ຕົ້ມຮາກ ຫຼື ລຳຕົ້ນ ແລະ ດື່ມນ້ຳຕົ້ມທີ່ເຍັນແວ້ວຕາມຕ້ອງການ). Soejarto et al. 15019; 15124.

Lasianthus lucidus Bl.

ຄັນແຮ້ວນົກຂໍ່ – Khan Heo Nok Kho. Savannakhet MBP. Small shrub 1–2 m high, leaves opposite and hairy. Root to prepare a tonic medicine (ໃຊ້ຮາກບົ່ວຢາບຳລຸງສຸຂະພາບ). Cut the root into small pieces and dry, then take a handful and boil in two liters of water. Drink the solution as needed (ຜັ້ກ ຈອຍຫມົດຕົ້ມເປັນຕ່ອນນ້ອຍໆ ແລະ ຕາກແຫ້ງ, ຕົ້ມປະມານຫນຶ່ງກຳມື ໃນນ້ຳສອງລິດ. ດື່ມນ້ຳຕົ້ມຕາມຕ້ອງການ). Sydara et al. 19023.

Lasianthus verticillatus (Lour.) Merr.

ຄັນແຫ້ວນົກຂໍ່ – Khan Heo Nok Kho. Sekong MBP. Small tree, 2 m in height, stem greenish brown, leaves opposite, the underside tomentose, the fruits white, arising from leaf nodes. Root, stem to treat fever, stomachache, body weakness (as a tonic) — ໃຊ້ຮາກ, ລຳຕົ້ນ ບົ່ວໄຂ້, ເຈັບກະເພາະ, ສຸຂະພາບ

ອ່ອນເພຍ (ເປັນຢາບໍາລຸງກໍາລັງ). Boil a handful of small, dried pieces of root or stem in one liter of water and drink the decoction (ຕົ້ມຮາກ ຫຼື ລຳຕົ້ນທີ່ ຜັກ ຈຸອຍ ເປັນຕ່ອນນ້ອຍໆ ປະມານຫນຶ່ງກໍາມືໃນນ້ຳຫນຶ່ງລິດ ແລະ ດື່ມນ້ຳຕົ້ມ). Elkington et al. 296.

Mussaenda cf. aptera Pit.
ຕັງເບິເຄືອ – Tang Beu Kheua. Bolikhamxay MBP. Vine, bracts white, flower buds white. Root to treat malaria (ໃຊ້ຮາກບໍ່ໄຂ້ມາລາເຣຍ). Cut the root into small pieces, boil in water, and drink the decoction as needed (ຜັກ ຈຸອຍ ຮາກເປັນ ຕ່ອນນ້ອຍໆ, ຕົ້ມປະມານຫນຶ່ງກໍາມື ແລະ ດື່ມນ້ຳຕົ້ມຕາມຕ້ອງການ). Soejarto et al. 15027.

Mussaenda densiflora H.L. Li
ຕັງເບິເຄືອ – Tang Beu Kheua. Xieng Khouang MBP. Vine, flowers with white corolla tube, yellow corolla lobes, reddish orange on the surface. Vine is used to treat coughs (ໃຊ້ເຄືອບໍ່ໄອ). Boil a handful of small pieces of vine (stem) in a bowl of water and drink the decoction (ຕົ້ມເຄືອທີ່ຜັກ ຈຸອຍ ເປັນຕ່ອນນ້ອຍໆ ປະມານຫນຶ່ງກໍາມືໃນນ້ຳຫນຶ່ງຖ້ວຍແລະ ດື່ມນ້ຳຕົ້ມ). Soejarto et al. 14979.

Mussaenda longipetala Li. Figure 5.58
ຕັງເບິເຄືອ – Tang Beu Kheua. Savannakhet MBP. Vine, flowers with deep yellow corolla lobes, light dull green tube, and large white bract. The whole plant is used to prepare a medicine to treat wounds to accelerate the skin regeneration (ໃຊ້ໝົດຕົ້ນ ປິ່ນປວດແຜ ເພື່ອເລັ່ງໃຫ້ໜັງປົ່ງໄວ). Boil the plant and drink the decoction (ຕົ້ມພືດໃນນ້ຳ ແລະ ດື່ມນ້ຳຕົ້ມ). Soejarto et al. 14918.

Mussaenda cf. sanderiana Ridl.
ຕັງເບິເຄືອ – Tang Beu Kheua. Savannakhet MBP. A vine with opposite leaves, white flowers, and globular fruits. Root, stem to treat gonorrhea (ໃຊ້ຮາກ, ລຳ ຕົ້ນບໍ່ໂລກທານອງໃນ). Cut the root or stem into small pieces and dry, then take a handful and boil in 2 liters of water. Drink the decoction as needed (ຜັກ ຈຸອຍ ຮາກ ຫຼື ລຳຕົ້ນເປັນຕ່ອນນ້ອຍໆ, ຕົ້ມປະມານຫນຶ່ງກໍາມືໃນນ້ຳສອງລິດ ແລະ ດື່ມ ນ້ຳຕົ້ມຕາມຕ້ອງການ). Sydara et al. 19049.

Mycetia sp.
ຈຳຄາມ – Jam Kham. Bolikhamxay MPP. Shrub, flowers white, set on a loose, spreading panicle. Root to treat postnatal weakness (as a tonic) (ໃຊ້ຮາກບໍ່ສຸຂະພາບອ່ອນເພຍ ຫຼັງຈາກເກີດລູກ (ເປັນຢາບຳລຸງແມ່ຍິງ)). Cut the root into small pieces, boil in water, and drink the decoction (ຜັກ ຈຸອຍ ຮາກເປັນ ຕ່ອນນ້ອຍໆ, ຕົ້ມ ໃນນ້ຳສອງລິດ ແລະ ດື່ມນ້ຳຕົ້ມ). Soejarto et al. 15013.

FIGURE 5.58 The terminal end of a flowering stem of *Mussaenda longipetala* vine, bearing flowers set on a terminal cluster, with each of the lower flower subtended by a large white bract. Photo credit: D. Soejarto.

Oldenlandia microcephala Pierre ex Pit.
ມຸ້ງກະຕ່າຍຕົ້ນ – Moung Ka Tai Ton. Bolikhamxay MPP. Free-standing herb on forest floor, flowers set in axillary glomerules. The leaf is used to prepare a remedy to treat hemorrhoids and cases of fever (ໃຊ້ໃບບໍ່ວິດສິດຂອງຫະວານ ແລະ ບໍ່ໄຂ້). Take 1 handful of fresh leaves and pound to make a paste (ເອົາໃບສົດປະມານໜຶ່ງ ກໍານົື ຕໍາ ແລະ ເຮັດເປັນຍາກ້ອນບຸງກຽງ). Apply on the anus areas for hemorrhoid (ເອົາ ຍາກ້ອນບຸງກຽງນັ້ນແປະໃສ່ຮູຫະວານເພື່ອບໍ່ວິດສິດຂອງ). For fever, apply on the body (ສໍາລັບບໍ່ໄຂ້ແມ່ນແປະໃສ່ຕົນໂຕ). Soejarto et al. 15007.

Oldenlandia sp.
ຫຍ້າມຸ້ງກະຕ່າຍ – Nya Moung Ka Tai. Savannakhet MBP. Herb, flowers pale green, in globular cluster at nodes, axillary. Whole plant to treat gastritis, yellow urine, yellow eyes (ໃຊ້ໝົດຕົ້ນບໍ່ອກະເພາະອັກເສບ, ຍ່ຽວເຫຼືອງ, ຕາເຫຼືອງ). In all

cases, boil a handful of small, dried pieces of the whole plant in 1 liter of water and drink the decoction (ໃບທຸກກຳລະບິ, ຕົ້ມໝົດຕົ້ນທີ່ພັກ ຊອຍ ເປັນຕ່ອນນ້ອຍ ໆ ໃນນ້ຳທັງ1ລິດ ແລະ ດື່ມນ້ຳຕົ້ມ). Soejarto et al. 14936.

Oxyceros horridus Lour.

ເຄືອຂັດເຄົ້າ – Kheua Khat Khao. Oudomxay MBP. Liana with thorny hooks, fruits green, finely hairy. The liana (stem) is used to treat leucorrhea (ໃຊ້ເຄືອ (ວ່າຕົ້ນ) ບໍ່ລົງຂາວ). Cut the vine (stem) into small pieces and dry, then take a handful and boil in 2 liters of water and drink the solution as needed (ຜ່າ ຊອຍ ເຄືອ (ວ່າ) ເປັນຕ່ອນນ້ອຍໆ ແລະ ຕາກແຫ້ງ, ຕົ້ມບະມານທັງງກຳມືໃນນ້ຳສອງລິດ ແລະ ດື່ມນ້ຳຕົ້ມຕາມຕ້ອງການ). Soejarto et al. 15110.

ເຄືອຂັດເຄົ້າ, ຂັດເຄົ້າໜາມຊຶ – Kheua Khat Khao; Khat Khao Nam Seu. Savannakhet MBP. Vine-like shrub, with hard, curving spines, fruits green. The root or stem is used to treat many health ailments, including bellyache. Boil a handful of small pieces of the root or stem in a bowl of water and drink the decoction as needed (ໃຊ້ຮາກ ຫຼື ວ່າຕົ້ນບໍ່ຫຼາຍພະຍາດ ເຊັ່ນ ເຈັບທ້ອງ ໂດຍການຕົ້ມ ພົດທີ່ຕັດເປັນຕ່ອນນ້ອຍໆປະມານ ທັງງກຳມືໃນນ້ຳທັງງຖ້ວຍ ແລະ ດື່ມນ້ຳຕົ້ມ). Soejarto et al. 14910.

ຂັດເຄົ້າ – Khat Khao. Sekong MBP. Liana with thorns, leaves opposite, glabrous, inflorescence a rounded umbel. Root, stem to treat many health complaints (ໃຊ້ຮາກ, ວ່າຕົ້ນ ບໍ່ການເຈັບປ່ວຍຫຼາຍຢ່າງ). Boil a handful of small, dried pieces of root or stem in 1 liter of water and drink the decoction (ຕົ້ມຮາກ ຫຼື ວ່າຕົ້ນທີ່ພັກ ຊອຍ ເປັນຕ່ອນນ້ອຍໆປະມານທັງງກຳມືໃນນ້ຳທັງງລິດ ແລະ ດື່ມນ້ຳ ຕົ້ມ). Elkington et al. 298.

Paederia foetida L. Figure 5.59

ເຄືອຕົດໝາກ້ຽງ – Kheua Tot Ma Kieng. Oudomxay MBP. Vine, flowers dull pink, with a violet throat. The whole plant is used to make a remedy for stomachache and leucorrhea by boiling and drinking the decoction (ໃຊ້ໝົດຕົ້ນ ບໍ່ເຈັບກະເພາະ ໂດຍການຕົ້ມ ແລະ ດື່ມນ້ຳຕົ້ມ). It is also used to kill the bitter taste of a substance by chewing it (ຍັງໃຊ້ເພື່ອລະງັບລົດຂົມ ໂດຍການຄ້ຽວພົດນີ້). Soejarto et al. 15085.

ຕົດໝາໃບກາງ – Tot Ma Bai Kang. Luang Prabang MBP. Vine, young fruits green, mature fruits green. Whole plant to treat heart disease (ໃຊ້ໝົດຕົ້ນບໍ່ ພະຍາດທ່ຽວໃຈ). Cut the whole plant into small pieces and dry, then take a handful and boil in 2 liters of water (ຜ່າ ຊອຍ ໝົດຕົ້ນ ເປັນຕ່ອນນ້ອຍໆ ແລະ ຕາກ ແຫ້ງ, ຕົ້ມບະມານທັງງກຳມືໃນນ້ຳສອງລິດ). Drink the decoction as needed (ດື່ມ ນ້ຳຕົ້ມຕາມຕ້ອງການ). Soejarto et al. 15437.

ເຄືອຕົດໝາ – Kheua Tot Ma. Xieng Khouang MBP. Vine, flowers with gray tube and purplish red corolla cup. Aerial parts and root of the plant are used to treat stomachache and to kill a bitter taste (ໃຊ້ສ່ວນເໜືອດິນ ແລະ ຮາກ ບໍ່ເຈັບ

FIGURE 5.59 The terminal end of a flowering stem of *Paederia foetida* vine, bearing flowers set on a spreading terminal paniculate cluster. Photo credit: D. Soejarto.

ກະເພາະ ແລະ ກຳຈັດລົດຂົມ). Boil the plant and drink the decoction. For killing a bitter taste of a substance, chew the fresh plant (ຕົ້ມພົດ ແລະ ດື່ມນ້ຳຕົ້ມ. ສຳລັບ ກຳຈັດລົດຂົມ, ຫຍ້າພົດສົດ). Soejarto et al. 14978.

Pavetta indica L.
ຕ້າງເບິ່ນ້ອຍ – Tang Beu Noy. Oudomxay MBP. Treelet, fruits in dense, terminal paniculate cluster, dark green, finely pubescent. The root is used to make a medicine to treat cases of body swellings and edema (ໃຊ້ຮາກເພື່ອບໍ່ອຕົ້ນໂຕໄຄ່ບວມ ແລະ ພິກບວມ). Boil the root and drink the decoction (ຕົ້ມຮາກ ແລະ ດື່ມນ້ຳ ຕົ້ມ). Soejarto et al. 15120.

Psychotria cf. *fleuryi* Pit.

ປ້ອງຂຽວ – Pong Kiao. Bolikhamxay MPP. Shrub 2 m tall, flower buds white, fruits green set on a dense corymb. Whole plant to treat postpartum weakness (as a tonic) (ໃຊ້ໝົດຕົ້ນບໍ່ສຸຂະພາບອ່ອນເພຍຫຼັງຈາກເກີດລູກ (ບຳລຸງແມ່ກຳ)). Cut the whole plant into small pieces, boil in water, and drink the decoction (ພັກ ຊອຍ ໝົດຕົ້ນເປັນຕ່ອນນ້ອຍໆ, ຕົ້ມໃນນ້ຳ ແລະ ດື່ມນ້ຳຕົ້ມ). Soejarto et al. 15008.

Psychotria cf. *manillensis* Bartl. ex DC.

ຈວງຄານ, ເຕີ່ມກາວອອງ – Joung Khan, Teum Kao Wang. Sekong MBP. Small tree, leaf glabrous, margins smooth, 10 cm long, 4 cm wide, fruits red, glabrous, set on a round umbel. Leaf to treat weakness (as a tonic) — ໃຊ້ໃບບໍ່ສຸຂະພາບ ອ່ອນເພຍ (ເປັນຢາບຳລຸງ). Boil 1 handful of the dried leaves with 1.5 liters of water and drink the liquid (ຕົ້ມໃບແຫ້ງປະມານໜຶ່ງກຳມື ໃນນ້ຳໜຶ່ງລິດເຄິ່ງ ແລະ ດື່ມນ້ຳຕົ້ມ). Elkington et al. 264.

Psychotria cf. *sarmentosa* Blume. Figure 5.60

ເຂັມຂາວເຄືອ – Khem Khao Kheua. Bolikhamxay MPP. Liana, fruits globular, dull bluish gray, glossy, disposed on a terminal panicle, with side branches arising perpendicular to the main axis. The root is used as a medicine in cases of menstrual disorders (ໃຊ້ຮາກບໍ່ປະຈຳເດືອນບໍ່ປົກກະຕິ). Boil 1 handful of dried small pieces of the root in 1 liter of water and drink the decoction (ຕົ້ມຮາກທີ່ຕັດເປັນ ຕ່ອນນ້ອຍໆປະມານໜຶ່ງກຳມື ໃນນ້ຳໜຶ່ງລິດ, ດື່ມນ້ຳຕົ້ມ). Soejarto et al. 15044.

ເຂັມຂາວ – Khem Khao. Savannakhet MBP. Shrubby, fruits green, profuse on axillary cluster. Root to treat body weakness (as a general tonic) — ໃຊ້ຮາກບໍ່ສຸຂະພາບອ່ອນເພຍ (ເປັນຢາບຳລຸງ). Boil a handful of small pieces of root in a bowl of water and drink the decoction (ຕົ້ມຮາກທີ່ພັກ ຊອຍ ເປັນຕ່ອນນ້ອຍ ໃນນ້ຳໜຶ່ງຖ້ວຍ ແລະ ດື່ມນ້ຳຕົ້ມ). Soejarto et al. 14912.

Psychotria sp.

ເຂັມຂາວ – Khem Khao. Savannakhet MBP. Shrub, fruits green, profuse on axillary cluster. Root to treat body weakness (as a general tonic) — ໃຊ້ຮາກບໍ່ສຸຂະພາບອ່ອນເພຍ (ເປັນຢາບຳລຸງ). Boil a handful of small pieces of root in a bowl of water and drink the decoction (ຕົ້ມຮາກທີ່ພັກ ຊອຍ ເປັນຕ່ອນ ນ້ອຍ ໃນນ້ຳໜຶ່ງຖ້ວຍ ແລະ ດື່ມນ້ຳຕົ້ມ). Soejarto et al. 14912.

Rothmannia cf. *vietnamensis* Tirv.

ໝາກໝໍ້ – Mak Mo. Savannakhet MBP. Tree 8 m tall, flower buds gray-green, fruits globular, green. Common. Root, stem to treat bone and nerve pain (ໃຊ້ຮາກ, ລຳຕົ້ນ ບໍ່ເຈັບກະດູກ ແລະ ເຈັບເສັ້ນປະສາດ). Boil a handful of small pieces of root or stem in a bowl of water and drink the decoction (ຕົ້ມຮາກ ຫຼື ລຳຕົ້ນທີ່ພັກ ຊອຍ ເປັນຕ່ອນນ້ອຍ ໃນນ້ຳໜຶ່ງຖ້ວຍ ແລະ ດື່ມນ້ຳຕົ້ມ). Soejarto et al. 14906.

FIGURE 5.60 The terminal end of a flowering stem of *Psychotria* cf. *sarmentosa* liana, bearing fruits and flower buds set on terminal umbels, in turn borne on the terminal ends of the perpendicularly positioned lateral branches of the panicle. Photo credit: D. Soejarto.

Rubia argyi (H. Lév. & Vaniot) Hara ex Lauener

ເຄືອລິ້ນໝາໃນ – Kheua Lin Ma Nai. Xieng Khouang MBP. Vine with fine spiny bristles, flowers small, pale green, fruits green. Not common. The whole plant is used to prepare a medicine to treat kidney stones, backaches, and tonsilitis (ໃຊ້ໝົດຕົ້ນບໍ່ເໝື້ອໝາກໄຂ່ຫຼັງ, ເຈັບແອວ ແລະ ອັກເສບວ່າຄໍ). Boil entire plant and drink the decoction (kidney stones, backaches), or use as a gargle with salt (in case of tonsilitis) (ຕົ້ມພິດທ້າງໝົດໃນນ້ຳ ແລະ ດື່ມນ້ຳ (ສຳລັບບໍ່ເໝື້ອໝາກ ໄຂ່ຫຼັງ, ເຈັບແອວ) ຫຼື ໃຊ້ກ້ວຄໍ ໂດຍປະສົມກັບເກືອເລັກໜ້ອຍ (ບໍ່ອັກເສບວ່າຄໍ)). Soejarto et al. 14961.

Saprosma inaequilongum Pierre ex Pit.

ຕົດໝາຕົ້ນ – Tot Mah Tohn. Sekong MBP. Small tree, 2 m tall, stem thin, leaf margins smooth, fruits globular, glabrous, blue. When the leaves are crushed, it produces a foul odor. Root to treat stomachache (ໃຊ້ຮາກບົ່ວເຈັບກະເພາະ). Boil the dried root and drink the decoction (ຕົ້ມຮາກແຫ້ງ ແລະ ດື່ມນ້ຳຕົ້ມ). Elkington et al. 275.

Tarenna cf. *thorelii* Pit.

ກົກຜ້າຮ້າຍນ້ອຍ – Kok Pa Hai Noy. Oudomxay MBP. Shrub 2 m tall, fruits green. Stem bark to treat cases of toothache (ໃຊ້ເປືອກຕົ້ນ ບົ່ວເຈັບແຂ້ວ). Cut the stem bark into small pieces and dry, then take a handful and boil in 2 liters of water. Use the liquid as a gargle and a drink (ຟັກ ຊອຍ ເປືອກຕົ້ນເປັນຕ່ອນ ນ້ອຍ ແລະ ຕາກໃຫ້ແຫ້ງ, ຕົ້ມໃນນ້ຳສອງລິດ ແລະ ດື່ມນ້ຳຕົ້ມ ກ້ວດ ແລະ ດື່ມ). Soejarto et al. 15122.

Tarennoidea wallichii (Hook.f.) Tirveng. & Sastre. Figure 5.61

ເຂັມຂາວ – Khem Khao. Savannakhet MBP. Shrub, fruits glossy-green, turning deep red and glossy. Common plant. The root is used to make a medicine as an antidote to intoxication (ໃຊ້ຮາກເພື່ອແກ້ພິດ ຖອນພິດ). Boil the root and drink the decoction (ຕົ້ມຮາກ ແລະ ດື່ມນ້ຳຕົ້ມ). Soejarto et al. 14939.

FIGURE 5.61 A leafy twig of *Tarrenoidea wallichii* bearing fruits set on a small, axillary paniculate cluster. Photo credit: D. Soejarto.

Uncaria cf. *sessilifructus* Roxb.

ເຄືອເລັບຮຸ້ງຍອດຂາວ – Kheua Lep Houng Yod Khao. Oudomxay MBP. Liana, hooks subcircular, plant sterile. Healer said the flower is white. Liana (stem), root to treat stomachache (ໃຊ້ເຄືອ (ວ່າ), ຮາກ ບໍ່ເຈັບກະເພາະ)). Boil 1 handful of dried small pieces of the liana/stem or root in 1 liter of water, then drink the decoction (ຕົ້ມເຄືອ ຫຼື ຮາກທີ່ຟັກ ຂອຍ ເປັນຕ່ອນນ້ອຍ ໃນນ້ຳທັ່ງ1ລິດ ແລະ ດື່ມ ນ້ຳຕົ້ມ). Soejarto et al. 15066.

Urophyllum sp.

ເນັ້ງດ້ຽວ – Neng Diao. Sekong MBP. Small tree, 1 m in height, leaves 5–6 cm in length, 3 cm in width, underside has a coarse texture. Stem, root to treat many kinds of afflictions (ໃຊ້ລຳຕົ້ນ, ຮາກບໍ່ວຫ່າຍພະຍາດ). Boil the dried root or stem and drink (ຕົ້ມຮາກແຫ້ງ ແລະ ດື່ມ). Elkington et al. 276.

Rutaceae

Acronychia pedunculata (L.) Miq.

ເປົ້າທອງ; ຂົມລະຫານຈໍ້ – Pao Tong; Khom La Wan Cho. Oudomxay MBP. Tree 10 m tall, dbh 30 cm, bark gray, fruits green. The stem or leaf is used to make a medicine to treat cases of nervous disorders by boiling the crushed stem and drinking the decoction (ໃຊ້ລຳຕົ້ນ ຫຼື ໃບ ບໍ່ພະຍາດເສັ້ນບະສາດ ໂດຍການຕົ້ມ ແລະ ດື່ມນ້ຳຕົ້ນ). Soejarto et al. 15092.

ເປົ້າທອງ; ເປົ້າຂີ້ເໝັນ; ຊົ້ມຊື້ນໃຫຍ່ – Pao Thong; Pao Khi Neh; Som Seun Nyai. Savannakhet MBP. Small tree, 3–4 m high with opposite leaves, fruits globular, yellow when ripe. Leaves, stem to treat red rashes on the body (ໃຊ້ໃບ, ລຳຕົ້ນ ບໍ່ຕຸ່ມແດງຕາມຕົນໂຕ). Macerate the leaves and stem in the water, soak the infected site, take bath with macerated liquid (ແຊ່ໃບ ແລະ ລຳຕົ້ນໃນນ້ຳ, ເອົາສ່ວນ ທີ່ມີອາການລົງແຊ່ ແລະ ອາບ ນ້ຳທີ່ໄດ້ຈາກການແຊ່ພິດນັ້ນ). Sydara et al. 19012.

ເຄືອຕາກວາງ – Kheua Ta Kuang. Savannakhet MBP. Treelet, fruits profuse, greenish yellow. Stem bark, stem wood to treat excessive flatulence (ໃຊ້ເປືອກ ຕົ້ນ, ແກ່ນລຳຕົ້ນ ບໍ່ອາການທ້ອງອືດ ທ້ອງເບັ່ງ). Cut the stem bark and stem wood into small pieces, boil in water, and drink the decoction (ຟັກ ຂອຍ ເປືອກ ຕົ້ນ ແລະ ແກ່ນລຳຕົ້ນເປັນຕ່ອນນ້ອຍໆ, ຕົ້ມໃນນ້ຳ ແລະ ດື່ມນ້ຳຕົ້ນ). Soejarto et al. 14942.

Clausena anisata (Willd.) Hook.f. ex Benth.

ເພ້ຍຟານ – Pia Fan. Oudomxay MBP. Tree 12 m tall, dbh 30 cm, bark gray, fissured, slash pale brown to white, fruits green, globular. Stem, leaves to treat nervous system disorders (ໃຊ້ລຳຕົ້ນ, ໃບ ບໍ່ອະບັບບະສາດ). Cut the root into small pieces and dry, then take a handful and boil in two liters of water and drink the solution as needed (ຟັກ ຢອຍ ຮາກເປັນຕ່ອນນ້ອຍໆ, ຕົ້ມບະມານທັ້ງໆກຳນົໃນນ້ຳ ສອງລິດ ແລະ ດື່ມນ້ຳຕົ້ມຕາມຕ້ອງການ). Soejarto et al. 15061.

ສະມັດຂົນ – Sa Mat Khon. Luang Prabang MBP. Treelet 2 m tall, flower buds green, pubescent, crushed leaves aromatic. Root, stem to treat fever and kidney stone (ໃຊ້ຮາກ, ລຳຕົ້ນ ບໍ່ໄຂ້ ແລະ ໜິ້ວໜາກໄຂ່ຫຼັ້ງ). Cut the root or stem into small pieces and dry, then take a handful and boil in two liters of water (ຟັກ ຢອຍ ຮາກ ຫຼື ລຳຕົ້ນເປັນຕ່ອນນ້ອຍໆ ແລະ ຕາກແຫ້ງ, ຕົ້ມບະມານທັ້ງໆກຳນົໃນນ້ຳ ສອງລິດ). Drink the decoction as needed (ດື່ມນ້ຳຕົ້ມຕາມຕ້ອງການ). Soejarto et al. 15427.

Clausena excavata Burm.f.
ກົກຂີ້ແບ້ – Kok Khi Beh. Luang Prabang MBP. Treelet, flowers greenish white, set on a dense terminal cluster. Crushed leaves very aromatic, citrus fragrance. The stem or the root is used to prepare a medicine to treat cases of kidney stones (ໃຊ້ລຳຕົ້ນ ຫຼື ຮາກ ບໍ່ໜິ້ວໜາກໄຂ່ຫຼັ້ງ). Boil the crushed stem or the root and drink the cooled liquid (ຕົ້ມຮາກ ແລະ ດື່ມນ້ຳຕົ້ມທີ່ເຢັນແລ້ວ). Soejarto et al. 15442.

Euodia simplicifolia Ridl.
ກໍຮ້າຍຂາວ; ຂົມລະທວານຈົ້ນ້ອຍ – Ko Ai Khao; Khom La Wan Cho Noy. Xieng Khouang MBP. Shrub, flowers white, in dense clusters, leaf simple and trifoliate, irregularly. Root to treat general body weakness (as a tonic medicine) (ໃຊ້ຮາກບໍ່ສຸຂະພາບອ່ອນເພຍ (ເປັນຍາບຳລຸງ)). Cut the root into small pieces, boil in water, and drink the decoction (ຟັກ ຢອຍ ຮາກເປັນຕ່ອນນ້ອຍໆ, ຕົ້ມບະມານທັ້ງໆກຳນົໃນນ້ຳ ແລະ ດື່ມນ້ຳຕົ້ມ). Soejarto et al. 14974.

Glycosmis parviflora (Sims) Little
ສົ້ມຊື້ນ – Som Seun. Luang Prabang MBP. Treelet 3.5 m tall, inflorescence terminal, flowers past anthesis, young fruits yellow, turning whitish to reddish, crushed leaves aromatic. Leaf to treat eye blindness; stem to treat asthma, fever, and rashes (ໃຊ້ໃບບໍ່ຕາມົ, ລຳຕົ້ນບໍ່ຫືດ, ບໍ່ໄຂ້ ແລະ ອອກຕຸ່ມຕາມຕົບໂຕ). In all cases, cut the stem or leaves into small pieces and dry, then take a handful and boil in two liters of water (ໃບຫຼຸກກໍລະນົ, ຟັກ ຢອຍ ລຳຕົ້ນ ຫຼື ໃບເປັນຕ່ອນນ້ອຍໆ ແລະ ຕາກ ແຫ້ງ, ຕົ້ມບະມານທັ້ງໆກຳນົໃນນ້ຳສອງລິດ). Drink the decoction as needed (ດື່ມ ນ້ຳຕົ້ມຕາມຕ້ອງການ). Soejarto et al. 15428; 15441; 15463.

Luvunga scandens (Roxb.) Buch-Ham.
ກຳລັງໜູ່ກ້າວ; ພະຍາຢາ; ພະຍາຮາກໄມ້ – Kam Lang Mou Kao; Pa Nya Ya; Pa Nya Hak Mai. Savannakhet MBP. Shrub 2–3 m high, fruits green in terminal clusters. Root to prepare a tonic medicine (ໃຊ້ຮາກເປັນຍາບຳລຸງ). Cut the root into

small pieces and dry, then take a handful and boil in 2 liters of water (ພັກ ຂອຍ ຮາກປັ່ນຕ່ອນນ້ອຍໆ ແລະ ຕາກແຫ້ງ, ຕົ້ມປະມານໜຶ່ງກຳມືໃນນ້ຳສອງລິດ). Drink as needed (ດື່ມນ້ຳຕົ້ມຕາມຕ້ອງການ). Sydara et al. 19053.

Micromelum minutum (G. Forst.) Wight & Arn. Figure 5.62
ກ່ອ້າຍກ່ານ; ສະມັດນ້ອຍ – Ko Ai Kan; Sa Mat Noy. Xieng Khouang MBP. Treelet 5 m tall, fruits green, crushed leaves aromatic. Common. Root to treat body weakness (as a tonic) (ໃຊ້ຮາກບໍ່ສຸຂະພາບອ່ອນເພຍ (ເປັນຢາບໍາລຸງ)). Cut the root into small pieces, boil in water, and drink the decoction (ພັກ ຂອຍ ຮາກປັ່ນຕ່ອນນ້ອຍໆ, ຕົ້ມໃນນ້ຳ ແລະ ດື່ມນ້ຳຕົ້ມ). Soejarto et al. 14954.

ສະມັດຂາວ – Sa Mat Khao. Bolikhamxay MBP. Treelet, fruits deep yellow, turning deep red, disposed on corymbose terminal clusters. The stem or the leaf is used to make a medicine to treat cases of kidney stones, poisoning, and dizziness (ໃຊ້ລຳຕົ້ນ ຫຼື ໃບບໍ່ອຫນ່ວາກໄຂ່ຫຼັງ, ຖອນພິດ ແລະ ວິນວຽນ). Cut the root into small pieces and dry, then take a handful and boil in 2 liters of water (ຕັດຮາກ ເປັນຕ່ອນນ້ອຍໆ ແລະ ຕາກໃຫ້ແຫ້ງ). Drink as needed in cases of kidney stones or as an antidote (ດື່ມນ້ຳຕົ້ມສຳລັບບໍ່ອຫນ່ວາກໄຂ່ຫຼັງ ແລະ ຖອນພິດ). Boil the leaves and use as sauna for dizziness (ຕົ້ມໃບເພື່ອຮົມອາຍນ້ຳ ສຳລັບບໍ່ອວິນ ວຽນ). Soejarto et al. 15055.

(a)   (b)

FIGURE 5.62   A corymbose cluster of the flowers at the terminal end of a flowering stem of *Micromelum minutum* treelet (a) and a close-up view of the flowers of *Micromelum minutum* (b). Photo credit: D. Soejarto.

ສະມັດຂາວ; ສະມັດນ້ອຍ – Sa Mat Khao; Sa Mat Noy. Savannakhet MBP. Small shrub 2–3 m high, leaves compound, flowers white; round fruit red or black when ripe. Stem, root to prepare a tonic and to treat fever, kidney stone (ໃຊ້ລຳຕົ້ນ, ຮາກ ເປັນຢາບຳລຸງ, ບໍ່ໄຂ້, ບໍ່ເຫງື່ອໝາກໄຂ່ຫຼັງ). Cut the stem or root into small pieces, dry, then boil in water and drink the decoction (ຟັກ ຊອຍ ຮາກເປັນຕ່ອນ ນ້ອຍໆ ແລະ ຕາກແຫ້ງ, ຕົ້ມໃນນ້ຳ ແລະ ດື່ມນ້ຳຕົ້ນ). Sydara et al. 19009.

ສະມັດຂາວ – Summat Khao. Sekong MBP. Small tree, 1.5–2 m in height, stem 4 cm in diam, leaves glabrous, margins smooth, fruits small, green, globular, set in an umbel. Stem to treat rashes and heart disease (ໃຊ້ລຳຕົ້ນບໍ່ໄຂ້ອອກ ຕຸ່ມຕາມຜີວໜັງໂຕ ແລະ ພະຍາດທ້ອໃຈ). Boil the dried stem and drink the decoction (ຕົ້ມລຳຕົ້ນທີ່ຟັກ ຊອຍ ເປັນຕ່ອນນ້ອຍໆ ແລະ ຕາກແຫ້ງ, ດື່ມນ້ຳຕົ້ນ). Elkington et al. 282.

Salicaceae

Xylosma controversa Clos. Figure 5.63

ກົກໝາກຫວດ – Kok Mak Huat. Xieng Khouang MBP. Tree 12 m tall, trunk slender, straight, 5 m to first branch, bark smooth, gray, flowers white, profuse, turning yellowish. Common. The stem or root is used to make a medicine to treat pain (unspecified) and as a tonic in nursing mother (ໃຊ້ລຳຕົ້ນ ຫຼື ຮາກ ບໍ່ເຈັບປອດ (ບໍ່ໄດ້ເຈາະຈົງໃສ່ສ່ວນໃດຂອງຮ່າງກາຍ) ແລະ ບຳລຸງແມ່ກ໌າ)). Boil small pieces of stem or root, cool the decoction, and drink as needed (ຕົ້ມລຳຕົ້ນ ຫຼື ຮາກ ທີ່ຕັດ ເປັນຕ່ອນນ້ອຍໆ, ປະໃຫ້ເຢັນ ແລະ ດື່ມ). Soejarto et al. 14952; 14996.

Sapindaceae

Dimocarpus longan Lour. Figure 5.64

ໝາກຄຳແລນ; ຄຳແລນຫຼວງ – Mak Kho Len; Kho Len Luang. Bolikhamxay MBP. Tree 20 m tall, fruits deep red with sweet tasting pulp. The stem bark is used to treat cases of poisoning (as an antidote). Cut the stem bark into small pieces and dry, then decoct 1 handful in a bowl of water; drink the decoction as needed. The stem wood is used to prepare a medicine to treat diseases of the thyroid gland (ໃຊ້ລຳຕົ້ນບໍ່ຄຳໜຸງ). A handful of small dry pieces of the stemwood is boiled in about 2 liters of water (ຕົ້ມລຳຕົ້ນທີ່ຕັດເປັນຕ່ອນນ້ອຍໆປະມານໜຶ່ງກຳມື ໃນນ້ຳ 2 ລິດ). Drink the cooled decoction as needed (ດື່ມນ້ຳຕົ້ນທີ່ເຢັນແລ້ວຕາມ ຕ້ອງການ). Soejarto et al. 15005; 15040.

Lepisanthes rubiginosa (Roxb.) Leenh.

ຫວດເຕ້ຍ – Huat Tia. Savannakhet MBP. Small tree, 1–2 m tall. The root is used to prepare a tonic and treat stomachache (ໃຊ້ຮາກເປັນຢາບຳລຸງກຳລັງ). Boil the root and drink the liquid as needed (ຕົ້ມຮາກ ແລະ ດື່ມນ້ຳຕົ້ນຕາມຕ້ອງການ). Sydara et al. 19005.

(a)

(b)

FIGURE 5.63   A flowering branch of a *Xylosma controversa* tree bearing profuse white small flowers (a) each set on racemose clusters and a close-up view of the flower of *Xylosma controversa* (b). Photo credit: D. Soejarto.

Saururaceae

Houttuynia cordata Thunb.

ຜັກຄາວທອງ – Pak Khao Tong. Luang Prabang MBP. Small herb on waterlogged terrains at base of slope, sterile, in stands. Whole plant to treat hot fever (ໄຂ້ໜັດ

FIGURE 5.64   A fruiting twig of *Dimocarpus longan* tree, bearing red, tubercled fruits.
Photo credit: D. Soejarto.

ຕົ້ນບໍ່ໄຂ້ຮ້ອນ). Macerate the fresh whole plant in the water and drink the macerated liquid as needed (ແຊ່ໝົດຕົ້ນໃນນ້ຳ ແລະ ດື່ມນ້ຳທີ່ໄດ້ຈາກການ ແຊ່ຕາມ
ຕ້ອງການ). Soejarto et al. 15467.

Simaroubaceae

Brucea javanica (L.) Merr.

ກົກບີຄົນ; ເພ້ຍຟານ – Kohk Bii Khon; Pia Fan. Savannakhet MBP. Small shrub
on forest floor, flowers disposed along narrow panicles. Leaves to treat rashes, itch
on body (ໃຊ້ໃບບໍ່ໄຂ້ອອກຕຸ່ມຕາມຕົນໂຕ, ຄັນຕາມຜິວໜັງ). Cut the leaves
into small pieces, dry, then boil in water and drink the decoction or use the liquid
to wash the affected part (ຟັກ ຽອຍ ໃບເປັນຕ່ອນນ້ອຍໆ, ຕາກ, ຕົ້ມໃນນ້ຳ ແລະ
ໃຊ້ນ້ຳຕົ້ມນີ້ລ້າງບ່ອນທີ່ມີອາການ). Sydara et al. 19010.

ກົກບີຄົນ – Kohk Bii Khon. Sekong MBP. Tree, approx. 3 m high, trunk 20 cm
dbh; the leaves pilose, alternate, pinnate, with leaflets opposite, crenate; axillary
inflorescence a spike, flowers minuscule, brown to orange. The root or stem is
used to make a medicine to treat cases of diabetes and malaria (ໃຊ້ຮາກ ຫຼື ລຳຕົ້ນ
ບໍ່ເບົາທານ ແລະ ໄຂ້ມາລາເຣຍ). Boil the root or the stem and drink the decoction (as needed) (ຕົ້ມຮາກ ຫຼື ລຳຕົ້ນ ແລະ ດື່ມນ້ຳຕົ້ມ). Elkington et al. 315.

Eurycoma longifolia Jack

ຍິກບໍ່ຕຍໆ – Yik Bo Tong. Savannakhet MBP. Treelet with slender and straight stem, fruits green, semi-glossy. Rather common. Root to prepare a medicine to treat body weakness (as a general tonic) (ໃຊ້ຮາກບໍ່ສຸຂະພາບອ່ອນເພຍ (ເປັນຍາບຳລຸງ)). Cut the root into small pieces, boil in water, and drink the decoction (ຜ້າ ຂອຍ ຮາກເປັນຕ່ອນນ້ອຍໆ, ຕາກ, ຕົ້ມໃນນ້ຳ ແລະ ດື່ມນ້ຳຕົ້ມ). Soejarto et al. 14924.

Harrisonia perforata (Blanco) Merr. Figure 5.65

ກົກກົນທາ – Kok Konh Ta. Savannakhet MBP. Small shrub with pinnately compound leaves, flowers with red petals outside, and white stamens. The fruit is used to prepare a medicine for itching rashes (ໃຊ້ໝາກເພື່ອບໍ່ຕຸ່ມຄັນ). Grill the

FIGURE 5.65 The terminal end of a branch of *Harrisonia perforata* shrub, bearing flowers set on a terminal paniculate cluster on each twig. Photo credit: K. Sydara.

fruits, then crush and apply on the itching skin (ເອົາໝາກຂາງໄພ, ຕຳ ແລະ ຖູໃສ່ ບ່ອນຄັນ). Sydara et al. 19008.

Staphyleaceae

Dalrympelea pomifera Roxb.

ກົກດູກ; ຕ້າງຕໍ່; ສາມສີບສອງໂຕສັດ – Kok Douk; Tang Toh, Sam Sip Song Toh Sat. Oudomxay MBP Tree 7 m tall, fruits green, globular. The leaf and stem are used to make a medicine to treat broken bones (leaf) and rheumatism (stem) by boiling and drinking (in both cases) the cooled solution as needed (ໃຊ້ໃບ ແລະ ລ່າຕົ້ນບໍ່ອກະດູກຫັກ (ໃບ) ແລະ ບໍ່ປະດົງ (ລ່າຕົ້ນ) ໂດຍການຕົ້ມ ແລະ ດື່ມນ້ຳຕົ້ມທີ່ເຍັນ (ທັ້ງສອງກໍລະນີ)). Soejarto et al. 15113.

ສິບສອງລາສີ – Sip Song Lassi. Sekong MBP. Large tree, 10–15 m in height, 50 cm dbh; leaves simple, glabrous, slightly crenate, flowers small borne on paniculate inflorescence. Fruit, stem, or root to treat liver diseases, skin diseases, and "12 other" diseases (ໃຊ້ໝາກ, ລ່າຕົ້ນ, ຮາກ ບໍ່ພະຍາດຕັບ, ໂລກຜິວໜັງ ແລະ ອີກ 12 ພະຍາດ). Boil the fruit or the root or the stem in water and drink the decoction (ຕົ້ມໝາກ ຫຼື ຮາກ ຫຼື ລ່າຕົ້ນໃນນ້ຳ ແລະ ດື່ມນ້ຳ). Elkington et al. 280.

Theaceae

Schima wallichii (DC.) Korth.

ກົກໝີຄາຍ; ໄມ້ໂຫ້; ບົກຄາຍໂຄ້ – Kok Mii Khai; Mai Toh; Bok Khai Koh. Xieng Khouang MBP. Tree 25 m tall, dbh 50 cm, bark gray, coarsely fissured, flowers and fruits collected from under the tree. Leaves to treat pain, stomach worms (ໃຊ້ໃບບໍ່ອາການເຈັບປວດ, ຂ້າແມ່ທ້ອງ). Cut the leaves into small pieces, boil in water, and drink the decoction (ຟັກ ຂອຍ ໃບເປັນຕ່ອນນ້ອຍໆ, ຕົ້ມໃນນ້ຳ ແລະ ດື່ມນ້ຳຕົ້ມ). Soejarto et al. 15002.

Urticaceae

Debregeasia longifolia (Burm.f.) Wedd.

ຊ້າປ່ານ, ປ່ານປ່າ – Xa Pan, Pan Pa. Oudomxay MBP. Shrub, upper leaf surface green, lower surface grayish white; flowers minute, greenish, in globular glomerules, in turn arranged in branched clusters. The whole plant is used to make a medicine to treat cases of fever and red rashes on the skin (ໃຊ້ໝົດຕົ້ນເພື່ອບໍ່ໄຂ້ ແລະ ມີຕຸ່ມແດງເທິງຜິວໜັງ). In either case, boil the root and drink the decoction as needed (ຕົ້ມຮາກ ແລະ ດື່ມນ້ຳຕົ້ມຕາມຕ້ອງການ). Soejarto et al. 15076.

Debregeasia squamata King ex Hook.f.

ກົກສະບູ; ຊ້າປ່ານຂົນ – Kok Sa Bou; Sa Pan Khon. Oudomxay MBP. Shrub, flowers minute, greenish white, set in a dense, globular, axillary clusters. The leaves are used to wash dirty clothes and body parts—ໃຊ້ໃບລ້າງສິ່ງສົກກະປົກອອກຈາກ ເຄື່ອງນຸ່ງ ແລະ ຮ່າງກາຍ (ເປັນສະບູ). Crush the fresh leaves in the water to make

the foam. Use this liquid to wash the dirty clothes and body parts (ຍ່ອງໃບສົດ
ໃນນ້ຳໃຫ້ເກີດພອງ. ໃຊ້ນ້ຳນີ້ລ້າງສິ່ງສົກກະປົກ). Soejarto et al. 15134.

Elatostema cyrtandrifolium (Zoll. & Moritzi) Miq.
ໂສມດິນ; ຫ້ອມຫ້ວຍ – Som Dinh; Hom Houay. Oudomxay MBP. Semi-succulent
herb on forest floor, gregarious, flowers minute, white, in pale green axillary
clusters, stipules prominent, green. Leaves, flowers to treat cases of swollen leg
and other edemas (ໃຊ້ໃບ, ດອກ ບໍ່ຂາໄຕ່ ແລະ ອາການບວມອື່ນໆ). Boil 1 handful of dried small pieces of the flowers or leaves in 1 liter of water and drink
the decoction (ຕົ້ມໃບ ຫຼື ດອກທີ່ຫ້ຽກ ຂອຍເປັນຕ່ອນນ້ອຍໆ ປະມານຫຶ່ງກຳມື
ໃນນ້ຳຫຶ່ງລິດ, ດື່ມນ້ຳຕົ້ມ). Soejarto et al. 15072.

Oreocnide integrifolia (Gaud.) Miq. Figure 5.66
ກົກຮ້ອຍປາ, ສ້ານນ້ຳ – Kok Hoy Pa; San Nam. Oudomxay MBP. Tree 6 m tall,
branches pendulous, liana-like, flowers minute, whitish green, in dense clusters
along branches and twigs, the clusters in turn arranged in a cymose clusters. The
stem is used to make a medicine to treat cases of fever (ໃຊ້ລຳຕົ້ນບໍ່ໄຂ້). Boil and
drink the decoction (ຕົ້ມນ້ຳ ແລະ ດື່ມນ້ຳຕົ້ມ). Soejarto et al. 15117.

Vitaceae
Cayratia japonica (Thunb.) Gagn.
ເຄືອເມືອກ, ຫຍ້າປົກຕໍ – Kheua Muak; Nya Pok Toh. Xieng Khouang MBP.
Vine, flowers yellow, fruits green, cups green. The vine (stem) is used to make
a medicine to treat intestinal infection (disease) by boiling and drinking the
cooled decoction (ໃຊ້ເຄືອບໍ່ພະຍາດລຳໄສ້ (ລຳໄສ້ຕິດເຊື້ອ) ໂດຍການຕົ້ມ ແລະ
ດື່ມນ້ຳຕົ້ມທີ່ເຍັນແລ້ວ)). Soejarto et al. 14990.

Leea aequata L.
ຄອນແຄນ, ຕ້າງໄກ່ຂາວ – Khone Khen; Tang Kai Khao. Oudomxay MBP. Large
liana-like shrub, leaf pinnate to bipinnate, fruits dull green, in dense clusters. Root
to treat uterus infection (ໃຊ້ຮາກບໍ່ມົດລູກ). Cut the root into small pieces and
dry, then take a handful and boil in 2 liters of water (ຟັກ ຂອຍ ຮາກເປັນຕ່ອນ
ນ້ອຍ, ຕາກ, ຕົ້ມປະມານຫຶ່ງກຳມືໃນນ້ຳສອງລິດ). Drink the liquid as needed
(ດື່ມນ້ຳຕົ້ມຕາມຕ້ອງການ). Soejarto et al. 15131.

Leea indica (Burm.f.) Merr.
ຕ້າງໄກ່ຂາວ, ກໍຂ້າງຂໍ – Tang Kai Khao; Ko Khang Kho. Xieng Khouang MBP.
Large vine, flower buds dull green, leaves pinnately compound. The root is used
to treat rheumatism (ໃຊ້ຮາກບໍ່ປະດົງ). Cut the root into small pieces, boil in
water, and drink the decoction (ຟັກ ຂອຍ ຮາກເປັນຕ່ອນນ້ອຍ, ຕົ້ມປະມານຫຶ່ງ
ກຳມືໃນນ້ຳ ແລະ ດື່ມນ້ຳຕົ້ມ). Soejarto et al. 14995.

(a)

(b)

FIGURE 5.66 A stem of *Oreocnide integrifolia* treelet with the lateral branches each coated by tiny brownish green fruitlets (a) and a close-up view of the fruitlet clusters of *Oreocnide integrifolia* infructescence (b). Photo credit: D. Soejarto.

Tetrastigma crassipes Pl.

ເຄືອຂົ້າແປ້ນ – Kheua Khao Pen. Oudomxay MBP. Large liana, stem flattish, segmented, fruits profuse along stem, globular, dark brownish green, granular on the surface, in globular clusters. The stem is used to make a medicine to treat gastric pain (stomachache) by boiling small pieces of dried stem and drinking the decoction (ໃຊ້ລຳບົດຈົບກະເພາະລຳໄສ້ ໂດຍການຕົ້ມ ເຄືອທີ່ຕັດເປັນຕ່ອນ ນ້ອຍໆ ແລະ ດື່ມນ້ຳຕົ້ມ). Soejarto et al. 15077.

Tetrastigma erubescens Pl. Figure 5.67
ເຄືອຫຸນ – Kheua Houn. Savannakhet MBP. Vine, fruits green to greenish yellow. The root is used to prepare medicine to treat boils and other abscesses (ໃຊ້ຮາກບໍ່ຕຸມຝີ ແລະ ຝີໜອງອື່ນໆ). Cut the root into small pieces, boil in water, and drink the decoction (ຟັກ ຊອຍ ຮາກເປັນຕ່ອນນ້ອຍ, ຕົ້ມປະມານເຄິ່ງກ່ຳມືໃນນ້ຳ ແລະ ດື່ມນ້ຳຕົ້ມ). Soejarto et al. 14947.

ເຄືອຫຸນ; ເຄືອຂີ້ເຂັບ; ເຄືອໄຂ່ເຕົ່າ – Kheua Houn; Kheua Khi Khep; Kheua Khai Tao. Xieng Khouang MBP. Vine, fruits deep red. Common. The whole plant is used to make a medicine for boils and other abscesses, skin disease, and rashes (ໃຊ້ໝົດເຄືອ ບໍ່ຕຸມຝີ ແລະ ຝີໜອງ, ໂລກຜິວໜັງ ແລະ ຄັນ ຄາຍ). Boil the whole plant in plenty of water and take a bath with the warm

FIGURE 5.67 *Tetrastigma erubescens* vine bearing an axillary racemose cluster of red fruits. Photo credit: D. Soejarto.

decoction (ຕົ້ມໝົດຕົ້ນຕົ້ມໃບນ້ຳຫຼາຍໆ ເພື່ອອາບ ເມື່ອເວລານ້ຳຕົ້ມອຸ່ນໆ). Soejarto et al. 15000.

Tetrastigma harmandii Pl.
ເຂົາແປ່ນ – Khao Pen. Luang Prabang MBP. Vine, fruits green, set on a dense, axillary, umbellate cluster. The vine (stem) is used to make a medicine to treat rheumatism by boiling and drinking the decoction (ໃຊ້ເຄືອ ບໍ່ວະດົງຂໍ່ ໂດຍການ ຕົ້ມນ້ຳ ແລະ ດື່ມນ້ຳຕົ້ມ). Soejarto et al. 15469.

Tetrastigma cf. *lanceolarium* (Roxb.) Pl.
ເຄືອໝາກໄຂ່ເຕົ່າ – Kheua Mak Khai Tao. Oudomxay MBP. Liana, stem flat, fruits profuse, dull green, in hanging clusters. Liana (stem) to treat intestinal infection (ໃຊ້ເຄືອ (ວ່າຕົ້ນ) ບໍ່ພະຍາດກະເພາະວ່າໄສ້ເປັນບາດ). Cut the liana (stem) into small pieces and dry (if not dry, may cause itching), then take a handful and boil in 2 liters of water (ຟັກ ຊອຍ ເຄືອ (ວ່າຕົ້ນ) ເປັນຕ່ອນນ້ອຍໆ, ຕາກໃຫ້ແຫ້ງ (ຖ້າບໍ່ແຫ້ງຈະຄັນຄໍ), ຕົ້ມປະມານນຶ່ງກຳມືໃນນ້ຳສອງລິດ). Drink the decoction as needed (ດື່ມນ້ຳຕົ້ມຕາມຕ້ອງການ). Soejarto et al. 15105.

Tetrastigma quadrangulum Gagn. & Craib Figure 5.68
ເຄືອເຂົາຣໍແດງ – Kheua Khao Ho Deng. Oudomxay MBP. Large liana, stem thick, fissured, gray, flowers white on axillary clusters. Liquid flowing from the vine when cut is used to treat red eyes or conjunctivitis (ນ້ຳທີ່ໄຫຼອອກຈາກເຄືອທີ່ຖືກຕັດ ໃຊ້ເພື່ອບໍ່ຕາແດງ). Cut the stem and aim the liquid that drips from the cut over the affected eyes (ຕັດເຄືອ ແລ້ວໂຕ່ງເອົານ້ຳທີ່ໄຫຼອອກເພື່ອຍອດໃສ່ຕາທີ່ແດງ). Soejarto et al. 15082.

(a) (b)

FIGURE 5.68 A portion of the vine of *Tetrastigma quadrangulum* showing the quadrangular stem bearing dense axillary clusters of flowers (a) and a close-up view of a dense umbellate cluster of greenish white flowers of *Tetrastigma quadrangulum* (b). Photo credit: D. Soejarto.

GROUP 2 ANGIOSPERMAE – MONOCOTYLEDONAE

Araceae

Aglaonema simplex Bl. Figure 5.69

ຫາງອິ້ນ; ຄ້ອນເທົ້າພະຍາອິນ – Hang Onh; Khon Tao Pa Nya Iin. Oudomxay MBP. Semi-succulent herb on forest floor, leaves ovate-lanceolate, fruits green, globular, on a subglobular cluster, borne on a long pedicel. The whole plant is used to prepare a medicine to treat cases of thermal burns (ໃຊ້ໝົດຕົ້ນເພື່ອບໍ່ ໄພໃໝ້). Crush the whole plant and apply it on the injured skin (ຢ່ອງພົດໝົດຕົ້ນ ແລະ ແປະໃສ່ບ່ອນໄພໃໝ້). Soejarto et al. 15118.

Aglaonema ovatum Engler

ຊາຍນອນດຽວ – Sai Nohn Diao; Sekong MBP. Herbaceous plant 20 cm high, leaves glabrous, fruits red, glabrous, in a round umbel. Root to treat fatigue for a "man who sleeps alone" (but not as a tonic) (ໃຊ້ຮາກບໍ່ອສຸຂະພາບອ່ອນເພຍ

FIGURE 5.69   An *Aglaonema simplex* herb in its natural habitat, bearing dense clusters of green fruits. Photo credit: D. Soejarto.

(ເບັນຢາບຳລຸງ). Cut the root into small pieces and boil in water, then drink the decoction as needed (ພັກ ຈອຍ ຮາກເບັນຕ່ອນນ້ອຍໆ, ຕົ້ມໃນນ້ຳ ແລະ ດື່ມນ້ຳ ຕົ້ມຕາມຕ້ອງການ). Elkington et al. 267.

Pothos chinensis (Raf.) Merr.

ຫວາຍສະນອຍ; ລິ້ນມັ້ງກອນ; ຫນາມຕຳຫານ – Wai Sah Noy; Lin Mang Khon; Nam Tam Tan. Sekong MBP. Small herb, growing on a rock; leaves consist of two parts, with smooth margins. Sterile. Whole plant to prepare a tonic to treat fever (ໃຊ້ໝົດຕົ້ນເບັນຢາບຳລຸງກຳວັງ ເພື່ອບໍ່ໄຂ້). Boil the dried pieces of the whole plant and drink the decoction (ຕົ້ມຕ່ອນນ້ອຍໆທີ່ພັກ ຈອຍມາຈາກໝົດຕົ້ນໃນນ້ຳ ແລະ ດື່ມນ້ຳຕົ້ມ). Elkington et al. 263.

Pothos scandens L.

ເຄືອວຸ່ມນ້ອຍ; ຫວາຍດິນ – Kheua Voum Noy; Houai Diin. Oudomxay MBP. Epiphytic climber, hugging on a tree, petiole winged. Plant sterile. Large vine, leaves to treat rheumatism, joint pain (ໃຊ້ເຄືອ, ໃບ ບໍ່ປະດງຂໍ້, ເຈັບປວດຂໍ້). Boil 1 handful of dried small pieces of the liana/stem or the leaves in 1 liter of water and drink the decoction (ຕົ້ມຕ່ອນນ້ອຍໆທີ່ພັກ ຈອຍມາຈາກເຄືອ (ລ່າຕົ້ນ) ຫຼື ໃບ ບະມານຫນຶ່ງກຳມືໃນນ້ຳຫນຶ່ງລິດ ແລະ ດື່ມນ້ຳຕົ້ມ). Soejarto et al. 15067.

Rhaphidophora decursiva (Roxb.) Schott

ເຄືອວຸ່ມໃຫຍ່ – Kheua Voum Nyai. Oudomxay MBP. Large, semi-succulent herbaceous epiphyte, climbing on a tree with its root system. Sterile. The vine is a medicine for rheumatism (ໃຊ້ເຄືອ ບໍ່ປະດງຂໍ້). Pieces of vine are boiled in water and the decoction drunk (ຕົ້ມເຄືອທີ່ ພັກ ຈອຍ ເບັນຕ່ອນນ້ອຍໃນນ້ຳ ແລະ ດື່ມນ້ຳ ຕົ້ມ). Soejarto et al. 15078.

Arecaceae

Arenga caudata (Lour.) H.E. Moore

ຕາວຮ້າງນ້ອຍ – Tao Hang Noy. Bolikhamxay MBP. Liana-like shrub with a slender stem, fruits green, profuse. Root to prepare a tonic for body weakness (ໃຊ້ ຮາກບໍ່ສຸຂະພາບອ່ອນເພຍ). Cut the root into small pieces, boil in water, and drink the decoction (ພັກ ຈອຍຮາກເບັນຕ່ອນນ້ອຍໆ, ຕົ້ມໃນນ້ຳ ແລະ ດື່ມນ້ຳ ຕົ້ມ). Soejarto et al. 15011.

ຕາວໄກ່ນ້ອຍ – Tao Kai Noy. Savannakhet MBP. Small shrub, leaflets narrowly obtriangular, caudate at apex. The root is used to make a medicine to treat cases of fever with rashes (ໃຊ້ຮາກບໍ່ໄຂ້ອອກຕຸ່ມ). Boil the root and drink the liquid (ຕົ້ມຮາກ, ດື່ມນ້ຳຕົ້ມ). Soejarto et al. 14926.

Asparagaceae

Disporopsis longifolia Craib

ຫົວເຂົ້າເຢັນ; ຫວ້ານຫົວຕໍ່ – Hua Khao Yen; Wan Hua Toh. Oudomxay MBP. Herb with gnarled rhizomes, white surface when slashed. The rhizome is used to prepare a medicine (tonic) to be given in cases of poor appetite (ໃຊ້ຮາກເປັນຢາບຳລຸງສຸຂະພາບ ໃນກໍລະນີທີ່ກິນເຂົ້າບໍ່ແຊບ). Grill the rhizome and then boil and drink the decoction (with new rice aroma) (ເອົາຫົວເຜົາໄຟ ແລ້ວເອົາໄປຕົ້ມ, ນ້ຳຕົ້ມຈະມີກິ່ນຫອມຄືເຂົ້າໃໝ່, ດື່ມນ້ຳຕົ້ມ). Soejarto et al. 15071.

Dracaena cambodiana Pierre ex Gagn. Figure 5.70

ກົກຄອນແຄນ; ກົກອ້ອຍລິງ – Kok Khon Khen; Kok Oy Ling. Bolikhamxay MPP. Small tree, 3 m tall, fruits pale greenish yellow, set on loose, spreading, long panicles. The stem or the root is used to prepare a medicine to treat rheumatism and muscle pain (ໃຊ້ລຳຕົ້ນ ຫຼື ຮາກ ບໍ່ປະດົງຂໍ້ ແລະ ເຈັບກ້າມຊີ້ນ). Cut the root into small pieces and dry (ຕັດຮາກ ຫຼື ລຳຕົ້ນເປັນຕ່ອນນ້ອຍ, ຕາກແດດ

FIGURE 5.70 A specimen of *Dracaena cambodiana* shrub in its natural habitat, showing the foliage and hanging inflorescences bearing yellowish green fruits. Photo credit: D. Soejarto.

ໃຫ້ແຫ້ງ). Decoct one handful of the root by boiling in approximately 2 liters of water (ຕົ້ມຍາງແຫ້ງປະມານ ໜຶ່ງກຳມືໃນນ້ຳສອງລິດ). Cool it down and drink as needed (ປະໃຫ້ນ້ຳຕົ້ມເຢັນ, ດື່ມຕາມຕ້ອງການ). Soejarto et al. 15035.

Dracaena elliptica Thunb.

ກົກຄອນແຄນ – Kok Khon Khen. Savannakhet MBP. A large shrub with long pendulous inflorescences. The root is used to make a medicine to treat scorpion stings and snakebites (ໃຊ້ຮາກບົວງຂົບ ແມງງອດຕອດ). Rub the root with some water over a rock and place the sludge on the affected part (ຝົນຮາກໃສ່ຫີນຝົນ ແລະ ເອົານ້ຳຍາທີ່ຝົນໄດ້ແປະໃສ່ບ່ອນສັດຂົບ ສັດຕອດ). Sydara et al. 19022.

Peliosanthes teta Andrews

ແໜງ ດຽບ – Neng Diap. Sekong MBP. Herbaceous plant 10 cm high, leaves lanceolate, 20–30 cm in length, fruits white and blue, ca. 0.5 cm across. Leaf, root to treat depression, uterine cancer (ໃຊ້ໃບ, ຮາກບົວອາການຊຶມເສົ້າ, ມະເຮັງມົດລູກ). Boil the leaves and root, then drink the decoction (ຕົ້ມໃບ ແລະ ຮາກ, ດື່ມນ້ຳຕົ້ມ). Elkington et al. 278.

Commelinaceae

Amischotolype hispida (A. Rich.) D.Y. Hong

ກາບປີຊ້າງ – Kap Pii Xang. Sekong MBP. Herbaceous plant 2 m in height, leaves alternate, ca. 40 cm long, fruits oblong, purple and pink. Stem to treat hiccups, broken bones (ໃຊ້ລຳຕົ້ນບໍ່ໄອໄກ່, ກະດູກຫັກ). Boil a handful of small, dried pieces of stem in 1 liter of water and drink the decoction (ຕົ້ມລຳຕົ້ນທີ່ຫຍ້າກ ຫຼວຍ ເປັນຕ່ອນນ້ອຍໆປະມານໜຶ່ງກຳມືໃນນ້ຳໜຶ່ງລິດ ແລະ ດື່ມນ້ຳຕົ້ມ). Elkington et al. 308.

Commelina communis L.

ເຄືອກາບປີນ້ອຍ – Khcua Kab Pii Noy. Oudomxay MBP. Small creeping herb, flowers blue. Herb to treat fever (ໃຊ້ເຄືອ (ເຄືອນ້ອຍໆ) ບໍ່ໄຂ້ຮ້ອນ). Pieces of herb are boiled in water and the decoction drunk (ຕົ້ມເຄືອທີ່ຫຍ້າກ ຫຼວຍ ເປັນຕ່ອນນ້ອຍໆ ໃນນ້ຳ ແລະ ດື່ມນ້ຳຕົ້ມ). Soejarto et al. 15075.

Pollia secundiflora (Bl.) Bakh.f.

ກາບປີໃຫຍ່ – Kab Pii Nyai. Oudomxay MBP. Creeping herb on dark forest floor, flowers white, set in a paniculate cluster, fruits black. Vine (stem) and root to treat hot fever (ໃຊ້ເຄືອ ແລະ ຮາກບໍ່ໄຂ້ຮ້ອນ). Pieces of vine/stem and root are boiled in water and the decoction taken by mouth (ຕົ້ມເຄືອ ແລະ ຮາກ ທີ່ຫຍ້າກ ຫຼວຍ ເປັນ ຕ່ອນນ້ອຍໆ ໃນນ້ຳ ແລະ ດື່ມນ້ຳຕົ້ມ). Soejarto et al. 15074.

ກາບປີໃຫຍ່ – Kab Pii Nyai. Sekong MBP. Herbaceous plant, approximately 20 cm tall, leaves 8–10 cm in length and 2–3 cm in width. Fruits purple.

Whole plant to treat back pain (ໃຊ້ໝົດຕົ້ນບົວເຈັບແອວ). Cut the whole plant into small pieces and boil in water, then drink the decoction as needed (ຟ້ກ ຊອຍໝົດຕົ້ນເບັນຕ່ອນນ້ອຍໆ, ຕົ້ມໃນນ້ຳ ແລະ ດື່ມນ້ຳຕົ້ມຕາມຕ້ອງການ). Elkington 265.

Cyperaceae
Cyperus cf. *diffusus* Vahl
ຫຍ້າຄົມປາວດົງ; ຄົມປາວໃຫຍ່; ຄົມປາວຜູ້ – Nya Khom Pao Dong; Khom Pao Nyai; Khom Pao Pou. Savannakhet MBP. Herb; inflorescences brown. Root to treat asthma (ໃຊ້ຮາກບໍ່ຫືດ). Cut the root into small pieces, boil in water and drink the decoction (ຟ້ກ ຊອຍ ຮາກເປັນຕ່ອນນ້ອຍ, ຕົ້ມໃນນ້ຳ ແລະ ດື່ມນ້ຳຕົ້ມ). Soejarto et al. 14931.

Cyperus sp.
ຫຍ້າກາບປີ – Nya Kab Pii. Bolikhamxay MBP. Herb on forest floor in a patch of forest opening, inflorescences dull green to blackish green. Whole plant to treat cases of high fever (ໃຊ້ໝົດຕົ້ນບໍ່ໄຂ້ຮ້ອນ). Pound the whole plant with water into a paste. Use the wet paste to cover the important part of the body to help reduce temperature (ຕຳໝົດຕົ້ນໃຫ້ເປັນແປ້ງບຼຼກ ແລ້ວເອົາມາໂປະໃສ່ຈຸດສຳຄັນ ຂອງຮ່າງກາຍ ຊ່ວຍຫຼຸດຜ່ອນຄວາມຮ້ອນ). Soejarto et al. 15012.

Dioscoreaceae
Dioscorea cirrhosa Lour. Figure 5.71
ຫົວເປົ້າເລືອດ; ຫົວຕ້ອມເລືອດ – Hua Pao Leuat; Hua Tom Leuat – Qos Yaj Thawj (Hmong). Oudomxay MBP. Large liana with a spiny black, cylindrical stem; tubers gray-brown, coarsely scaly, subglobular, exposed, in dense cluster on ground surface; the stem arises from the cluster. Cut surface of tuber dark red with scanty red sap. Leaves 3-veined. Sterile. The rhizome is used to make a medicine to treat cases of diarrhea and dysentery (ໃຊ້ຫົວບໍ່ທ້ອງຖອກ, ທ້ອງບິດ). Slice the rhizome into small pieces, then dry them (ຊອຍຫົວເປັນຕ່ອນ ນ້ອຍໆ, ຕາກແດດໃຫ້ແຫ້ງ). Take 1 handful of the dried pieces and bring to a boil in about 1 liter of water for 10–15 minutes (ເອົາຢາແຫ້ງປະມານໜຶ່ງກຳມື ຕົ້ມໃນນ້ຳປະມານໜຶ່ງລິດ ໃນເວລາ 10–15 ນາທີ). Drink the cooled decoction as needed (ດື່ມນ້ຳຕົ້ມທີ່ເຢັນແລ້ວຕາມຕ້ອງການ). Soejarto et al. 15126.

Marantaceae
Stachyphrynium placentarium (Lour.) Clausager & Borchs.
ຕອງຈິງ – Tong Ching. Oudomxay MBP. Large, tall herbaceous plant, leaf blade banana-like, on a long petiole, with flower cluster at base of petiole. Leaf, root to treat cases of intoxication from allergic foods during postpartum (puerperium) (ໃຊ້ໃບ, ຮາກ ແກ້ເບື່ອໃບກໍລະບີຜິດກຳ ຈາກການກິນອາຫານ). Cut the root or leaf into small pieces and dry, then take a handful and boil in 2 liters of

FIGURE 5.71 A cluster of stems of *Dioscorea cirrhosa*, each bearing globular tubers at its base (a) and a close-up view of blood red sap seeping out from the cut surface of the tuber of *Dioscorea cirrhosa* (b) and some leaves of *Dioscorea cirrhosa* showing their upper surface view with three primary veins running from the leaf base (c). Photo credit: D. Soejarto.

water. Drink the decoction as needed (ຝັກ ຸອຍ ຮາກ ຫຼື ໃບເປັນຕ່ອນນ້ອຍໆ, ຕົ້ມປະມານຫນຶ່ງກຳມືໃນນ້ຳສອງລິດ. ດື່ມນ້ຳຕົ້ມຕາມຕ້ອງການ). Soejarto et al. 15128.

Orchidaceae

Cymbidium aloifolium (L.) Sw.

ເອື້ອງສົບເປັດ; ເອື້ອງນາງກາຍ; ກະດ້າມຜີ – Euang Sop Pet; Euang Nang Kai; Ka Dam Pii. Luang Prabang MBP. Succulent epiphyte on a tree trunk, 4 m above ground. Fruits dry, splitting open. Leaf to treat cases of infection in the ears (otitis) (ໃຊ້ໃບບໍ່ ທູຫນວງ). Grill the leaves, press to collect the liquid. Use that liquid as a drop in the ear to treat ear infection (otitis) (ເອົາໃບລົນໄຟ ຫຼື ບີ້ງໄຟ ແລ້ວບີບເອົາ ນ້ຳ ຍອດໃສ່ຫູທີ່ເປັນຫນວງ). Soejarto et al. 15424.

Poaceae

Lophatherum gracile Brongn. Figure 5.72

ຫຍ້າແຟດນົກປິດ, ຊ້າຮົກ – Nha Fed Nok Piid; Xa Hok. Oudomxay MBP. Herb with cylindrical to ellipsoid root tubers, inflorescences green. The root and tubers

(a)   (b)

FIGURE 5.72   A cluster of the herbaceous grass specimens *Lophatherum gracile* (a) and a close-up view of the tuberous roots of the herbaceous grass *Lophatherum gracile* (b). Photo credit: D. Soejarto.

are used to make a medicine to treat kidney stones and kidney inflammation. Collect a handful of roots and tubers and dry them, then bring to a boil in plenty of water; cool down the decoction and drink as needed (ເກັບເອົາທົ່ວ ແລະ ເຫງົ້າ ໃຫ້ໄດ້ບະມານຫນຶ່ງກຳມື ແລະ ຕາກໃຫ້ແຫ້ງ, ຈາກນັ້ນຕົ້ມໃນນ້ຳຫຼາຍໆ, ປະໃຫ້ນ້ຳ ຕົ້ມເຢັນລົງ ແລະ ດຶ່ມຕາມຕ້ອງການ). Soejarto et al. 15088.

Smilacaceae
Smilax glabra Roxb.
ເຄືອງໃບຍາວ – Kheuang Bai Yao. Savannakhet MBP. Vine, flowers greenish white on umbellate, axillary clusters. Common. Liana (stem) to treat cases of gastritis (ໃຊ້ເຄືອ (ວ່າຕົ້ນ) ບໍ່ວກະເພາະອັກເສບ). Boil a handful of small pieces of the stem in a bowl of water and drink the decoction (ຕົ້ມເຄືອທີ່ຫັ່ນ ຊອຍ ເປັນ ຕ່ອນນ້ອຍໆ ໃນນ້ຳຫນຶ່ງຖ້ວຍ ແລະ ດຶ່ມນ້ຳຕົ້ມ). Soejarto et al. 14915.

Zingiberaceae
Alpinia sp.
ຂ່າໂຄມ – Kha Khome. Bolikhamxay MBP. Large herb, leaves banana-like, blade 60 cm long, fruits globular, green, on a terminal racemose cluster. The rhizome is used to treat cases of pain in the uterus (ໃຊ້ຫົວຂ່າບໍ່ວເຈັບມົດລູກ). Cut the

rhizome into small pieces, boil in water, and drink the decoction as needed (ຂຸດ ຫົວຂ່າເປັນຕ່ອນນ້ອຍໆ, ຕົ້ມໃນນ້ຳ ແລະ ດື່ມນ້ຳຕົ້ມຕາມຕ້ອງການ). Soejarto et al. 15033.

Amomum cf. *thyrsoideum* Wight

ໝາກແໜ່ງ – Mak Neng. Oudomxay MBP. Herb 2 m tall, fruits brown-haired, flowers white with a yellow spot. Whole plant to treat nervous system disorders; the root to discharge blood for nursing mother (ໃຊ້ໝົດຕົ້ນບໍ່ອະນັບນະຊາດ. ໃຊ້ ຮາກເພື່ອໄລ່ເລືອດແມ່ລູກອ່ອນ). Use the whole plant to make a sauna for the nervous system. For a nursing mother, decoct the roots and drink the decoction (ໃຊ້ໝົດຕົ້ນຮົມເພື່ອບໍ່ອະນັບເສັ້ນນະຊາດ. ຕົ້ມຮາກ ເພື່ອໄລ່ເລືອດສາລັບແມ່ ລູກອ່ອນ, ດື່ມນ້ຳຕົ້ມ). Soejarto et al. 15062.

Costus tonkinensis Gagn.

ຄ້ອນເກົ້າຜີຍະວາຍ; ເອື້ອງອ້ອຍ – Khon Tao Pii Nya Wai; Euang Oy. Oudomxay MBP. Large herb, single-stemmed, with rhizome, 2½ m tall; infructescence grows separately from the stem, subglobular, with the persistent calyx spiny, greenish to pinkish. The stem is used to treat conjunctivitis (red eyes), otitis, uterus pains (ໃຊ້ລຳບໍ່ອຕາແດງ, ຫູໜວອງ, ເຈັບມົດລູກ). For conjunctivitis and otitis, grill the stem, press to collect the liquid, filter, and put the filtered liquid into the eyes and the ears. For uterus pain, decoct the stem and drink the decoction as needed (ສຳລັບບໍ່ອຕາແດງ ແລະ ຫູໜວອງ, ເອົາລຳຕົ້ນມາລົນໄຟ (ປີ້ງ) ບີບເອົານ້ຳ, ຕອງ, ເອົານ້ຳຕອງທີ່ສະອາດຫຍອດໃສ່ຕາ ແລະ ຫູ. ສຳລັບບໍ່ອເຈັບມົດລູກ, ຕົ້ມລຳ ຕົ້ນໃນນ້ຳ ແລະ ດື່ມນ້ຳຕົ້ມ). Soejarto et al. 15115.

GROUP 3: GYMNOSPERMAE

Gnetaceae

Gnetum latifolium Bl. Figure 5.73

ເຄືອມ້ວຍແດງ – Kheua Muay Deng. Savannakhet MBP. Liana climbing on a tree, with round fruits, when ripe red in color, on dense spikes along stem. The liana (stem) is used to make a tonic for treating blood circulation problems. Cut the stem into small pieces, boil a handful in plenty of water, and drink the cooled decoction as needed (ຕັດເຄືອໃຫ້ເປັນຕ່ອນນ້ອຍໆ, ຕົ້ມບະມານໜຶ່ງກຳມືໃນນ້ຳ, ດື່ມນ້ຳຕົ້ມທີ່ເຢັນແລ້ວຕາມຕ້ອງການ). Women also take a drink of the decoction in cases of absence of menstrual flow (ຜູ້ຍິງທີ່ປະຈຳເດືອນບໍ່ມາສາມາດກິນ ນ້ຳຕົ້ມນີ້ເພື່ອປິ່ນບໍ່ອພະຍາດດັ່ງກ່າວ). Sydara et al. 19024.

Gnetum leptostachyum Bl.

ມ້ວຍຂາວ – Muay Khao. Bolikhamxay MPP. Large liana, fruits dark green, with shiny and silvery, membranous scales. The root is used to treat diarrhea, dysentery

FIGURE 5.73 A portion of the leafy stem of the *Gnetum latifolium* liana, bearing clusters of fruits set on dense lateral racemes along the stem. Photo credit: K. Sydara.

(ໃຊ້ຮາກບໍ່ທ້ອງຕອກ, ທ້ອງບິດ). Cut the root into small pieces, boil in water, and drink the decoction (ຟັກ ຉອຍ ຮາກບັ່ນຕ່ອນນ້ອຍໆ, ຕົ້ມໃນນ້ຳ ແລະ ດື່ມ ນ້ຳຕົ້ມ). Soejarto et al. 15026.

Gnetum montanum Markgr.
ມ້ວຍຂາວ – Muay Khao. Savannakhet MBP. Liana resting on a tree, 10–15 m in length, the fruits round. The liana (stem) is used to treat stomachache (ໃຊ້ເຄືອ (ວ່າ ຕົ້ນ) ບໍ່ເຈັບກະເພາະ). Cut the liana into small pieces, dry, then boil in water and drink the decoction (ຟັກ ຉອຍ ເຄືອບັ່ນຕ່ອນນ້ອຍໆ, ຕາກແຫ້ງ, ຕົ້ມໃນນ້ຳ ແລະ ດື່ມນ້ຳຕົ້ມ). Sydara et al. 19016.

GROUP 4: PTERIDOPHYTA

Lygodiaceae
Lygodium flexuosum (L.) Swartz
ກູດງ້ອງ – Kout Ngong. Savannakhet MBP. A vine, sori black, common. The whole plant is used to treat cases of pneumonia (ໃຊ້ໝົດຕົ້ນບໍ່ອັກເສບປອດ). Boil a handful of small, dried pieces of the whole plant in 1 liter of water and drink the decoction (ຕົ້ມໝົດເຄືອທັ້ງຟັກ ຉອຍ ບັ່ນຕ່ອນນ້ອຍໆ ແລະ ຕາກແຫ້ງ ໃນນ້ຳທັ້ງລິດ ແລະ ດື່ມນ້ຳຕົ້ມ). Soejarto et al. 14916.

Selaginellaceae

Selaginella sp.

ຫຍ້າຂ້າຫນອນ, ກູດມຸ້ງເຕົ່າ – Nya Kha Nohn; Kout Mong Tao. Xieng Khouang MBP. Herb, on forest floor, creeping on ground surface, sporangia light green. The whole plant is used to treat stomach worms (ໃຊ້ໝົດຕົ້ນບໍ່ວເມ່ທ້ອງ). Boil a handful of small, dried pieces of the whole plant in 1 liter of water and drink the decoction (ຕົ້ນທັ້ງໆກຳມື ພົດໝົດຕົ້ນທີ່ພັກ ຊອຍເປັນຕ່ອນນ້ອຍ ແລະຕາກແຫ້ງ ໃນນ້ຳ 1 ລິດ ແລະ ດື່ມນ້ຳຕົ້ນ). Soejarto et al. 1496.

REFERENCES

Chase, M.W., M.J.M. Christenhusz, M.F. Fay, J.W. Byng, W.S. Judd, D.E. Soltis, D.J. Mabberley, et al. 2016. "An update of the Angiosperm Phylogeny Group classification for the orders and families of flowering plants: APG IV." *Botanical Journal of the Linnean Society* 181 (1): 1–20. doi:10.1111/BOJ.12385.

Delang, C. 2007. "The role of medicinal plants in the provision of health care in Lao PDR." *Journal of Medicinal Plants Research* 1 (3): 50–59. doi:10.5897/JMPR.9001235.

Elkington, B.G., D.D. Soejarto, and K. Sydara. 2014. "Ethnobotany of tuberculosis in Laos." *SpringerBriefs in Plant Science*. Accessed on January 8, 2023. https://www.springer.com/gb/book/9783319106557.

Elkington, B.G., K. Sydara, J.F. Hartmann, B. Southavong, and D.D. Soejarto. 2012. "Folk epidemiology recorded in palm leaf manuscripts of Laos." *Journal of Lao Studies* 3 (1): 1–14. Accessed on January 8, 2023. http://laostudies.org/system/files/subscription/JLS-v3-i1-Oct2012-elkington.pdf.

Fay, Michael F. n.d. APG – classification by consensus. *Royal Botanic Gardens Kew*. Accessed on January 8, 2023. https://www.kew.org/read-and-watch/apg-classification-consensus.

Field Museum. 2022. "Botanical collections." Accessed on January 8, 2023. https://collections-botany.fieldmuseum.org/.

Ho, P.H. 2000. *Cây cỏ Việt Nam – An Illustrated Flora of Viet Nam*. Montreal: Mekong Printing.

Inthakoun, L., and C.O. Delang. 2011. *Lao Flora: A Checklist of Plants Found in Lao PDR with Scientific and Vernacular Names*. Morrisville, NC: Lulu Enterprises, Inc.

JSTOR®. 2022. "Global plants on JSTOR." Accessed on January 8, 2023. https://plants.jstor.org/.

Kew. 2022a. "The Plant List: A Working List of All Plant Species. Version 1." *Published on the Internet*. Accessed on January 8, 2023. http://www.theplantlist.org/.

———. 2022b. "Plants of the World Online." *Plants of the World Online. Facilitated by the Royal Botanic Gardens, Kew*. Accessed on January 8, 2023. https://powo.science.kew.org/.

Lao Statistics Bureau, and World Bank. 2020. "Poverty profile in Lao PDR: Poverty report for the Lao expenditure and consumption survey 2018–2019." Accessed on January 8, 2023. https://thedocs.worldbank.org/en/doc/923031603135932002-0070022020/render/LaoPDRPovertyProfileReportENG.pdf.

Libman, A.S., B. Southavong, K. Sydara, S. Bouamanivong, C. Gyllenhaal, M.C. Riley, and D.D. Soejarto. 2009. "The influence of cultural tradition and geographic location on the level of medicinal plant knowledge held by various Cultural Groups in

Laos." In *Contemporary Lao Studies: Research on Development, Language and Culture, and Traditional Medicine*, edited by C.J. Compton, J.F. Hartmann, and V. Sysamouth, p. 339. DeKalb, San Fransisco: Southeast Asia Publications.

Muséum National d'Histoire Naturelle. 2023. "Consultation Des Collections." *Botanique*. Accessed on January 8, 2023. https://science.mnhn.fr/institution/mnhn/search.

Newman, M., S. Ketphanh, B. Svengsuksa, P. Thomas, K. Sengdala, V. Lamxay, and K. Armstrong. 2007. *A Checklist of the Vascular Plants of Lao PDR*. Edinburgh. Accessed on January 8, 2023. https://portals.iucn.org/library/efiles/documents/2007-014.pdf.

Smithsonian Institution. 2022. "Botany collections search." *Department of Botany Collections*. Accessed on January 8, 2023. https://collections.nmnh.si.edu/search/botany/.

Soejarto, D.D., C. Gyllenhaal, M.R. Kadushin, B. Southavong, K. Sydara, S. Bouamanivong, M. Xaiveu, et al. 2012. "An ethnobotanical survey of medicinal plants of Laos toward the discovery of bioactive compounds as potential candidates for pharmaceutical development." *Pharmaceutical Biology* 50 (1): 42–60.

Tagwerker, E. 2009. *Siho and Naga – Lao Textiles: Reflecting a People's Tradition and Change*. Peter Lang GmbH, Internationaler Verlag der Wissenschaften.

Thiên, T.Đ.T., and S. Ziegler. 2001. "Utilization of medicinal plants in Bach Ma National Park, Vietnam." *Medicinal Plant Conservation* 7 (August). Accessed on January 8, 2023. https://www.researchgate.net/publication/242282209_Utilization_of_medicinal_plants_in_Bach_Ma_National_Park_Vietnam.

Tropicos.org. 2023. "Tropicos, Botanical Information System at the Missouri Botanical Garden." St. Louis, Missouri. Accessed on January 8, 2023. http://www.tropicos.org/.

Vidal, J. 1959. "Noms vernaculaires de plantes en usage au Laos." In *Extrait Du Bulletin de l'Ecole Francaise d'Extrême-Orient, Tome XLIX, Fascicule 2*. Paris: l'Ecole Francaise d'Extrême-Orient (EFEO).

WFO. 2022. "Home." *The World Flora Online*. Accessed on January 8, 2023. http://www.worldfloraonline.org/ and the WFO *Plant List* (formerly The Plant List). https://wfoplantlist.org/plant-list.

Wittmer, H., and H. Gundimeda. 2012. *The Economics of Ecosystems and Biodiversity in Local and Regional Policy and Management*. Washington, DC: Earthscan.

6 Literature Review and Analysis of Select Medicinal Plants in the MBPs and MPPs

James G. Graham and Bethany Gwen Elkington
University of Illinois at Chicago, Chicago, IL, USA
Field Museum of Natural History, Chicago, IL, USA

Jonathan Bisson and Charlotte Gyllenhaal
University of Illinois at Chicago, Chicago, IL, USA

Yulin Ren and A. Douglas Kinghorn
The Ohio State University, Columbus, Ohio, USA

Kongmany Sydara
Institute of Traditional Medicine, Ministry of Health,
Vientiane, Laos
University of Health Sciences, Vientiane, Laos

Djaja Djendoel Soejarto
University of Illinois at Chicago, Chicago, IL, USA
Field Museum of Natural History, Chicago, IL, USA

CONTENTS

DOI: 10.1201/9781003216636-6

6.1 INTRODUCTION

Chapter 3 of this book demonstrated that bioactive compounds are found in many medicinal plants of Laos. These plants and their compounds possess potential to be developed further into medicines. For example, *Marsypopetalum modestum* is traditionally used to treat coughs, and follow-up research on this species has resulted in the isolation of compounds that show promising activity against virulent *Mycobacterium tuberculosis* (Elkington et al. 2014). Various plants traditionally used to treat various types of fevers have demonstrated activity against *Plasmodium falciparum* and *P. vivax*, causative agents of malaria (Libman et al. 2008; Elliott et al. 2020). This pattern is also seen in other regions, as many therapeutic drugs have been discovered from folk use of plants, such as precursors for morphine from *Papaver somniferum* L. (Papaveraceae), aspirin from *Salix* spp. (Salicaceae), and artemisinin from *Artemisia annua* L. (Asteraceae), to name only a few.

Chapter 5 presented a list of 279 medicinally useful plant species documented through interviews with traditional healers in Laos. From this species pool we selected a set of 99 species used in five medicinal categories, with the purpose of evaluating their medicinal uses and the presence of supporting evidence—either corroboration of traditional uses or convergence through experimental bioactivity testing of extracts or isolated compounds reported for these species in the global scientific literature. While laboratory-demonstrated bioactivity may not always be available to support traditional use, over time the use of certain plant species as medicine often implies bioactivity in a plant. There are examples of serendipitous discoveries of very important drugs from traditionally used plants. The discovery of vincristine and vinblastine from *Catharanthus roseus* (L.) G. Don (Apocynaceae) is an excellent example (Noble, Beer, and Cutts 1958; Duffin 2000; Cooper and Deakin 2016; Caputi 2018).

Selection of species for evaluation was guided by the relatively abundant use of all species reported in Chapter 5 for the treatment of pain, rheumatism, fever, and diarrhea. All these species present a set of relatively salient (objectively observed) set of symptoms with which to evaluate the efficacy of traditional uses of a plant. Liver complaints, which are less directly observable, were included, due to the prevalence of liver disease reported in Laos (Sitbounlang et al. 2021; Black et al. 2022). Several "tonics" and treatments for "weakness" that appear relatively frequently in the list of uses are included, as are a respectable number of treatments for "stomachache." But these were deprioritized, in the interest of identifying the clearest evidence, particularly for "best translation" of categories of traditional use with biomedical evidence from basic science experimental bioactivity testing.

The plant species being employed in the regions surrounding the MBPs and MPPs in Laos highlight the critical relationship between maintenance of cultural traditions, the conservation of biodiversity, and the sustainable utilization of botanical resources in the country. As mentioned in Chapter 4, the use and continuing use of a medicinal plant species in a particular region of the world, such as Laos, may affect the very existence of the species. This chapter also briefly notes possible conservation issues of these plants through conservation-related databases (IUCN Red List 2022; CITES 2022).

The unique and diverse practices of folkloric medicine are an invaluable component of traditional ecological knowledge systems that have very deep roots in the many ethnic groups that currently live in Laos. Traditional beliefs about health and wellness, the nature of diseases and disorders, and the botanical resources that form the basis of their *materia medica* represent an as-yet largely untapped resource, guiding future research efforts that may prove to be of great benefit to the people of Laos and the world.

6.2   METHODS OF LITERATURE SEARCH

6.2.1   Sources of Data

Following the identification of each medicinal plant species, as outlined in Chapter 5, those species used in the treatment of diarrhea, fever, liver complaints, pain, and rheumatism were evaluated through a series of literature searches, using the plant's scientific name as the initial search term. The following online text aggregators were queried: Google Scholar (2022), PubMed (2022), EMBASE (2022), along with databases NAPRALERT (2022), and LOTUS (2021). The uniform resource locator (URL) address for each of these websites is listed in the References section for this chapter. For each of the five databases, the Latin binomial or the species name, using Global Names Verifier (2022) and Plants of the World Online (Kew 2022) for each plant, was queried as the initial search term, and the number of citations was noted. If the name was listed as "ambiguous," the current accepted name in Plants of the World Online (Kew 2022) was used.

There are important qualitative differences between the information presented in the columns of Table 6.1. The NAPRALERT and LOTUS columns represent queries of *databases*. These represent curated, human-annotated knowledge bases documenting specific experiments directly involving a species evaluated in an experiment, with a record of the literature source that generated the data. LOTUS provides chemical data for verified compound-species pairs, and NAPRALERT provides data on folkloric use, experimental bioactivity testing of extracts, and the compounds isolated from those species. Data from these two *databases* are already confirmed, in contrast to citations generated by keyword searches of *online text-aggregators*, which have the disadvantage of requiring a confirmation of relevance for each citation and is impractical for large numbers of citations. The process requires manual review, namely cleaning, organizing, and classifying information in each citation for meaningful data. Often, a citation may be suggested that simply contains text reflecting the name of a parent genus and the specific epithet of a species from an entirely different genus. Nevertheless, the overall number of citations yielded by these online text aggregators is useful to ascertain the relative amount of data available for species being queried. The number of citations recorded in the "Google Scholar," "PubMed," and "EMBASE" columns in Table 6.1 reflect the total (uncleaned) number of citations generated by the respective text-based searches of a specific epithet. This allows the comparative ranking of species with a relatively large number of literature citations along with those for which there are very few or no literature records. In contrast, extensive review of the recommended citations identified in keyword-generated searches (methodology presented below) for convergent use of species ("corroborated?" column in Table 6.1) required extensive review of each relevant citation to confirm purported bioactivity or corroborative folkloric use.

6.2.2 QUERY METHODOLOGY

The term corroborative has Latin roots (Harper 2022); from com "with, together" and robare "make strong," and defined as: "to support with evidence or authority; make more certain." (Merriam-Webster Dictionary 2022). The Latin roots of convergent (Etymology online 2022), from com "with, together" and vergere, meaning "to bend, turn, tend toward," defined as (Merriam-Webster online 2022): "to tend or move toward one point or one another; come together; meet converging paths."

Table 6.1 contains information acquired through a process that reflects both primary and secondary datatypes. *Primary data* were derived from extensive field research involving ethnobotanical inventory in Laos, with interviews and documentation of plant-use information by local healers, subsequent collection of plants, and followed-up with species determination through detailed herbarium work. *Secondary data* found in Table 6.1 are derived from heterogeneous work published by other researchers in two different forms: (1) Databases (NAPRALERT 2022; LOTUS 2021), which provide ethnobotanical uses and phytochemical data (and citations), along with informatics data on conservation status (IUCN Red List 2022; CITES 2022) and global distribution (Kew 2022) data for the species found in the table; and (2) Keyword searching of several popular online text-based aggregators (Google Scholar 2022; PubMed 2022; EMBASE 2022), producing a set of literature citations, which require, as stated above, additional confirmatory evidence of accuracy.

Keyword searching methods used for the online text aggregators, including Google Scholar, PubMed, and EMBASE, started with the species name ("Genus-species"), enclosed in parentheses, used to generate the total number of citations found in the columns labeled "Google Scholar," "PubMed," and "EMBASE" in Table 6.1. The number of citations generated by each species was then entered into the three corresponding columns in Table 6.1. For each species in Table 6.1 a second series of queries was conducted using the species name of each species, along with another disease-specific set of keywords: for diarrhea, the keywords "diarrhea" and "dysentery" were applied; for fever, the keywords "fever" and "antipyretic" were applied; for liver complaints, the keywords "liver," "jaundice," and "hepatoprotective" were applied; for pain, the keywords "pain," "analgesia," and "nociceptive" were applied; while for rheumatism, the keywords "rheumatism," "arthritis," and "inflammation" were applied. Both corroborative (ethnomedical) evidence as well as convergent (positive bioassay data) evidence, if confirmed, were used to identify supporting data from the literature, which is represented by the letter "Y" (for yes) in the confirmed uses column in Table 6.1. Species found with shared uses between sites in Laos, but without confirmation of use (for example, a fever treatment) in the global literature, are represented by the letter "L" (Laos; for Lack of confirmation) in the "corroborated?" column in Table 6.1.

Note that in Figure 6.1 the relative ratio of corroboration of uses (height of column) in the global literature is highly correlated with the overall number of citations generated for that species in Google Scholar. Of all the 30 species returning less than 50 citations per query, none (0%) had ethnomedical uses corroborated in the literature, while all 17 species yielding more than 1000 citations per query (right side of image) had 100% corroboration of ethnomedical uses in the literature.

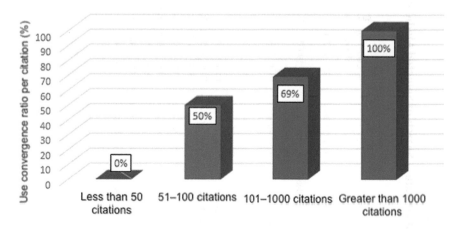

FIGURE 6.1 Pattern of corroborative evidence from literature citations reported in Google Scholar (accessed June 30, 2022) for selected ethnomedical uses (diarrhea, fevers, pain, rheumatism, and liver complaints) for the 99 species listed in Table 6.1.

TABLE 6.1

Results of Literature Search of Selected Medicinal Plant Species with Reported Uses[a]

1. Disease	2. Family	3. Species	4. Collector coll no	5. Corroborated?	6. Google Scholar	7. PubMed	8. EMBASE	9. LOTUS cit	10. LOTUS cmpds	11. Napra E cit	12. Napra Ethno	13. Napra Bio cit	14. Napra Bio	15. IUCN Red List	16. CITES Check-list	17. Range Code
F	Acanthaceae	Phlogacanthus curviflorus	DDS_15081	Y	119	2	4	2	39	3	5	0	0	0	0	w
F	Ancistrocladaceae	Ancistrocladus tectorius	KS_19035	Y	395	14	23	17	26	2	2	8	112	0	0	w
L	Annonaceae	Monocarpia kalimantanensis	BGE_274		29	0	0	0	0	0	0	0	0	LC	0	1
P	Annonaceae	Monocarpia maingayi	BGE_303		5	0	0	0	0	0	0	0	0	LC	0	2
F	Annonaceae	Monoon thorelii	BGE_277		2	0	0	0	0	0	0	0	0	0	0	2
D	Annonaceae	Polyalthia evecta	DDS_14925, KS_19056	Y, L	182	5	9	2	8	1	1	0	0	0	0	4
P	Annonaceae	Polyalthia evecta	DDS_15006		182	5	9	2	8	1	1	0	0	0	0	4
R	Annonaceae	Polyalthia evecta	DDS_15006, KS_19056	Y, L	182	5	9	2	8	1	1	0	0	0	0	4
L	Apocynaceae	Alstonia scholaris	DDS_15462	Y	10400	33	501	43	128	51	190	32	424	LC	0	w
L	Apocynaceae	Hoya cf. parasitica	DDS_15457		32	0	0	0	0	0	0	0	0	0	0	5
L	Apocynaceae	Hoya oblongacutifolia	KS_19039		2	0	0	0	0	0	0	0	0	0	0	3
P	Apocynaceae	Hoya oblongacutifolia	KS_19039		2	0	0	0	0	0	0	0	0	0	0	3
D	Apocynaceae	Kibatalia macrophylla	KS_19028		21	1	0	1	1	0	0	0	0	0	0	6
D	Apocynaceae	Tabernaemontana peduncularis	DDS_15048		20	1	1	1	7	0	0	0	0	0	0	5

	Family	Species	Code													
F	Araceae	*Pothos chinensis*	BGE_263		201	3	8	0	0	3	4	0	0	0	0	w
P	Araceae	*Pothos scandens*	DDS_15067	Y	915	7	25	0	0	6	7	1	2	0	0	w
R	Araceae	*Pothos scandens*	DDS_15067,	Y	915	7	25	0	0	6	7	1	2	0	0	w
R	Araceae	*Rhaphidophora decursiva*	DDS_15078	Y	479	4	10	3	34	2	1	3	314	0	0	w
P	Araliaceae	*Heteropanax fragrans*	DDS_15083		472	2	7	2	8	3	5	1	4	0	0	w
R	Araliaceae	*Macropanax schmidii*	DDS_15069		3	0	0	0	0	0	0	0	0	0	0	1
L	Araliaceae	*Schefflera tunkinensis*	DDS_15064		1	0	0	0	0	0	0	0	0	0	0	1
F	Arecaceae	*Arenga caudata*	DDS_14926		67	0	0	0	0	0	0	0	0	0	0	7
P	Asparagaceae	*Dracaena cambodiana*	DDS_15035	Y	690	29	40	5	44	1	0	0	0	0	0	6
R	Asparagaceae	*Dracaena cambodiana*	DDS_15035		690	29	40	5	44	1	1	0	0	0	0	6
F	Asteraceae	*Cyanthillium cinereum*	BGE_312	Y	832	10	163	18	43	0	43	0	0	0	0	6
L	Asteraceae	*Elephantopus mollis*	DDS_14991	Y	2110	22	42	14	64	10	25	9	578	0	0	w
L	Capparaceae	*Capparis acutifolia*	DDS_14938	Y	74	0	0	0	0	0	0	0	0	0	0	5
R	Capparaceae	*Capparis trinervia*	DDS_15053		6	0	0	0	0	0	0	0	0	0	0	6
R	Celastraceae	*Salacia verrucosa*	KS_19027		65	1	4	1	24	0	0	0	0	0	0	w
R	Chloranthaceae	*Chloranthus spicatus*	DDS_14997, 15004	Y	385	10	17	9	89	2	2	0	0	0	0	2
D	Clusiaceae	*Garcinia xanthochymus*	DDS_15121	Y	1600	37	67	14	98	2	5	3	20	LC	0	w
L	Clusiaceae	*Garcinia xanthochymus*	DDS_15042	Y	1600	37	67	14	98	2	5	3	20	LC	0	w

(Continued)

TABLE 6.1 (CONTINUED)
Results of Literature Search of Selected Medicinal Plant Species with Reported Uses[a]

1. Disease	2. Family	3. Species	4. Collector coll no	5. Corroborated?	6. Google Scholar	7. PubMed	8. EMBASE	9. LOTUS cit	10. LOTUS cmpds	11. Napra E cit	12. Napra Ethno	13. Napra Bio cit	14. Napra Bio	15. IUCN Red List	16. CITES Checklist	17. Range Code
R	Clusiaceae	*Garcinia xanthochymus*	DDS_15042	Y	1600	37	67	14	98	2	5	3	20	LC	0	w
F	Commelinaceae	*Commelina communis*	DDS_15075	Y	10800	177	153	6	21	5	17	18	129	0	0	w
F	Commelinaceae	*Pollia secundiflora*	DDS_15074		58	0	0	0	0	0	0	0	0	0	0	w
P	Commelinaceae	*Pollia secundiflora*	BGE_265		58	0	0	0	0	0	0	0	0	0	0	w
P	Connaraceae	*Connarus paniculatus*	DDS_15043	Y	66	1	2	1	1	0	0	0	0	0	0	w
R	Connaraceae	*Connarus paniculatus*	DDS_15043, KS_19011	Y, L	66	1	2	1	1	0	0	0	0	0	0	w
D	Dioscoreaceae	*Dioscorea cirrhosa*	DDS_15126		295	5	8	5	24	3	3	1	1	0	0	w
L	Elaeocarpaceae	*Elaeocarpus ovalis*	DDS_14985		11	0	0	0	0	0	0	0	0	0	0	4
D	Fabaceae	*Afzelia xylocarpa*	DDS_15472		741	4	2	0	0	1	1	0	0	EN	0	5
L	Fabaceae	*Albizia lucidior*	DDS_15445		23	1	2	1	1	2	4	0	0	0	0	w
D	Fabaceae	*Campylotropis pinetorum*	DDS_15444		11	0	0	0	0	0	0	0	0	0	0	6
R	Fabaceae	*Dalbergia rimosa*	SDDS_15051	Y	97	0	1	0	0	0	0	0	0	LC	II	w
P	Fabaceae	*Dialium cochinchinense*	KS_19043		130	0	1	0	0	0	0	0	0	NT	0	6

	Family	Species	Code													
R	Fabaceae	Fordia cauliflora	DDS_15454	Y	89	5	7	1	5	0	0	0	0	LC	0	w
F	Fabaceae	Piliostigma malabaricum	DDS_15422	Y	820	7	25	1	8	1	2	1	36	LC	0	w
R	Fabaceae	Piliostigma malabaricum	DDS_14903	Y	820	7	25	1	8	1	2	1	36	LC	0	w
L	Fabaceae	Tadehagi triquetrum	DDS_14992, BGE_293	Y, L	371	15	21	1	36	1	2	1	1	0	0	w
D	Fabaceae-Caesal.	Caesalpinia pulcherrima	KS_19007	Y	6460	64	188	22	115	36	81	30	537	0	0	z
F	Fabaceae-Caesal.	Peltophorum dasyrhachis	DDS_14951		101	0	1	0	0	0	0	0	0	0	0	w
R	Fabaceae-Caesal.	Senna sophera	DDS_15448	Y	407	3	8	54	83	8	12	7	27	0	0	z
D	Gnetaceae	Gnetum leptostachyum	DDS_15026		29	0	1	0	0	0	0	0	0	LC	0	4
L	Hydrangeaceae	Dichroa febrifuga	DDS_14972	Y	763	0	26	8	16	7	9	20	1442	0	0	w
P	Lamiaceae	Callicarpa rubella	DDS_14986	Y	129	5	3	0	0	2	4	0	0	LC	0	w
P	Lamiaceae	Clerodendrum chinense	DDS_14955	Y	286	3	10	3	31	0	0	0	0	LC	0	w
F	Lamiaceae	Congea tomentosa	BGE_291	Y	192	1	1	0	0	1	1	0	0	0	0	w
L	Lamiaceae	Rotheca serrata	DDS_15135, BGE_289	Y, L	210	3	6	2	4	0	0	0	0	0	0	w
F	Lamiaceae	Sphenodesme griffithiana	KS_19034		1	0	0	0	0	0	0	0	0	0	0	2
L	Lamiaceae	Vitex quinata	DDS_14989	Y	370	2	4	2	36	2	2	4	16	LC	0	w
L	Lauraceae	Actinodaphne rehderiana	DDS_14968		5	0	0	0	0	0	0	0	0	0	0	1
L	Lauraceae	Litsea cambodiana	KS_19020		31	24	30	0	0	0	0	0	0	0	0	4
P	Loganiaceae	Strychnos nux-blanda	DDS_15419		135	4	4	2	1	0	0	0	0	0	0	6

(Continued)

TABLE 6.1 (CONTINUED)
Results of Literature Search of Selected Medicinal Plant Species with Reported Uses[a]

1. Disease	2. Family	3. Species	4. Collector coll no	5. Corroborated?	6. Google Scholar	7. PubMed	8. EMBASE	9. LOTUS cit	10. LOTUS cmpds	11. Napra E cit	12. Napra Ethno	13. Napra Bio cit	14. Napra Bio	15. IUCN Red List	16. CITES Checklist	17. Range Code
R	Loranthaceae	*Macrosolen cochinchinensis*	DDS_15050, KS_19038	Y, L	373	6	11	0	0	0	0	0	0	0	0	w
P	Malvaceae	*Microcos paniculata*	DDS_15085	Y	939	19	49	3	19	1	1	1	3	0	0	w
P	Malvaceae	*Pterospermum argenteum*	DDS_15017		5	0	0	0	0	0	0	0	0	0	0	1
R	Malvaceae	*Pterospermum argenteum*	DDS_15017		5	0	0	0	0	0	0	0	0	0	0	1
R	Malvaceae	*Pterospermum semisagittatum*	DDS_15473	Y	132	1	4	1	2	0	0	1	3	0	0	w
R	Melastomataceae	*Pseudodissochaeta septentrionalis*	DDS_14966		27	0	1	0	0	0	0	0	0	0	0	7
F	Meliaceae	*Cipadessa baccifera*	DDS_15459	Y	706	17	38	18	157	3	3	0	0	LC	0	w
L	Moraceae	*Ficus triloba*	DDS_15089		10	1	1	0	0	0	0	0	0	0	0	w
R	Moraceae	*Maclura tricuspidata*	KS_19018	Y	230	16	14	37	113	0	0	0	0	LC	0	5
D	Myristicaceae	*Knema furfuracea*	DDS_15111	Y	183	3	2	3	12	0	0	1	50	LC	0	3
P	Oleaceae	*Ligustrum robustum*	DDS_15107	Y	879	23	29	2	57	0	0	4	235	0	0	w
F	Phyllanthaceae	*Antidesma comptum*	BGE_292		1	0	0	0	0	0	0	0	0	0	0	w
F	Phyllanthaceae	*Antidesma japonicum*	DDS_14913		156	1	2	0	0	0	0	0	0	LC	0	w

	Family	Species	Specimen														
F	Phyllanthaceae	Antidesma tonkinense	BGE_295		3	0	0	0	0	0	0	0	0	0	0	0	1
P	Phyllanthaceae	Aporosa ficifolia	DDS_15015		18	0	0	0	0	0	0	0	0	0	0	0	3
R	Phyllanthaceae	Aporosa ficifolia	DDS_15015		18	0	0	0	0	0	0	0	0	0	0	0	3
P	Phyllanthaceae	Aporosa octandra	DDS_15014	Y	158	5	5	0	0	0	0	0	0	LC	0	0	w
R	Phyllanthaceae	Aporosa octandra	DDS_15014	Y	158	5	5	0	0	0	0	0	0	LC	0	0	w
L	Phyllanthaceae	Baccaurea ramiflora	DDS_ KS_19025	Y	1150	13	26	2	14	1	1	0	0	LC	0	0	w
L	Phyllanthaceae	Phyllanthus urinaria	BGE_287	Y	5800	137	267	17	61	12	18	34	184	0	0	0	w
F	Primulaceae	Ardisia helferiana	DDS_14928, BGE_268	L	13	0	0	0	0	0	0	0	0	0	0	0	6
R	Primulaceae	Ardisia polysticta	DDS_14993		77	1	2	1	17	0	17	0	0	LC	0	0	w
P	Primulaceae	Ardisia villosa	DDS_15108	Y	96	1	4	0	0	0	0	1	8	0	0	0	w
D	Primulaceae	Embelia ribes	DDS_14976	Y	5130	133	340	11	36	17	31	71	429	0	0	0	w
F	Primulaceae	Maesa japonica	DDS_15101		912	7	7	4	20	0	0	1	174	0	0	0	7
P	Rhamnaceae	Gouania leptostachya	BGE_284	Y	806	3	5	1	2	0	0	0	0	0	0	0	w
D	Rhamnaceae	Ventilago denticulata	DDS_15461		176	7	15	10	50	2	2	0	0	0	0	0	w
L	Rhizophoraceae	Carallia brachiata	DDS_14937	Y	1250	5	9	2	45	2	4	5	151	0	0	0	w
R	Rhizophoraceae	Carallia brachiata	DDS_14937	Y	1250	5	9	2	45	2	4	5	151	0	0	0	w
R	Rubiaceae	Diplospora dubia	DDS_15109		149	0	0	8	110	0	0	0	0	LC	0	0	7
F	Rubiaceae	Exallage microcephala	DDS_15007		1	0	0	0	0	0	0	0	0	0	0	0	2
P	Rubiaceae	Gardenia stenophylla	DDS_15018		36	1	1	0	0	0	0	0	0	0	0	0	3
D	Rubiaceae	Ixora chinensis	DDS_15423	Y	678	3	6	2	8	2	2	0	0	0	0	0	7
F	Rubiaceae	Ixora delpyana	DDS_15080		3	0	0	0	0	0	0	0	0	0	0	0	1

(Continued)

TABLE 6.1 (CONTINUED)
Results of Literature Search of Selected Medicinal Plant Species with Reported Uses[a]

1. Disease	2. Family	3. Species	4. Collector coll no	5. Corroborated?	6. Google Scholar	7. PubMed	8. EMBASE	9. LOTUS cit	10. LOTUS cmpds	11. Napra E cit	12. Napra Ethno	13. Napra Bio cit	14. Napra Bio	15. IUCN Red List	16. CITES Check-list	17. Range Code
D	Rubiaceae	Lasianthus attenuatus	DDS_15098		45	1	1	1	1	0	0	0	0	0	0	w
P	Rubiaceae	Lasianthus chinensis	DDS_15124		1	2	1	0	0	0	0	0	0	0	0	w
F	Rubiaceae	Lasianthus verticillatus	BGE_296		51	2	3	0	0	0	0	0	0	0	0	w
P	Rubiaceae	Rubia argyi	DDS_14961		151	1	1	13	27	0	0	0	0	0	0	6
P	Rubiaceae	Tarenna thorelii	DDS_15122		1	0	0	0	0	0	0	0	0	0	0	1
F	Rutaceae	Clausena anisata	DDS_15427	Y	3470	53	111	29	141	13	27	19	412	LC	0	w
F	Rutaceae	Glycosmis parviflora	DDS_15463	Y	107	2	3	17	68	0	0	1	50	LC	0	w
F	Rutaceae	Micromelum minutum var. minutum	KS_19009	Y	158	7	11	7	37	2	2	1	26	0	0	w
P	Rutaceae	Xylosma controversa	DDS_14952, 14996		9	0	0	0	0	0	0	0	0	LC	0	w
F	Saururaceae	Houttuynia cordata	DDS_15467	Y	16100	303	552	20	156	17	41	81	2204	0	0	w
F	Simaroubaceae	Brucea javanica	BGE_315	Y	4560	211	345	61	435	13	26	83	3778	LC	0	w
L	Staphyleaceae	Dalrympelea pomifera	BGE_280		176	1	2	0	0	2	2	1	10	LC	0	w
R	Staphyleaceae	Dalrympelea pomifera	DDS_15113		176	1	2	0	0	2	2	1	10	LC	0	w

P	Theaceae	*Schima wallichii*	DDS_15002	Y	5360	15	35	1	13	11	13	9	86	LC	0	w
F	Urticaceae	*Debregeasia longifolia*	DDS_15076	Y	424	1	6	0	0	1	1	1	1	LC	0	w
R	Vitaceae	*Leea indica*	DDS_14995	Y	2080	18	48	0	0	5	9	6	243	LC	0	w
R	Vitaceae	*Tetrastigma harmandii*	DDS_15469		39	0	0	0	0	0	0	0	0	0	0	4

Note: URL of Table 6.1 posted at MPoL website: https://mpol.fieldmuseum.org/collection-analyses. Also supplied as a Table 6.1 in Word and PDF with the text of Chapter 6.

ᵃ **Explanation for Table 6.1.**

Column 1: Disease/disorder codes – diarrhea (D), fever (F), liver effects (L), pain (P), and rheumatism (R). **Column 2**: Plant Family. **Column 3**: species Latin binomial. **Column 4**: Voucher collection number (deposited in herbaria). **Column 5**: Codes for corroborative uses for specific treatments (ethnobotanical or bioassay results) found in global literature, namely Present (Y), Not found (blank), and shared use within Laos (L). **Column 6**: Google Scholar citation counts (per species). **Column 7**: PubMed citation counts (per species). **Column 8**: EMBASE citation counts (per species). **Column 9**: LOTUS citation counts (number of citations per species). **Column 10**: LOTUS compound counts (number of distinct chemical isolates, per species). **Column 11**: NAPRALERT ethnomedical citation count (number of citations recording relevant ethnomedical uses, per disease, per species). **Column 12**: NAPRALERT ethnomedical uses count (number of ethnomedical uses reported, per disease, per species). **Column 13**: NAPRALERT bioassay citation count (number of bioassays reported, per disease, per species). **Column 14**: NAPRALERT bioassay experiment count (number of bioassay experiments reported, per disease, per species). **Column 15**: codes for IUCN RedList-Endangered (EN); Least concern (LC); not listed (0). **Column 16**: codes for CITES (Convention on International Trade in Endangered Species of Wild Fauna and Flora): not listed (0). **Column 17**: species distribution code: alpha-numerical code reflecting the number of countries in which the species has been reported as native, namely w (widespread); z (introduced from western hemisphere).

6.3 OVERALL RESULTS

Of the species listed in Table 6.1 (https://mpol.fieldmuseum.org/collection-analyses), each had citations in at least one of the columns, especially in the largest text aggregator, Google Scholar, which accounted for the only citations recorded for more than a quarter of the species in the table. The overall pattern of citations in Figure 6.1 illustrates an interesting pattern in the data, where a relatively limited number of species with widespread distribution is comparatively well represented in the literature, while species of more limited geographic distribution ("rare" species) tend to be very poorly represented in the literature. These relatively rarer species are in consequence very incompletely studied. The overall lack of supporting evidence for ethnomedical use, biological activity data, or chemical constituents for these native species signals that there remains a vast amount of biological, ethnomedical, and chemical knowledge to be gained from investigating the traditional plant medicines of Laos.

6.4 DISCUSSION

6.4.1 CORROBORATION OF MEDICINAL USES BETWEEN COMMUNITIES SURROUNDING MBPs-MPPs

One approach to identification of efficacious medicinal plants is to compare similar usage of a species between different cultures. This approach infers that independent discovery by different cultures for treatment of similar symptoms provides supportive evidence for the medicinal use of a species. Of the species presented in Table 6.1, a total of six plant species were identified by healers in different communities in different provinces as treatment for similar disorders: (1) *Ardisia helferiana* was used as a fever treatment by two different healers in communities in adjacent provinces in the south of Laos (Savannakhet and Sekong); (2) *Connarus paniculatus*, *Macrosolen cochinchinensis*, and *Polyalthia evecta* were used as a remedy for rheumatism by two different healers in the communities. These corroborative uses involved a convergence of use across much wider distances (Bolikhamxay province in the central part of the country, and Savannakhet province in the south, about 235 km apart); (3) *Rotheca serrata* was reported as a fever treatment by healers in two different provinces (Sekong in the south and Oudomxay in the north, about 650 km apart); and (4) *Tadehagi triquetrum* was used independently for liver complaints in two different protected provinces (Sekong in the south and Xieng Khouang in the north), around 480 km apart, and with 700 meters difference in elevation between locations.

Of the above six species, only *Ardisia helferiana* had no corroborative data found in the global literature. In contrast, around half of the 99 species in Table 6.1 were found to contain some inference of similar biological effect when a search of the global literature was conducted, either in published reports of experimental bioassay testing, or in reports of ethnomedical usage of the same species for treatment of a similar disorder. Regardless of the overall pattern of shared uses found among the different provincial floras represented, five of the six species with shared usage from Table 6.1 were found to have shared uses in the global literature. Even if one might initially consider that 6% of species convergence was small, in comparison with the global convergence rate of 50% for all 99 species, an 83% convergence rate for shared local uses and corroborative evidence from the global literature provides strong support for folkloric use patterns in guiding future research effort.

It will be important to also remember that similarities or divergences in patterns of medicinal use of the species documented in Table 6.1 and illustrated in Figure 6.2 reflect a complex pattern embedded in a broader context of geography (topography, forest types, and regional conservation status, including prevalence of primary vs. secondary vegetation cover), cultural diversity (language, traditional knowledge, and ethnomedical belief systems), and biodiversity gradients (species distributions, as driven by local climatic and edaphic conditions, as well as patterns of species densities, namely the relative commonality or rarity of each individual species).

The Flora of Laos is one of the least-known in Asia (Chapter 3; Lim et al. 2016), with one vascular plant checklist documenting 4850 species (Newman et al. 2007); it is expected that further botanical exploration will reveal an extensive number of new distribution records for Laos. Efforts associated with the botanical fieldwork recorded in this book have already contributed new species records to the country checklist.

6.4.2   HABITAT HETEROGENEITY EFFECTS ON THE MEDICINAL FLORA OF LAOS

In terms of geography, the central and southern regions of Laos are primarily found at below 1000 meters elevation, with gently sloping terrain in the southern provinces dominated by the alluvial plains and terraces of the Mekong River and its tributaries, at a base elevation of about 200 meters. Along the eastern border with Vietnam, the Annamite Mountain range rises to higher elevations. The northern region of Laos is mountainous, a terminus of the Himalayan range, reaching elevations above 2000 meters. Laos has a tropical monsoon climate (with a pronounced rainy season from May through October), a cool dry season from November through February, and a hot dry season in March and April. Generally, monsoons occur at the same time across the country, although that time may vary significantly from one year to the next. Rainfalls vary regionally, with highest annual rainfalls recorded in the south (Whitaker 1979). (Also, see Chapter 1.)

6.4.3   EFFECT OF CULTURAL DIVERSITY ON THE MEDICINAL FLORA OF LAOS

In terms of cultural diversity, 49 distinct ethnic groups in Laos are officially recognized, although that number may be significantly higher. Major ethnic groups in Laos include the Lao (53.2%), Khmu (11%), Hmong (9.2%), Phouthay (3.4%),

Tai (3.1%), Makong (2.5%), Katong (2.2%), Lue (2%), and Akha (1.8%), among others (The World Fact book, 2021). Laos, in relation to neighboring countries, has an index of cultural diversity of 0.449, with less culturally diverse neighbors China (0.138 index), Cambodia (0.205 index), and Vietnam (0.235 index), and with similar, but slightly higher diversity indices noted for neighbors Myanmar (0.489) and Thailand (0.496) (WPR 2023; Gören 2013).

Due to the noteworthy biological and cultural diversity found in Laos, its geographic and habitat diversity, as well as the preliminary nature of our investigation of traditional knowledge of medicinal plants in the MBPs-MPPs, it is perhaps not surprising that a relatively limited number of shared uses for the species of selected medicinal plants of Table 6.1 is recorded. Ongoing progress in ethnobotanical inventory, and the generation of much larger sample sizes, is expected to better identify emergent patterns in the utility of medicinal plant resources in Laos.

6.4.4 Exploring Use Pattern between Provinces

Six plant species in Table 6.1 were used to treat different diseases between locations, while six species were found to have shared uses. The other 87 species in Table 6.1 were reported from only a single community. Given the floristic and cultural diversity found in Laos, the prioritization of a relatively different set of medicinal species by different healers adjacent to the six MBPs/MPPs—in different locations spreading across the entire country, north to south—is not surprising. Data are insufficient, however, to account for interesting patterns of variation between communities in the study, resulting in the heterogeneous nature of the data. Differences in useful species reported in interviews, as well as the relative ranking or prevalence of disorders being treated, present a great deal of heterogeneity on a community-to-community and province-to-province basis. This is also consistent with the fact that Lao traditional medicinal knowledge is folk medicine (see Chapter 2) and more heterogeneous in form, in contrast to formal systems of medicine, which have centralized knowledge sources and require greater degrees of specialization and training (Elliott 2021). The data illustrated in Figure 6.2 reveals heterogeneity of specific uses for plant species, both within and between communities.

6.4.5 Indicators for Future Study

Divergence in the use of plant species, referred to by different healers in different locations spread across the entire country, north to south, is not surprising. Data are insufficient, however, to account for interesting patterns of variation between communities in the study. Differences in useful species reported in interviews, as well as the relative ranking or prevalence of disorders being treated, present a great deal of heterogeneity on a community-to-community basis.

Those species in Table 6.1 that show a significant absence of literature data (about a third of all species found in the table) present a great opportunity for further investigation to establish a baseline of scientific knowledge. Prioritizing efforts to study both the safety and efficacy of these poorly studied plant remedies

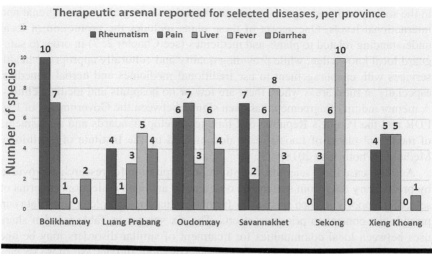

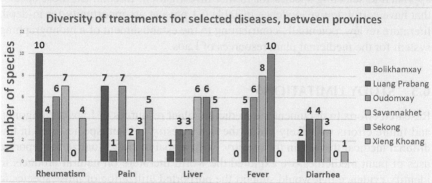

FIGURE 6.2 This figure provides a comparison of use pattern for selected medicinal plant species, both within a single village, and between provinces.

will be useful in primary health care to promote the rational use of plant medicines, as well as to foster training and capacity building in the basic sciences. Care must be taken to ensure that all medicinal plant species, particularly those poorly studied species that are of limited distribution, are not subject to over-exploitation and are properly protected. Chapters 2 and 4 cover conservation of medicinal species of Laos in more detail.

6.4.6 MEDICINAL PLANTS, ECONOMIC DEVELOPMENT, AND CAPACITY-BUILDING

The Government of Lao PDR has long recognized the role of local culture, knowledge, and skills as a means to improve economic development. Culture is seen as the main foundation for sustainable development of the nation, leading to solidarity in the society and pushing the society to expand and integrate at the regional and international levels. One aspect of these plans includes the promotion of local understanding related to plants and medicines (see Chapter 2; 7) in order to safeguard local knowledge, while providing quality and culturally appropriate health services with encouragement to use traditional medicines and herbal remedies, especially in rural areas where there are few or no hospitals and medical clinics. A memorandum of agreement has been signed between the Governments of Lao PDR and the People's Republic of China to develop standards and assessments of medicinal plants of Laos already documented by the Institute of Traditional Medicine (Phothisane 2018).

As is the case in all scientific investigation, inspiration for research is provided by preliminary data from systematic observation and data collection. In terms of selecting a poorly studied species for further evaluation, complementary data can provide the context to prioritize efforts. The six species in Table 6.1 that share uses between local communities for treatment of similar disorders may be one approach to selecting species for further investigation. In contrast, those species that have corroborative evidence in Table 6.1 present an opportunity for in-depth literature review, potentially contributing to the establishment of a monographing system for the medicinal plant resources of Laos.

6.5   STUDY LIMITATIONS

Research efforts to document the medicinal plant resources of Laos are ongoing, and future efforts will likely add to the rich tapestry of plant species used in traditional medical therapy in the country. With a goal to contextualize the reported uses of plant medicines, we explored the world literature on natural products to identify evidence that would support the purported utilization of those species in Laos. We were able to identify a significant lack of scientific literature for some of the more geographically range-limited plant species in our list, as well as to find some convergent published evidence in support of some species. It is important to take a moment to consider some of the challenges to conducting such a literature search strategy, including identifying some notable limitations for using online search engines to highlight the utility of these plants as medicines. Each database was queried individually using the species names provided in Chapter 5. There is overlap (redundancy) in many of the citations between each column in Table 6.1, as some articles turned up in more than one of the five databases queried.

Another important element of this research is careful attention to taxonomic identity of medicinal plant species, taxonomic synonymy, as well as local names (folk taxonomy) for medicinal plant species. Synonyms, a scientific name that applies to a taxon that (now) goes by a different scientific name, may reveal additional research articles that were not identified in our literature screening methods. Adding to the confusion of taxonomic synonymy, common names used locally to refer to different medicinal plant species often overlap, with one common name referring to multiple different species and one single species having multiple common names.

REFERENCES

Black, A.P., V. Khounvisith, K. Xaydalasouk, K. Sayasinh, A. Sausy, C.P. Muller, and J.M. Hübschen. 2022. "Unexpectedly high prevalence of hepatitis C virus infection, Southern Laos." *Emerging Infectious Diseases* 28 (1): 256–59. doi:10.3201/EID2801.211307.

Caputi, L. 2018. "Vinblastine and vincristine: Life-saving drugs from a periwinkle." *John Innes Centre - Excellence in Plant and Microbial Science - Blog.* Accessed on January 8, 2023. https://www.jic.ac.uk/blog/vinblastine-and-vincristine-life-saving-drugs-from-a-periwinkle/.

CITES. 2022. "Checklist of CITES Species." *Convention on International Trade in Endangered Species of Wild Fauna and Flora.* Accessed on January 8, 2023. https://checklist.cites.org/#/en.

Cooper, R., and J.J. Deakin. 2016. *Botanical Miracles: Chemistry of Plants That Changed the World.* 1st ed. Boca Raton, FL: CRC Press. doi:10.1201/B19538.

Duffin, J. 2000. "Poisoning the spindle: Serendipity and discovery of the anti-tumor properties of the vinca alkaloids." *Canadian Bulletin of Medical History* 17 (1–2): 155–92. doi:10.3138/CBMH.17.1.155.

Elkington, B.G., K. Sydara, A. Newsome, C.H. Hwang, D.C. Lankin, C. Simmler, J.G. Napolitano, et al. 2014. "New finding of an anti-TB compound in the genus *Marsypopetalum* (Annonaceae) from a traditional herbal remedy of Laos." *Journal of Ethnopharmacology* 151 (2): 903–11. doi:10.1016/j.jep.2013.11.057.

Elliott, E.M. 2021. "Potent plants, cool hearts: A landscape of healing in Laos." (PhD thesis, University College London). Accessed on January 8, 2023. https://discovery.ucl.ac.uk/id/eprint/10126896/.

Elliott, E.M., F. Chassagne, A. Aubouy, E. Deharo, O. Souvanasy, P. Sythamala, K. Sydara, et al. 2020. "Forest fevers: Traditional treatment of malaria in the southern lowlands of Laos." *Journal of Ethnopharmacology* 249 (March). doi:10.1016/J.JEP.2019.112187.

Embase. 2022. "Embase." Elsevier Limited. Accessed on January 8, 2023. https://www.embase.com/search/quick.

Global Names Verifier. 2022. "Global names verifier." *Global Names.* Accessed on January 8, 2023. https://verifier.globalnames.org/.

Etymology Online. 2022. "covergent (adj)." Accessed on January 8, 2023. https://www.etymonline.com/word/convergent.

Google Scholar. 2022. "About Google Scholar." Accessed on January 8, 2023. https://scholar.google.com/intl/en/scholar/about.html.

Gören, E. 2013. "Economic effects of domestic and neighbouring countries' cultural diversity." *SSRN Electronic Journal* 16 (April). doi:10.2139/SSRN.2255492.

Harper, D. 2022. "Online Etymology dictionary." *Etymonline.* Accessed on January 8, 2023. https://www.etymonline.com/.

IUCN Red List. 2022. "The IUCN red list of threatened species." *International Union for the Conservation of Nature.* Accessed on January 8, 2023. https://www.iucnredlist.org/.

Kew 2022. "Plants of the world online." *Plants of the World Online. Facilitated by the Royal Botanic Gardens, Kew.* Accessed on January 8, 2023. https://powo.science.kew.org/.

Libman, A, H. Zhang, C. Ma, B. Southavong, K. Sydara, S. Bouamanivong, G.T. Tan, H.H.S. Fong, and D.D. Soejarto. 2008. "A first new antimalarial pregnane glycoside from *Gongronema napalense.*" *Asian Journal of Traditional Medicines* 3 (6): 203–10.

Lim, C.-K., J. Kim, V. Saysavanh, and H. Won. 2016. New records of flowering plants from Lao PDR. *Korean Journal of Plant Taxonomy* 46(4): 348–55. Accessed January 11, 2023. doi:https://doi.org/10.11110/kjpt.2016.46.4.348.

LOTUS. 2021. "LOTUS: Natural products online." *Natural Products Occurrence Database*. Accessed on January 8, 2023. https://lotus.naturalproducts.net/.

Merriam-Webster Dictionary. 2022. "Dictionary by Merriam-Webster: America's most-trusted online dictionary." Accessed on January 8, 2023. https://www.merriam-webster.com/.

NAPRALERT. 2022. "The NAPRALERT database of natural products, ethnomedicine, pharmacology, and botany." *Pharmacognosy Institute*. Accessed on January 8, 2023. https://pharmacognosy.pharmacy.uic.edu/pharmacognosy/napralert/.

Newman, M., S. Ketphanh, B. Svengsuksa, P. Thomas, K. Sengdala, V. Lamxay, and K. Armstrong. 2007. *A Checklist of the Vascular Plants of Lao PDR*. Royal Botanic Garden Edinburgh. A paperback. Accessed on January 8, 2023. https://portals.iucn.org/library/efiles/documents/2007-014.pdf.

Noble, R.L., C.T. Beer, and J.H. Cutts. 1958. "Role of chance observations in chemo-therapy: *Vinca Rosea*." *Annals of the New York Academy of Sciences* 76 (3): 882–94. doi:10.1111/J.1749-6632.1958.TB54906.X.

Phothisane, T. 2018. "Lao People's Democratic Republic 2018 report." *UNESCO: Diversity of Cultural Expressions*. Accessed on January 8, 2023. https://en.unesco.org/creativity/node/16028.

PubMed. 2022. "PubMed." *National Library of Medicine | National Center for Biotech-nology Information*. Accessed on January 8, 2023. https://pubmed.ncbi.nlm.nih.gov/.

Sitbounlang, P., A. Marchio, E. Deharo, P. Paboriboune, and P. Pineau. 2021. "The threat of multiple liver carcinogens in the population of Laos: A review." *Livers* 1 (1): 49–59. doi:10.3390/LIVERS1010005.

Whitaker, D.P. 1979. *Laos, a Country Study*. 2nd ed. Supt. of Docs., U.S. Govt. Print. Accessed on January 8, 2023. https://openlibrary.org/books/OL22919531M/Laos_a_country_study.

The World Factbook. 2021. *The World Factbook*. Accessed on January 8, 2023. https://www.cia.gov/the-world-factbook/.

WPR. 2023. "Laos population 2023 (demographics, maps, graphs)." *World Population Review*. Accessed on January 8, 2023. https://worldpopulationreview.com/countries/laos-population.

7 Importance of Engagement with the Communities

Impact on Preserving Medicinal Plants for Future Collections

Djaja Djendoel Soejarto
University of Illinois at Chicago, Chicago, IL, USA
Field Museum of Natural History, Chicago, IL, USA

Kongmany Sydara
Institute of Traditional Medicine, Ministry of Health,
Vientiane, Laos
University of Health Sciences, Vientiane, Laos

Mouachanh Xayvue and Onevilay Souliya
Institute of Traditional Medicine, Ministry of Health,
Vientiane, Laos

Bethany Gwen Elkington
University of Illinois at Chicago, Chicago, IL, USA
Field Museum of Natural History, Chicago, IL, USA

Mary Riley and Charlotte Gyllenhaal
University of Illinois at Chicago, Chicago, IL, USA

DOI: 10.1201/9781003216636-7

CONTENTS

7.1   INTRODUCTION

Protecting forests situated in the vicinity and under the jurisdiction of a community will be successful only if members of the surrounding population help to protect them. Therefore, it is important to engage the communities in monitoring and managing the MPPs and MBPs, which include the medicinal plants therein. As the coordinating body of the MPP/MBP network, the ITM has the responsibility to engage the local population to participate in the monitoring and management of these preserves. Meanwhile, efforts to establish additional MPPs and MBPs continue throughout Laos. As of July 2022, 27 MPPs and MBPs have been established (see Chapter 4).

7.2   SIGNIFICANCE OF THE INVOLVEMENT
OF THE COMMUNITY

The ITM has a decades-long commitment of working with communities throughout Laos to promote the knowledge, utilization, and conservation of medicinal plants through the network of Traditional Medicine Stations/Units (referred to here as TMS), which the ITM has established throughout the country. Through this TMS network, the ITM has conducted educational seminars, carried out research on medicinal plants, and performed community outreach efforts for many years. This work reflects the strong support for the use of traditional medicines at all levels of the government and has led to the establishment of close connections between the ITM and community leaders and healers. These close links are necessary to protect the tracts of land declared MPPs and MBPs. Such links must be maintained by a continuing process of engagement with the community.

A review of several international efforts at community engagement, which focus on medicinal plants and forest conservation, allows us to point out some relevant features of successful programs that may help our present and future efforts in Laos.

Perhaps the most closely analogous model of community engagement efforts focused on MPPs and MBPs is the project "Conservation of living pharmacies of Tasek Bera," conducted by Wetlands International (Wetlands International 2006).

This project proposed to set up an herbal garden in a village forest in Malaysia inhabited by the Semelai tribe. Efforts at community engagement took place in several different phases.

The first important phase was site selection. Community members were consulted about which sites were traditionally used for gathering herbs, thus indicating availability of needed species. Proposed sites took into consideration accessibility to the village, physical characteristics such as availability of trails in the area, difficulty of the terrain, and whether it was prone to flooding.

After choosing a site, outreach programs were conducted to explain the rationale for preserving territory to both community leaders and villagers so that an agreement could be signed with the community's consent. Identification of medicinal plants and their uses was performed in collaboration with local healers. Once the preserve was set up and plants identified, further outreach efforts were conducted, including attempts to train tour guides for ecotourism, produce signage of important species for educational purposes, and produce books with photos of medicinal plants for use by the local community, schools, local governments, and NGOs. Subsequent reporting on community engagement emphasized active participation by the Semelais as being critical to the success of conservation management of Tasek Bera (Haidar 2014).

In Puerto Rico, three forest preserves that implemented community engagement reported a wide variety of techniques, many of which rely on much higher levels of monetary investment (Schadler 2011) than is possible in Laos. One interesting activity mentioned in this report is engagement of local communities in forest preserve service projects that include general cleanup and boundary maintenance.

A project through Conservation International in a Cambodian forest preserve draws attention to the production of books and videos for use in outreach events. It stresses the importance of specifically evaluating community needs and wants relative to protected areas and negotiation of signed agreements regarding preserves (Seng and McCallum 2007).

For a larger-scale example of community engagement, it is notable to cite CGIAR (Consortium of International Agricultural Research Centers). CGIAR is an international organization that has specialized in supporting research partnerships for reducing rural poverty, increasing food security, and improving sustainability of natural resources since the 1970s. Their global impact in more than 80 countries includes a network of genebanks with more than 700,000 germplasm accessions of food plants, 15 research centers, and more than 3,000 partnerships with governments, academic institutions, private companies, and NGOs (CGIAR 2023). CGIAR has already been involved with some projects in Laos, and it may be worthwhile to seek a collaboration focusing on medicinal plant conservation.

7.3 ENGAGEMENT BETWEEN THE ITM AND THE SOMSAVATH COMMUNITY

The ITM-UIC team has used many similar methods of community engagement in the set-up and management of the MPPs and MBPs (see Chapter 4). First, it should be noted that villages exist in a multilayered regional and provincial governmental framework that integrates into their communities. Obtaining consent to set up preserves takes place at multiple governmental layers, down to the level of the village heads and boards. Community leaders, including the Village Head and Council, are some of the first to be involved. District and provincial officials attend events such as inaugurations and award ceremonies for LBF awards and the inauguration of the preserves. Meetings with village boards (councils) and heads to formalize agreements are important steps in the initiation of MPPs and MBPs (Chapter 4).

Early in the process the local expertise of both healers and, where available, heads of TMS, should be engaged to identify suitable tracts of land, as well as to identify and assess the medicinal plants present in each of the preserves. The TMS heads are important links between local village healers and the ITM. Once a preserve has been agreed upon and inaugurated (Chapter 4), the local Village Board, under the auspices of the Food and Drug Division of the Provincial Health Department, takes over management of the preserve with input and expertise from the ITM.

In the case of the Somsavath MPP, the initial engagement was to inventory medicinal plants and mark select medicinal trees using blue painted aluminum plates with the local name of the plant, family name, and species painted white on the plate. In addition, two other community engagement projects were also implemented:

- To reciprocate the generosity of the Somsavath community and their collaboration in the enactment and inauguration of the Somsavath MPP, in 2006 the ITM-UIC team (the ICBG Project team) contributed funds to the village to assist with building a primary school on the village grounds, including benches and blackboards. Once completed, the ICBG team donated textbooks to the school.
- A more significant community engagement act was an informal joint-venture project between the ITM and the Somsavath village in a large-scale cultivation of Javanese turmeric (*Curcuma xanthorrhiza* Roxb., Zingiberaceae) (Kew 2022) in the village terrain. Funding came from the ICBG project and submission of the harvested product (rhizomes) to the ITM as part of its medicinal plant product manufacturing process. This was intended to initiate a process that would build a revolving account in the village. Unfortunately, the outcome was not cost effective, and the project was discontinued after two years.

7.4 IMPLEMENTING A FORMAL COMMUNITY ENGAGEMENT AT THE BOLIKHAMXAY MPP

Formal community events, such as inaugurations and anniversary celebrations, provide opportunities for recognizing a community's commitment to a preserve. In 2014 we had an opportunity to hold a community engagement event at the Bolikhamxay (= Somsavath) MPP to celebrate the 10-year anniversary of the founding of the Somsavath preserve. Participants in the event were the Village Head, the Village Council, the Women's Association, and members of the Somsavath community.

As part of a project funded by the CCF, a joint plan was made between the Village Head, staff of the ITM, and a UIC investigator to hold multiple events. The planned events were announced to the entire Somsavath population by the Village Head several days in advance, resulting in a big turnout of community members, including primary school students and their teachers (Figure 7.1). A major focus of the event was the creation of a banner to be displayed inside the visitor shelter at the preserve where the event was held. It read: "COMMUNITY POWER. Protecting the forest is protecting its medicinal plants for the health of present and future generations."

By way of building community awareness, the ITM and UIC have collaborated on educational posters. One shows the "Medicinal Plant Network," and one shows "Rare and Threatened Medicinal Plants of Laos" (see Chapter 4). At the anniversary celebration, a copy of each poster was unrolled and exhibited to the audience for explanation.

FIGURE 7.1 Some members of the gathering and students and teacher of Somsavath primary school posed in front of the street sign of the Somsavath preserve at the start of the Community Engagement event. Photo credit: M. Xayvue.

7.5 CONTINUING ENGAGEMENT WITH THE SOMSAVATH COMMUNITY BEYOND 2014

As a follow-up to the community engagement at Somsavath, periodic visits were made by the ITM staff to meet with the Village Head and some members of the community. These visits were intended to maintain the connection between the ITM and the community. Activities during such visits included: a walk in the forest to inspect the condition of vegetation in the preserve; collecting herbarium specimens for study purposes; fixing any broken parts of the preserve boundary (with the assistance of community members); cleaning broken and fallen trees caused by storms; and monitoring the effects of occasional use of the preserve by the members of the community—activities that included collection of mushrooms, bamboo shoots, firewood, and medicinal plants. Further, name labels on some of the more outstanding trees around the visitor shelter were renewed or upgraded.

More significant engagement interactions also took place after the 2014 formal engagement celebration. In 2015, a plant collection trip to the Bolikhamxay MPP was undertaken as part of a collaborative project to discover anticancer agents from plants based at the OSU and in collaboration with UIC (P01 project) (see Chapters 1 and 3). This collection trip is a component of the activities as defined in the UIC-ITM Memorandum of Agreement (Amendment IV, 2015–2017; Amendment V, 2017–2020) (Henkin et al. 2017). During this trip, a formal engagement with the Village Head took place, while the healer of Somsavath and several male members of the village participated and assisted the researchers in the collection effort. During this collection visit, the UIC-ITM researchers had ample opportunity to observe the excellent condition of the preserve, which enabled them to gather a set of plant materials for the P01 project.

In 2016, staff of the ITM made a visit to recollect some plant species (original collection in 2015) that had shown good initial anticancer test results for further analyses. The Village Head, together with some members of the Somsavath community, joined the collection effort.

In 2017, as part of the P01 project, joint plant collection expeditions between UIC and ITM were undertaken in the Bokeo MBP, Savannakhet MBP, and Attapeu MBP. During the trips to Savannakhet and Attapeu (in the southern part of the country), a brief side trip was made to Somsavath Village to meet with the village leaders and to see the condition of the MPP. As luck had it, the expedition team spotted a red-fruited plant in the preserve and collected it. Follow-up taxonomic identification indicated this plant belonged to a new species, *Garcinia nuntasaenii* Ngerns. & Suddee (Clusiaceae) (Common name: Kam Lang Mou Thok or Ian Done Don), described just one year earlier (in 2016) (Ngernsaengsaruay and Suddee 2016). This finding represents a new record for the flora of Laos and demonstrates the significant scientific value of the Somsavath MPP.

The Somsavath MPP is the first of its kind in Laos, and the ITM will maintain its normal and continuing engagement with the community. To date, the community has maintained its commitment to help monitor and protect the preserve.

In the words of the former Director of the ITM, Somsavath MPP is the "MODEL," or the "MOTHER," for all the other medicinal plant and medicinal biodiversity preserves in Laos to follow (Kongmany Sydara, personal communication, 2022).

7.6 FUTURE EFFORT IN COMMUNITY ENGAGEMENT

Laos has 27 MPPs and MBPs (Chapter 4). Although the continuing and long-term protection of these preserves rests in the hands of the Ministry of Health authorities, namely the ITM and the Provincial FDD, in terms of day-to-day protection, these preserves are entrusted to the local communities, which form a part of the preserve. These communities, the real stewards of the forest, the plants, medicinal plants, and other resources, have a direct link with the preserves. Their participation in the monitoring and protection of these biodiversity preserves is imperative.

The community engagement activities established and initiated at the Somsavath preserve act as a template for work at other MPPs and MBPs. Small group meetings to maintain community awareness, and interaction by the ITM with local experts, will help motivate community members to care for the preserves. The ITM experts discuss management challenges with village boards at these meetings. Funds for celebrations, gifts for local school children, and for improvement of projects as preserves age may be secured from the government, from NGOs, or other sources.

In the context of community engagement, data presented in Chapters 5 and 6 will serve as a scientific platform for continuing explorations of and new discoveries in other preserves, as well as for the preparation of educational materials for communities that will be based on scientific evaluations of medicinal species. These activities are planned to expand around the full breadth of the Medicinal Plant Network.

REFERENCES

CGIAR. 2023. "Our Assets." Accessed on January 8, 2023. https://www.cgiar.org/.

Henkin, J.M., K. Sydara, M. Xayvue, O. Souliya, A.D. Kinghorn, J.E. Burdette, W.-L. Chen, B.G. Elkington, and D.D. Soejarto. 2017. "Revisiting the linkage between ethnomedical use and development of new medicines: A novel plant collection strategy towards the discovery of anticancer agents." *Journal of Medicinal Plants Research* 11 (40): 621–34. doi:10.5897/jmpr2017.6485.

Kew. 2022. "*Curcuma zanthorrhiza* Roxb." *Kew Science.* Accessed on January 8, 2023. https://powo.science.kew.org/taxon/urn:lsid:ipni.org:names:796485-1.

Mokbolhassen, H.K. 2014. "Managing Tasek Bera: Department of wildlife and national parks and local community participation." *Journal of Wildlife and Parks* 27: 121–27. Accessed on January 11, 2023. http://myagric.upm.edu.my/id/eprint/14440/.

Ngernsaengsaruay, C., and S. Suddee. 2016. "*Garcinia Nuntasaenii* (Clusiaceae), a new species from Thailand." *Thai Forest Bulletin (Botany)* 44 (2): 134–39. doi:10.20531/TFB.2016.44.2.09.

Schadler, E. 2011. "A tale of two forests: Community engagement in Vermont and Puerto Rico." Accessed on January 8, 2023. https://blog.uvm.edu/place-twoforests/files/2011/04/CommunityEngagementinPRandVT.pdf.

Seng, B., and W. McCallum. 2007. "Conservation International Cambodia: Community Engagement in the Central Cardamom Protected Forest." Accessed on January 8, 2023. https://toolkits.knowledgesuccess.org/toolkits/phe/conservation-international-cambodia-community-engagement-central-cardamom-protected.

Wetlands International. 2006. "Conservation of living pharmacies in Tasek Bera, a wetland of international importance in Malaysia." Accessed on January 8, 2023. https://studyres.com/doc/4605951/conservation-of-living-pharmacies-in-tasek-bera--a-wetlan.

8 Summary and Future Prospects

Djaja Djendoel Soejarto
University of Illinois at Chicago, Chicago, IL, USA
Field Museum of Natural History, Chicago, IL, USA

Kongmany Sydara
Institute of Traditional Medicine, Ministry of Health,
Vientiane, Laos
University of Health Sciences, Vientiane, Laos

Bethany Gwen Elkington
University of Illinois at Chicago, Chicago, IL, USA
Field Museum of Natural History, Chicago, IL, USA

Chun-Tao Che and Guido F. Pauli
University of Illinois at Chicago, Chicago, IL, USA

Bounleuane Douangdeuane
Institute of Traditional Medicine, Ministry of Health,
Vientiane, Laos

Charlotte Gyllenhaal and Mary Riley
University of Illinois at Chicago, Chicago, IL, USA

CONTENTS

DOI: 10.1201/9781003216636-8

8.1   INTRODUCTION: THE VALUE OF MEDICINAL PLANTS

Medicinal plants are an essential component of traditional medicine. They play an important role in the health maintenance of people in many countries, including those on the Asian continent. In Laos, medicinal plants serve as the backbone of primary health care for the population, especially in rural areas (see Chapter 2). In this respect, healers play a central role in health management and disease treatment throughout the country. Old medical documents attest to the centuries-old use of medicinal plants by the people of Laos for the treatment of various diseases (Elkington et al. 2012).

The perceived value of "medicinal plants" lies in their actual use to treat ailments. In practice, if a treatment is efficacious, the patient will be cured. A chemical constituent in a plant, or more often a mixture of multiple constituents, is likely responsible for the preventative or therapeutic effect. In the case where the patient is not cured, the healer would try to use an alternative approach by selecting a different plant or a formulated mixture until the disease is cured. However, in most cases, the healer will continue to perform such a healing practice without knowing the underlying biochemical mechanisms.

In recent years, an important effort to document medicinal plants of Laos through field interviews with healers was undertaken by a collaborative team of investigators from the ITM and UIC under the International Cooperative Biodiversity Group (ICBG) Program (Chapters 1 and 3). From this effort, 990 medicinal plant collections were made, of which 725 were identified to the species level, comprising 573 unique species of medicinal plants (Soejarto et al. 2012). A complete set of documentation of these collections has been deposited at the ITM Herbarium (in the capital of Vientiane). A duplicate set is at the John G. Searle Herbarium of the Field Museum in Chicago. The ITM also has a database of the medicinal plant collection (Sydara et al. 2014).

Medicinal plants, however, may have additional value beyond what healers use in their day-to-day practices. Through the isolation of bioactive compounds, new medications may be developed for use in Laos and other parts of the world. In 1985, 74% of the 119 plant-derived drugs of great importance were discovered through studies of "medicinal plants" based on traditional medicine knowledge (Farnsworth et al. 1985). Efforts continue for the discovery of new drug candidates from plants through many other ethnobotany studies (Heinrich 2000; McClatchey et al. 2009; Salmerón-Manzano et al. 2020).

Our past efforts to elucidate the biochemical basis of medicinal plants have led to the discovery of new bioactive natural products (see Chapter 3). However, isolated bioactive compounds may not necessarily correlate with the traditional use of the plant by healers. The results of our research demonstrate the presence of ingredients that would exert biological activity (see Chapter 6), providing scientific support for the ethnomedicinal applications of the plant. In addition, modern biotechnology has shown that a chemical entity isolated from a medicinal plant may serve as a precursor to the process of drug discovery and development.

Biochemical analysis holds potential to provide scientific support to the practices of healers. As shown in the previous chapters, several steps are involved to achieve this goal. The first is to inquire about a healer's knowledge of medicinal plants (see Chapter 5), preferably where they keep their medicinal plant armamentarium (home garden or forest). This is followed by scientific/taxonomic identification of the plant species, which allows for a review of the scientific/ experimental literature records of biochemical or biomedical studies of the plants (see Chapter 6). Simultaneously, laboratory research can be performed on the plant to isolate and identify the bioactive components of the plants. The next step is to inform the healers of the scientific findings. It is for this purpose that the work presented in this book has been undertaken.

8.2   AVAILABILITY OF MEDICINAL PLANTS TO HEALERS

Healers will be able to continue their healing practices if the medicinal plants they use remain available. Sadly, many forests containing medicinal plants are being eliminated for various reasons. Forest conversion to generate food (farming) is one such reason (see Chapter 4). As forests are eliminated, gone also are the medicinal genetic resources, which signify the loss of medicinal plant supplies for the healers. Indeed, many medicinal plants of Laos have become difficult to find in the wild.

The establishment of the MPPs and MBPs is a novel experiment in the preservation and sustainable utilization of medicinal plant genetic resources of the country. This system of medicinal plant protection will help preserve these genetic resources to safeguard the health of present and future generations, while also supporting the continuation of studies toward new discoveries. This experiment resulted in the establishment of eight medicinal MPPs/MBPs through the support of the LBF (2007–2013), as well as an additional 19 new MPPs/MBPs through the support of the Government of Laos and other independent groups, modeled after the Somsavath Medicinal Plant Preserve.

In the future, an evaluation of the importance and benefits of this system of medicinal plant conservation effort should be implemented through field surveys and interviews with healers (Chapter 5) and members of the communities surrounding the MPPs/MBPs, as well as with provincial government officials. Although there are medicinal plant conservation strategies in other countries (Weiss 1994; UNDP 2012; Botanical Dimensions 2022; Wetlands International 2006), this MPP/MBP system and the effort to evaluate medicinal plant protection systems are unique for Laos.

8.3   THE NEED FOR COMMUNITY ENGAGEMENT

The research described in this book illustrates a model of contemporary ethnobotanical (ethnomedical) research. The continuing need for—and the strength of—research plans that integrate multidisciplinary components, including biodiversity conservation, community-based stewardship, and the search for plants with medical potential, is highlighted. Central to these components, however, is community involvement in the overall research design. While it is important to know the specific interactions between people and plants, community involvement is key to understanding the use of traditional medicine and ensuring its preservation through the conservation and protection of forests (see Chapter 7). Any discussions of medicinal plants without the inclusion of community stewardship and the use of resources would overlook crucial linkages. Community planning and involvement serve as the foundation for maintaining plant populations that include medicinal plant species.

It is especially important to include local healers and community leaders in the research. Due to their excellent knowledge, experience, and expertise in using plants in traditional remedies, healers have a keen eye for the landscapes in which traditional medicinal plants grow. They can determine when resources are dwindling by visual inspection and by gauging the amount of time it takes for them to locate and collect specific plants. Community leaders—which in some cases also include healers—are well positioned to communicate, motivate, and educate the greater community to become active stewards in conserving medicinal plant resources.

We learned in Laos that when communities are actively involved in stewardship and conservation of local plant resources, regional and national authorities can devote their attention to the higher-level ("macro") parts of the conservation puzzle: (1) enact national laws to promote conservation and sustainable harvest practices; (2) plan additional land tracts to be granted to local communities for the establishment of medicinal plant gardens and preserves; (3) develop guidelines to regulate the growth of promising industries (ecotourism, handicrafts, organic farming, and intensive cultivation of exotic herbs/spices, etc.); and (4) facilitate communication channels among local, regional, and national authorities to decrease deforestation and unsustainable harvest practices.

8.4   WHAT REMAINS TO BE DONE

The healers we interviewed live near an MPP. They are fortunate that their source of medicinal plants will continue to be available, not only for themselves and other members of the community, but also for future generations of healers and for new discoveries. This assertion is made, however, with the belief that the MPPs will continue to exist and be protected.

We have initiated efforts to evaluate six of the 27 medicinal plant preserves of Laos through on-site interviews with healers (see Chapter 5), followed by reviews on the research being done on plants that healers use (see Chapter 6). We will keep the healers and members of the communities surrounding each preserve informed about our findings, with the goal of promoting and strengthening plant-based healing practices. This should benefit the community members where the healers live. We will also present the results of our research to the provincial government officials who are well positioned for the protection and continuing existence of the protected forests and the medicinal plants therein.

Much remains to be done. We cannot state strongly enough the necessity of engaging members of the local communities, as well as the entire population of Laos, in all efforts involving the protection and preservation of the forests where medicinal plants are located. We feel that the following actions are measures of priority:

- *Give back to the community*. It is of utmost importance to inform the healers and members of the community about our findings. This can be done through formal and informal interactions between the ITM staff scientists and the healers and members of the communities, in settings such as the community engagement undertaken in Somsavath Village (see Chapter 7). Such meetings provide opportunities to discuss the importance of community collaboration in the monitoring and management of protected forests and for the healer(s) to discuss the scientific studies on plants they use.
- *Promote continuing research*. The goal is to demonstrate the value of all MBPs and MPPs to the health maintenance of the population by performing studies similar to the work undertaken in Somsavath MPP. Collaborative efforts between the ITM and other institutions may be sought.
- *Increase public awareness*. The participation and cooperation of people in local communities must be part of every effort to protect the natural resources and ensure sustainable supplies of medicinal plants. In this regard, public education plays an important role in keeping laypeople informed and emphasizing the importance of traditional medicine for the surrounding communities. The aim is to motivate people to actively involve themselves in the protection and preservation of these precious natural resources. Educational programs should be in place to increase awareness about the important roles of healers and to encourage younger generations to acquire knowledge and training in traditional healing arts. At the same time, people need to acquire knowledge about the proper use of medicinal plants (Sydara et al. 2005; Chen et al. 2016 as a model).
- *Promote education about medicinal plants*. Short seminars for students in secondary schools near the MBPs and MPPs throughout Laos, and distribution of posters about medicinal plants and their protection, should foster a better understanding of the values of medicinal plants, so students can contribute to the protection of forests for their futures (Devkota and Watanabe 2020 as a model).
- *Promote continuing discoveries*. Foster research to elucidate the phytochemical, biochemical, and biological basis of medicinal plants used by the healers, especially those that have not been studied previously. This will involve the sampling and taxonomic documentation of plant species, broad biological evaluation of their extracts, and the characterization of their chemical constituents, especially those responsible for therapeutic effects.

8.5 CONCLUDING REMARKS

As Laos continues its national economic development and its participation in global markets and international trade increases, the concomitant issues of conservation and sustainable development will continue to intensify. Over the past several decades there has been increased interest in traditional medicinal plants—and in creating ways to prevent the disappearance of these valuable plant resources due to loss of forest/land to development. Models for "best practices," as well as models of what not to do, will continue to emerge during the course of empirical research in traditional medicine.

Although any country may look to other countries for examples of how to promote conservation, sustainable development, discoveries, and national economic development, every country has its own unique local circumstances and constraints that will require unique solutions. The Government of Laos has already started several national initiatives to encourage conservation and sustainable development of medicinal plants. Additional work to encourage local grassroots efforts to participate in these initiatives is needed to ensure that the national objectives are met without unduly exhausting natural resources or threatening the livelihoods of local communities. Meanwhile, research should continue in the effort to evaluate the medicinal plants of Laos as potential sources of beneficial medicines.

REFERENCES

Botanical Dimensions. 2022. "Ethnobotanical forest-garden on Hawaii." *Collect, Protect, Propagate, Understand.* Accessed on January 8, 2023. http://botanicaldimensions. org/ethnobotanical-forest-garden/.

Chen, S.-L., H. Yu, H.-M. Luo, Q. Wu, C.-F. Li, and A. Steinmetz. 2016. "Conservation and sustainable use of medicinal plants: problems, progress, and prospects." *Chinese Medicine* (United Kingdom) 11 (1): 1–10. doi:10.1186/s13020-016-0108-7.

Devkota, H.P., and M. Watanabe. 2020. "Role of medicinal plant gardens in pharmaceutical science education and research: An overview of medicinal plant garden at Kumamoto University, Japan." *Journal of Asian Association of Schools of Pharmacy* 9: 44–52. Accessed on January 8, 2023. https://www.researchgate.net/publication/344507830_Role_of_medicinal_plant_gardens_in_pharmaceutical_science_education_and_research_An_overview_of_medicinal_plant_garden_at_Kumamoto_University_Japan.

Elkington, B.G., K. Sydara, J.F. Hartmann, B. Southavong, and D.D. Soejarto. 2012. "Folk epidemiology recorded in palm leaf manuscripts of Laos." *Journal of Lao Studies* 3 (1): 1–14. Accessed on January 8, 2023. http://laostudies.org/system/files/subscription/JLS-v3-i1-Oct2012-elkington.pdf.

Farnsworth, N.R., O. Akerele, Audrey S. Bingel, D.D. Soejarto, and Z. Guo. 1985. "Medicinal plants in therapy." *Bulletin of the World Health Organization* 63 (6): 965–81. Accessed on January 8, 2023. https://www.ncbi.nlm.nih.gov/pmc/articles/PMC2536466/pdf/bullwho00089-0002.pdf.

Heinrich, M. 2000. "Ethnobotany and its role in drug development." *Phytotherapy Research* 14 (7): 479–88. doi:10.1002/1099-1573(200011)14:7<479::aid-ptr958>3.0.co;2-2.

McClatchey, W.C., G.B. Mahady, B.C. Bennett, L. Shiels, and V. Savo. 2009. "Ethnobotany as a pharmacological research tool and recent developments in CNS-active natural products from ethnobotanical sources." *Pharmacology & Therapeutics* 123 (2): 239–54. doi:10.1016/J.PHARMTHERA.2009.04.002.

Salmerón-Manzano, E., J.A. Garrido-Cardenas, and F. Manzano-Agugliaro. 2020. "Worldwide research trends on medicinal plants." *International Journal of Environmental Research and Public Health*, 17 (10): 3376. doi:10.3390/IJERPH 17103376.

Soejarto, D.D., C. Gyllenhaal, M.R. Kadushin, B. Southavong, K. Sydara, S. Bouamanivong, M. Xaiveu, et al. 2012. "An ethnobotanical survey of medicinal plants of Laos toward the discovery of bioactive compounds as potential candidates for pharmaceutical development." *Pharmaceutical Biology* 50 (1): 42–60.

Sydara, K., S. Gneunphonsavath, R. Wahlstrom, S. Freudenthal, K. Houamboun, G. Tomson, T. Falkenberg. 2005. "Use of traditional medicine in Lao PDR." *Complementary Therapies in Medicine* 13 (3): 199–205. https://doi.org/10.1016/j.ctim.2005.05.004.

Sydara, K., M. Xayvue, O. Souliya, B.G. Elkington, and D.D. Soejarto. 2014. "Inventory of medicinal plants of the Lao People's Democratic Republic: A mini review." *Journal of Medicinal Plants Research* 8 (43): 1262–74. Accessed on January 12, 2023. https://www.academia.edu/66268982/Inventory_of_medicinal_plants_of_the_Lao_Peoples_Democratic_Republic_A_mini_review.

UNDP. 2012. "Mainstreaming conservation and sustainable use of medicinal plant diversity in three Indian states: Project profile of Uttarakhand | United Nations Development Programme." Accessed on January 8, 2023. https://www.undp.org/india/publications/mainstreaming-conservation-and-sustainable-use-medicinal-plant-diversity-three-indian-states-project-profile-uttarakhand.

Weiss, R. 1994. "Learning traditional medicine in the jungles of Belize." *The Washington Post*, May 10. Accessed on January 8, 2023. https://www.washingtonpost.com/archive/1994/05/10/learning-traditional-medicine-in-the-jungles-of-belize/32f75f31-6d29-4e90-8bc5-e17379441a7d/.

Wetlands International. 2006. "Conservation of living pharmacies in Tasek Bera, a wetland of international importance in Malaysia." Accessed on January 8, 2023. https://studyres.com/doc/4605951/conservation-of-living-pharmacies-in-tasek-bera--a-wetlan.

Index

Pages in *italics* refer to figures and **bold** refer to tables.